Precalculus Mathematics

NEW IMPRESSION

Michael Payne

College of Alameda
Alameda, California

Julius Timbol 521-5865

W. B. SAUNDERS COMPANY
Philadelphia, London, Toronto

W. B. Saunders Company: West Washington Square
Philadelphia, Pa. 19105

1 St. Anne's Road
Eastbourne, East Sussex BN21 3UN, England

1 Goldthorne Avenue
Toronto, Ontario M8Z 5T9, Canada

Library of Congress Cataloging in Publication Data

Payne, Michael Noel.
    Precalculus mathematics.

    Includes index.
    1. Mathematics—1961-   I. Title.
QA39.2.P394 1978   512'.1   78-60160
ISBN 0-7216-7126-8

Precalculus Mathematics—*New Impression*   ISBN 0-7216-7126-8

Last digit is the print number: 9   8   7   6   5   4   3   2

To the Memory of

Otis S. Payne, Jr.

PREFACE

A textbook should be written for the student.

In recent years, there have been significant changes in the teaching of algebra and trigonometry, leading to a new type of course, precalculus. In such courses the emphasis has been to present the subject from a more theoretical and rigorous approach rather than the more established computational and intuitive point of view. The purpose of *Pre-Calculus Mathematics* is to achieve a balance between the two approaches.

As students work their way through this book, they will gain an understanding of the basic concepts and their applications, along with a proficiency in manipulative skills. To this end, many examples and illustrations stressing geometrical and physical intuition are given. It is hoped that this treatment of combining theory and problem-solving will open up the exciting world of mathematics to the reader.

The amount of material in the text is enough for a one-semester course which meets daily. The choice of topics and the pace will depend upon the individual instructor, the level of the class, and the academic calendar. Shorter courses can be constructed using this material. The following suggests some possible guidelines for one-quarter or one-semester courses.

In classes meeting daily for one semester, there should be sufficient time to progress through Chapters 1 to 7, in the order presented, although if additional time is needed for certain areas, Sections 3.3, 3.4, 3.7, and 3.8 in Chapter 3 may be omitted. In classes meeting daily for one quarter, it is recommended that the instructor work through the first six chapters; again, Sections 3.3, 3.4, 3.7, and 3.8 are optional. If the students are well prepared, and Chapter 1 can be reviewed quickly, Chapter 7 may be added to this schedule.

The material in this book was developed and used in courses at College of Alameda and San Francisco State University. I wish to express my thanks to those students for their suggestions and patience and in particular Mrs. Jan Smith, whose help in preparing this book was invaluable.

I am very grateful for the valuable suggestions for improving the manuscript that were made by Professors Craig Comstock of the Naval Postgraduate School, Harold B. Hackett, Jr. of State University of New York at Alfred, and Stanley M. Lukawecki of Clemson University, who reviewed the prepublication version. Thanks is also extended to Ms. Emilie Hance and Ms. Ruth Suzuki, who typed the manuscript. I would like to express my deep appreciation to my editor, Mr. John Snyder, Jr., and the staff of the W. B. Saunders Company, for their help and suggestions.

Finally, a very special thanks to Mr. Kim (HEY!) Fuller for his encouragement throughout this project.

MICHAEL PAYNE, PH.D.

CONTENTS

Chapter 1
REAL NUMBERS 1
   1.1 Rational Numbers and Irrational Numbers, **1**
   1.2 Order, **6**
   1.3 Inequalities and Intervals, **11**
   1.4 Absolute Value, **17**
   1.5 Cartesian Coordinate System, **20**
      Summary, **27**

Chapter 2
FUNCTIONS 30
   2.1 Set and Set Operations, **30**
   2.2 Relations, **40**
   2.3 Functions, **44**
   2.4 Algebra of Functions, **57**
   2.5 Types of Functions, **61**
   2.6 Composite Functions, **76**
   2.7 Inverse Functions, **79**
      Summary, **84**

Chapter 3
POLYNOMIALS AND RATIONAL FUNCTIONS 89
   3.1 Polynomials and Polynomial Functions, **89**
   3.2 Linear Functions, **103**
   3.3 Systems of Linear Equations, **120**
   3.4 Determinants, **142**
   3.5 Quadratic Functions, **156**
   3.6 Graphing Polynomials, **168**
   3.7 Complex Roots, **178**
   3.8 Rational Functions and Partial Fractions, **187**
      Summary, **198**

Chapter 4
EXPONENTIAL AND LOGARITHMIC FUNCTIONS 204
   4.1 Exponents, **204**
   4.2 The Exponential Function, **211**
   4.3 Applications of Exponential Functions, **217**
   4.4 Logarithmic Function, **223**
   4.5 Properties of Logarithmic Functions, **226**

4.6 Computation of Logarithms, **232**
4.7 Applications of Logarithmic Functions, **242**
Summary, **244**

Chapter 5
CIRCULAR FUNCTIONS **247**
5.1 The Winding Function, **247**
5.2 Sine and Cosine, **255**
5.3 The Graphs of Sine and Cosine, **262**
5.4 Inverse Sine and Inverse Cosine, **268**
5.5 Other Circular Functions, **273**
Summary, **283**

Chapter 6
TRIGONOMETRIC FUNCTIONS **287**
6.1 Angles, Radians, and Degrees, **287**
6.2 Trigonometric Functions, **294**
6.3 Trigonometric Identities, **299**
6.4 Trigonometric Equations, **308**
6.5 Triangles, Trigonometry and Applications, **313**
Summary, **325**

Chapter 7
INTRODUCTION TO ANALYTIC GEOMETRY **329**
7.1 Circle, **330**
7.2 Translation of Axes, **339**
7.3 Parabola, **344**
7.4 Ellipse, **355**
7.5 Hyperbola, **363**
Summary, **370**

Appendix I
BINOMIAL THEOREM **375**

Appendix II
MATHEMATICAL INDUCTION **379**

Appendix III
TABLES **382**
Common Logarithms, **383**
Natural Logarithms, **385**
Exponential Functions, **387**
Four-Place Values of Trigonometric Functions and Radians, **389**

Answers to Selected Odd-Numbered Problems **397**

Index **425**

CHAPTER 1

REAL NUMBERS

INTRODUCTION

What is algebra? Algebra is primarily the study of numbers and their general properties. In this chapter, we introduce the ideas of the real numbers and the real number line. We state some of the properties of the real numbers and what makes up the real numbers. We conclude the chapter with a discussion of the coordinate plane and the representation of "ordered pairs" of real numbers.

1.1 RATIONAL NUMBERS AND IRRATIONAL NUMBERS

One of the most important collections of numbers in mathematics is the collection of real numbers, denoted by R. Geometrically a real number can be represented by a **point** on a line called the **real line** and, conversely, every point on the real line represents a real number (see Fig. 1).

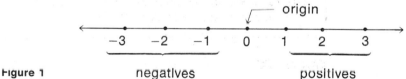

Figure 1

This correspondence between the real numbers and the real line can be described in the following way. We choose a point on the line to associate with the number 0 and call this point the **origin.** Points to the right of the origin correspond to the positive numbers, while points to the left correspond to the negative numbers.

Traditionally, the first set of numbers described were the **natural numbers,** denoted by the letter N. Thus, $N = \{1, 2, 3, 4, 5, 6, . . .\}$. This set is also called the set of **positive integers.** Therefore, the set

$$\{-1, -2, -3, -4, -5, . . .\}$$

is called the set of **negative integers.** The correspondence between these two sets and the real line is illustrated in Figure 2.

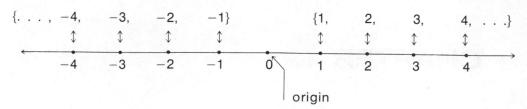

origin

Figure 2

The set which consists of the positive integers, negative integers, and the number zero is called the set of **integers** and is denoted by I. Thus,

$$I = \{\ldots, -3, -2, -1, 0, 1, 2, 3, \ldots\}$$

Looking at the real line we note that there are other points which do not correspond with the numbers in I. Examples of these are the points between 0 and 1 (see Fig. 3). We make use of the fractions to describe many of these points.

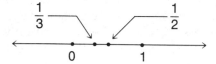

Figure 3

However, it can be shown that even the fractions are not enough to describe all the points that exist between the integers. We shall now discuss the two sets which together describe all the points on the real line.

Definition: Any number which can be expressed as the quotient of two integers, division by zero excluded, is called a **rational number.** The set of rational numbers is denoted by Q.

EXAMPLES

$\dfrac{4}{3}, \dfrac{1}{2}, \dfrac{16}{5}, \dfrac{-29}{3}$ are rational

---

PROBLEM

Are the integers rational?

Solution: Yes, for any integer can be written as the quotient of itself and 1.

$$4 = \frac{4}{1}, \; -2 = \frac{-2}{1}$$

Thus, the set of integers I is contained in the set of rational numbers Q.

There are real numbers, however, which *cannot* be expressed as the quotient of two integers. These numbers are called **irrational numbers.** It is generally difficult to show from the above statement that a number is irrational. However, we can show that numbers such as $\sqrt{2}$ and $\sqrt{12}$ are irrational without too much difficulty.

Theorem: The number $\sqrt{2}$ is an irrational number.

Proof (by contradiction). Suppose $\sqrt{2}$ were rational; that is, it could be expressed as the quotient of two integers, say $\frac{p}{q}$. Let us also assume that the fraction $\frac{p}{q}$ is reduced to its lowest form (that is, no common factors other than 1). Thus we have

$$\sqrt{2} = \frac{p}{q}$$

Squaring, we get $\qquad \frac{p^2}{q^2} = 2$ and $p^2 = 2q^2$ $\qquad$ **(1)**

This last equation implies that p^2 is even. Then if p^2 is even, p must be even. (Otherwise p would be odd and an odd $\cdot$ odd = odd, which would say p^2 would be odd.) Since p is even it can be written $p = 2n$ for some integer n. Substituting into Equation (1) we obtain

$$(2n)^2 = 2q^2 \qquad 4n^2 = 2q^2 \qquad 2n^2 = q^2 \qquad \textbf{(2)}$$

A similar argument as above shows that q is also even. Therefore, on the basis that $\sqrt{2}$ is rational we have shown that both p and q are even. If p and q are even, then they have 2 in common. This is contrary to our agreement that the fraction $\frac{p}{q}$ is in its lowest terms. Hence we conclude that the number $\sqrt{2}$ is an irrational number.

In the same way we could prove that $\sqrt{3}$, $\sqrt{12}$, for example, are irrational. There are other methods to determine if a number is rational or irrational.

Every rational number can be represented by an eventually repeating decimal or a terminating decimal. Conversely, every eventually repeating decimal or terminating decimal represents a rational number.

EXAMPLE 1

Consider the rational number $\frac{5}{7}$.

```
        .714285714285 . . .
    7)5.000000000000
      49
      ──
       10
        7
        ──
        30
        28
        ──
         20
         14
         ──
          60
          56
          ──
           40
           35
           ──
            50
             .
             .
             .
```

So $\frac{5}{7} = .714285714285\overline{714285}$ where the bar indicates the block of digits which repeats infinitely often.

EXAMPLE 2

$$4 = \frac{4}{1} = 4.\overline{0}$$

EXAMPLE 3

$$\frac{1}{3} = .33\overline{3}$$

EXAMPLE 4

$$7\frac{9}{14} = 7.6428571\overline{428571}$$

EXAMPLE 5

To illustrate the converse, consider the repeating decimal $4.76\overline{76}$.

Let $N = 4.76\overline{76}$
$N = 4.76767676 \ldots$

Since there are two numbers in the repeating block, we multiply N by 100.

$$100N = 476.767676 \ldots$$

Now subtract N,

$$100N = 476.767676 \ldots$$
$$N = 4.767676 \ldots$$
$$99N = 472$$
$$N = \frac{472}{99}$$

Thus $N = 4.76\overline{7676}$ represents the rational number $\frac{472}{99}$

Every irrational number is represented by a nonrepeating decimal.

EXAMPLE 1

We have shown previously that $\sqrt{2}$ is irrational. Its decimal representation

$$\sqrt{2} = 1.41421356 \ldots$$

is nonterminating and nonrepeating.

EXAMPLE 2

$$\pi = 3.1415926 \ldots$$

EXAMPLE 3

$$e = 2.7182818284 \ldots$$

In summary, the real numbers consist of the rational numbers and the irrational numbers. The following diagram shows the makeup of the real numbers.

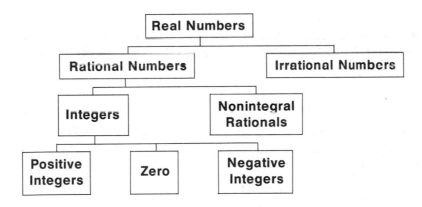

1.1 EXERCISES

Convert the following fractions to their decimal representations:

1. $\frac{1}{7}$ 2. $\frac{7}{8}$ 3. $\frac{3}{4}$ 4. $\frac{11}{12}$ 5. $\frac{9}{7}$

6. $\frac{7}{11}$ 7. $\frac{10}{8}$ 8. $\frac{29}{25}$ 9. $\frac{2}{9}$ 10. $\frac{23}{99}$

Convert the following decimals to a fraction in lowest terms:

11. .545454

12. 0.2

13. 1.002

14. 0.$\overline{5}$

15. 3.0007

16. 2.47474747 . . .

17. 0.$\overline{37}$

18. 12.$\overline{9}$

19. .625

20. .367436743$\overline{674}$

Determine which of the following represent rational numbers:

21. 3.3333 . . .

22. 1.6

23. 3 $\sqrt{7}$

24. 2.$\overline{0}$

25. 16.32$\overline{32}$

26. 4.9876543210117652 . . .

27. 1.742836$\overline{836}$

28. $\dfrac{1}{4}$

29. 2.02002000200002 . . .

30. 7

1.2 ORDER

Recall from the previous section that the real numbers are represented on the real line in such a way that as you go from left to right the numbers increase. Therefore, any real number whose corresponding point on the real number line lies to the right of the corresponding point of a second real number is said to be **greater than** the second number. The notation is ">." The second number is said to be **less than** the first number. The notation is "<."

EXAMPLE 1

5 < 7 (5 is "less than" 7) This can be interpreted as 7 > 5 (7 is "greater than" 5). Looking at the real number line we see that the point representing 7 lies to the right of the point representing 5 (see Fig. 4).

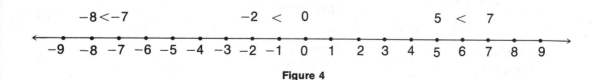

Figure 4

EXAMPLE 2

−2 < 0 (−2 is "less than" 0)

EXAMPLE 3

−7 > −8 (−7 is "greater than" −8)

> **Remark:** The point of the notation $\gg$ always indicates the smaller of the two numbers

EXAMPLE 4

$$-\frac{3}{4} < -\frac{1}{2} \left(-\frac{3}{4} \text{ is "less than" } -\frac{1}{2} \right)$$

The concepts of "greater than" and "less than" give **order** to the real numbers. Geometrically we understand what it means for $a > b$ or $a < b$, but how do we determine these order relations algebraically?

> **Definition:** Given any two real numbers a and b, then a is greater than b ($a > b$) or, equivalently, b is less than a ($b < a$) if $a - b$ is positive. That is, if $a - b > 0$, then $a > b$ or $b < a$.

EXAMPLE 1

$5 < 7$ because $7 - 5 = 2 > 0$

EXAMPLE 2

$-2 < 0$ because $0 - (-2) = 2 > 0$

EXAMPLE 3

$-7 > -8$ because $-7 - (-8) = 1 > 0$

EXAMPLE 4

$$-\frac{3}{4} < -\frac{1}{2} \text{ because } -\frac{1}{2} - \left(-\frac{3}{4} \right) = \frac{1}{4} > 0$$

Properties of Order Relations

Given two real numbers a and b and their point representation on the real number line, then only one of three possible cases can occur (see Fig. 5).

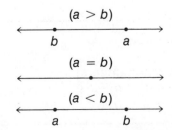

Figure 5

Property 1: The Trichotomy Law

If a and b are real numbers, then exactly one of the following is true: $a < b$ or $a = b$ or $a > b$.

	Algebraic Meaning	*Geometric Meaning*
1. $a > b$	a is greater than b	a is to the right of b
2. $a = b$	a is equal to b	a is the same point as b
3. $a < b$	a is less than b	a is to the left of b

EXAMPLE 1

Of the following statements, only the third one is true:

$$-4 > -2 \qquad -4 = -2 \qquad -4 < -2$$

Property 2: Transitive Law

For a, b, and c real numbers:
(i) If $a > b$ and $b > c$, then $a > c$.
(ii) If $a < b$ and $b < c$, then $a < c$.

EXAMPLE 1

Since $4 < 9$ and $9 < 11$, then $4 < 11$.
Geometrically:

Since 4 is to the left of 9 and 9 is to the left of 11, then 4 is to the left of 11.

EXAMPLE 2

Since $-1 > -2$ and $-2 > -5$, then $-1 > -5$.
Geometrically:

Since -1 is to the right of -2, and -2 is to the right of -5, then -1 is to the right of -5.

Property 3: Addition Law

For a, b, and c real numbers:
(i) If $a > b$, then $a + c > b + c$.
(ii) If $a < b$, then $a + c < b + c$.

EXAMPLE 1

Since $4 > 2$, then $4 + 1 > 2 + 1$.
$$5 > 3$$

Geometrically: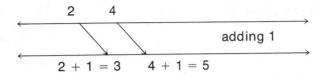

EXAMPLE 2

Since $-1 < 2$, then $-1 + 2 < 2 + 2$.
$$1 < 4$$

Geometrically: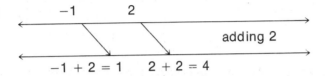

EXAMPLE 3

Since $3 > 1$, then $3 + (-2) > 1 + (-2)$.
$$1 > -1$$

Geometrically: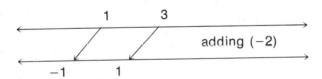

Property 4: Multiplication Law

For a, b, and c real numbers:
  (i) If $a > b$ and $c > 0$, then $ac > bc$.
 (ii) If $a < b$ and $c > 0$, then $ac < bc$.
(iii) If $a > b$ and $c < 0$, then $ac < bc$.
 (iv) If $a < b$ and $c < 0$, then $ac > bc$.

EXAMPLE 1

Since $3 > -1$, then $(3)(2) > (-1)(2)$.
$$6 > -2$$

Geometrically: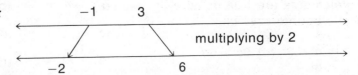

EXAMPLE 2

Since $-3 < 0$, then $(-3)(3) < (0)(3)$.
$$-9 < 0$$

Geometrically:

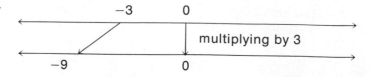

multiplying by 3

EXAMPLE 3

Since $4 > 1$, then $(4)(-1) < (1)(-1)$.
$$-4 < -1$$

Geometrically:

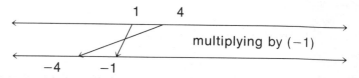

multiplying by (-1)

Note that multiplication by a negative number results in a change in the order.

EXAMPLE 4

Since $-1 < 2$, then $(-1)(-4) > (2)(-4)$.
$$4 > -8$$

Geometrically:

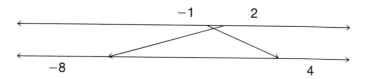

1.2 EXERCISES

Show that the following ordering relations are true:

1. $5 < 9$

2. $-4 > -5$

3. $0 < 6$

4. $\dfrac{1}{3} < \dfrac{5}{8}$

5. $3 > -2$

6. $-1 < 0$

7. $-2.3 > -4.5$

8. $3 < 5$

9. If $a < 0$ and $c > 0$, then $a < c$.

10. If $a > 0$ and $c < 0$, then $a > c$.

11. Given that $c > d$, determine if the following statements are true or false. If false, give a counterexample.
   (a) $-c < -d$ (b) $c^2 > d^2$
   (c) $c + 2 < d + 2$ (d) $6c > 6d$
   (e) $c - d < 0$

12. Insert $<$ or $>$ to make true statements.
   (a) $-7 \quad 6$ (b) $-7.6 \quad 2.1$
   (c) $\dfrac{1}{2} \quad \dfrac{1}{8}$ (d) $0 \quad 2$
   (e) $-4 \quad -5$ (f) $\dfrac{1}{3} \quad -3$

13. Indicate the ordering property which proves the following statements:

(a) If $x < y$, then $x - 2 < y - 2$.

(b) If $\frac{2}{3} > x$, then $-3x > -2$.

(c) If $6 < 9$, then $\frac{1}{3} < \frac{1}{2}$.

(d) If $y < x$ and $24 > x$, then $24 > y$.

1.3 INEQUALITIES AND INTERVALS

Inequalities

We now extend our ideas of ordering. A statement that one quantity is greater than or less than another quantity is called an **inequality.** Sometimes it is convenient to combine an inequality with an equality. Thus:

> $a \geq b$ means "a is greater than or equal to b."
> $a \leq b$ means "a is less than or equal to b."

EXAMPLE 1

The statement $4 \leq 10$ is true.

EXAMPLE 2

The statement $-6 \geq -6$ is true.

Intervals

Geometrically, inequalities can be described by means of line segments. Finite or infinite line segments are called **intervals.**

EXAMPLE 1

The picture of $x < 4$ looks like

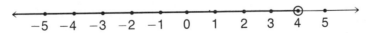

where $\bigcirc$ means the point 4 is not included and the left arrow indicates it is an infinite segment, in this case extending to the left.

EXAMPLE 2

The picture of $x \geq -1$ looks like

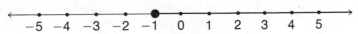

where $\bullet$ means the point -1 is included.

EXAMPLE 3

The picture of $-2 < x \leq 3$ looks like

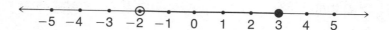

A more convenient notation can be used in describing intervals on the real line. This notation will also be helpful in discussing the collection of points between the real numbers *a* and *b*.

Definition: The **open interval** from *a* to *b*, where *a* < *b*, denoted by (*a*, *b*), is the collection of all points *x* such that *a* < *x* < *b*. The points *a* and *b* are called **end points** of the interval.

EXAMPLE

The open interval (2, 5) represents all the points *x* such that $2 < x < 5$.

Geometrically:

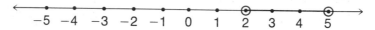

Note that the end points are not included in an open interval. If we do include the end points, we have a **closed interval.**

Definition: The **closed interval** from *a* to *b*, [*a*, *b*], is the collection of all points *x* such that $a \leq x \leq b$.

EXAMPLE

The closed interval [−2, 4] represents all points *x* such that $-2 \leq x \leq 4$.

Geometrically:

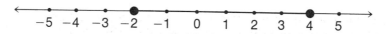

There are two other types of infinite intervals, those called **half-open** or **half-closed intervals.**

Definition: The **interval half-open on the right** from *a* to *b*, denoted by [*a*, *b*), is the collection of all points *x* such that $a \leq x < b$.

EXAMPLE

The interval [0, 7) is the collection of all points x such that $0 \leq x < 7$. *Geometrically:*

Definition: The **interval half-open on the left** from a to b, denoted by $(a, b]$, is the collection of all points x such that $a < x \leq b$.

EXAMPLE

The interval $(-6, -1]$ represents all points x such that $-6 < x \leq -1$.

Geometrically:

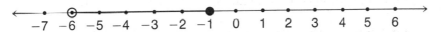

With this new notation we can describe any interval on the real number line. Not only the above finite intervals, but infinite intervals can be conveniently described.

EXAMPLE 1

The positive numbers can be written as $(0, +\infty)$.

EXAMPLE 2

The negative numbers can be written as $(-\infty, 0)$.

EXAMPLE 3

All the real numbers can be written as $(-\infty, +\infty)$.

Such infinite intervals are not referred to as open or closed intervals, although in Example 1 we could have $[0, \infty)$ which would include 0; in Example 2, $(-\infty, 0]$ would include 0.

Solving Inequalities

Now that we have the basic properties of inequalities, we can solve inequalities in one variable.

PROBLEM 1

Find the values of x such that $11x - 2 > 3x + 9$.

Solution: $11x - 2 > 3x + 9$
adding 2 to both sides gives $11x > 3x + 11$
subtracting $3x$ from both sides gives $8x > 11$

dividing by 8 $x > \dfrac{11}{8}$

Thus the solution consists of all x belonging to $\left(\dfrac{11}{8}, +\infty\right)$.

Geometrically: The solution is represented by the interval

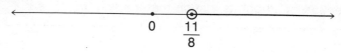

$$0 \qquad \frac{11}{8}$$

PROBLEM 2

Find all values of x that satisfy $-13 - 3x \geq 2x + 12$.

Solution: $-13 - 3x \geq 2x + 12$
$-2x - 3x \geq 12 + 13$
$-5x \geq 25$
$x \leq -5$ or $(-\infty, -5]$

Geometrically:

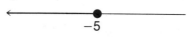

$$-5$$

PROBLEM 3

Solve $-4 \leq 3x + 8 < 23$.

Solution: $-4 \leq 3x + 8 < 23$
$-4 - 8 \leq 3x < 23 - 8$
$-12 \leq 3x < 15$
$-4 \leq x < 5$ or $[-4, 5)$

Geometrically:

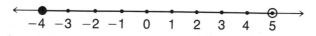

$$-4 \quad -3 \quad -2 \quad -1 \quad 0 \quad 1 \quad 2 \quad 3 \quad 4 \quad 5$$

PROBLEM 4

Solve for all values of x which satisfy

$x + 1 > 3$ and $3x - 2 \leq x$.

Solution: We are looking for the values of x that satisfy both inequalities.

$$x + 1 > -3$$
$$x > -4$$

and

$$3x - 2 \leq x$$
$$2x \leq 2$$
$$x \leq 1$$

Therefore the solution is the interval $(-4, 1]$.

Geometrically: The solution is represented by the interval where $x > -4$ overlaps $x \leq 1$.

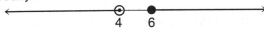

PROBLEM 5

Find all values of x which satisfy

$$2x - 5 < 3 \quad \text{or} \quad 4x - 2 \geq 22$$

Solution:

$2x - 5 < 3$		$4x - 2 \geq 22$
$2x < 8$	or	$4x \geq 24$
$x < 4$		$x \geq 6$

Therefore the solution consists of those values of x such that $x < 4$ or $x \geq 6$.

Geometrically:

The following problems are nonlinear inequalities. They require a different approach than the linear inequalities given previously.

Recall: Given any real numbers a and b:
1. $ab < 0$ if and only if $a < 0$ and $b > 0$ or $a > 0$ and $b < 0$.
2. $ab > 0$ if and only if $a > 0$ and $b > 0$ or $a < 0$ and $b < 0$.

We now illustrate how to solve nonlinear inequalities.

PROBLEM 6

$$x^2 - 1 > 0.$$

Solution: $x^2 - 1 > 0$

factoring $(x + 1)(x - 1) > 0$

If this were an equality, $(x + 1)(x - 1) = 0$, we could solve for the zeros and get $x = -1$ and $x = 1$. We use these values to divide the real number line into three intervals and describe each interval.

$$
\begin{array}{ccccc}
\text{Yes} & -1 & \text{No} & 1 & \text{Yes} \\
\underbrace{}_{x < -1} & & \underbrace{}_{-1 < x < 1} & & \underbrace{}_{x > 1}
\end{array}
$$

Each interval can be checked to see if it satisfies the inequality. Taking examples in each interval we see that the inequality is true only in the intervals $x < -1$ or $x > 1$. That is, let $x < -1$, then $x + 1 < 0$ and $x - 1 < 0$, and thus $(x + 1)(x - 1) > 0$. Now let $x > 1$, then $x + 1 > 0$ and $x - 1 > 0$, and thus $(x + 1)(x - 1) > 0$. Whereas if we let $-1 < x < 1$, then $x + 1 > 0$ but $x - 1 < 0$ and we have $(x + 1)(x - 1) < 0$, which does not satisfy the inequality.

PROBLEM 7

Solve $x^2 - 7x < -12$.

Solution: $x^2 - 7x < -12$
$x^2 - 7x + 12 < 0$
$(x - 3)(x - 4) < 0$

The zeros are $x = 3$ and $x = 4$.

$$
\begin{array}{ccccc}
\text{No} & 3 & \text{Yes} & 4 & \text{No} \\
\underbrace{}_{x < 3} & & \underbrace{}_{3 < x < 4} & & \underbrace{}_{x > 4}
\end{array}
$$

Checking each of the intervals, we see that $x^2 - 7 < -12$ is satisfied when $3 < x < 4$.

PROBLEM 8

Find values of x which satisfy $x^3 - 6x \le x^2$.

Solution: $x^3 - 6x \le x^2$
$x^3 - x^2 - 6x \le 0$
$x(x^2 - x - 6) \le 0$
$x(x + 2)(x - 3) \le 0$

The zeros are $x = 0$, $x = -2$, and $x = 3$.

$$
\begin{array}{ccccccc}
\text{Yes} & -2 & \text{No} & 0 & \text{Yes} & 3 & \text{No} \\
& & \text{Yes} & & \text{Yes} & & \text{Yes} \\
x < -2 & & -2 < x < 0 & & 0 < x < 3 & & x > 3
\end{array}
$$

Checking each interval and *the zeros,* because of $\leq$, we find that $x^3 - 6x \leq x^2$ is satisfied when $x \leq -2$ or $0 \leq x \leq 3$.

1.3 EXERCISES

Using the notation developed in this section, describe the following intervals:

1. The positive real numbers.

2. The real numbers between -1 and 7, including the number 7.

3. The nonnegative real numbers.

4. All real numbers greater than 4.

5. Real numbers less than or equal to -1.

Graph each of the following:

6. $x \geq 2$

7. $[0, 9)$

8. $-1 \leq x < 3$

9. $-4 < x < 2$

10. $(-2, 5)$

11. $x < -3$

12. $[-4, 4]$

13. $(2, 8]$

Solve the following inequalities:

14. $x + 8 < 14$

15. $2x + 4 < 3x + 9$

16. $2x + 7 \geq 15$

17. $-3(x + 5) \geq 5x + 2$

18. $4 < 3x + 5 < 1$

19. $5 - x < x < 7 - x$

20. $x^2 - 5x > 6$

21. $1 - x^2 > 0$

22. $x^2 \leq 16$

23. $(x + 4)(x - 2) \geq 0$

24. $(x + 1)(x + 2)(x - 3) < 0$

25. $x^3 - 6x^2 + 8x > 0$

Given that $a, b, c,$ and d are real numbers, determine whether Exercises 26 through 28 are true or false.

26. If $b > 1$, then $b^2 > b$.

27. If $a > 0$ and $b > 0$, then $\dfrac{1}{a} + \dfrac{1}{b} > \dfrac{2}{a + b}$.

28. If $a < b$ and $c < d$, then $a + c < b + d$.

29. Prove that $a^2 + b^2 \geq 2ab$. (Hint: Look at $(a - b)^2 \geq 0$.)

30. Prove that if $a > 0$ then $a + \dfrac{1}{a} \geq 2$. (Hint: Look at $(a - 1)^2 \geq 0$.)

1.4 ABSOLUTE VALUE

The absolute value of a real number can be thought of as its distance from 0 on the real number line. The absolute value of a real number x is denoted by $|x|$. The algebraic meaning is given on page 18.

> **Definition:** If x is a real number, the **absolute value** of x, denoted by $|x|$, is defined as follows:
>
> $$|x| = x \text{ if } x \geq 0.$$
> $$|x| = -x \text{ if } x < 0.$$

EXAMPLES

1. $|6| = 6$ because $6 \geq 0$
2. $|-6| = -(-6) = 6$ because $-6 < 0$
3. $|0| = 0$ because $0 \geq 0$
4. $|x + 4| = \begin{cases} x + 4 & \text{if } x + 4 \geq 0 \\ -(x + 4) & \text{if } x + 4 < 0 \end{cases}$

Properties of Absolute Value

> **Property 1:** Given any two real numbers a and b, then $|a - b|$ represents the distance between a and b.

EXAMPLE

If we let $a = 10$ and $b = 3$, then $|a - b| = |10 - 3| = |7| = 7$. Also, if we let $a = 3$ and $b = 10$ then $|a - b| = |3 - 10| = |-7| = 7$. This example illustrates a second property of absolute values.

> **Property 2:** For any real numbers a and b
>
> $$|a - b| = |b - a|$$

EXAMPLE

$|4 - 3| = 1$ and $|3 - 4| = |-1| = 1$.

> ✳ **Property 3:** If x is any real number and b is a positive real number, then $|x| < b$ if and only if $-b < x < b$.

EXAMPLE 1

$|x| < 7$ if and only if $-7 < x < 7$.

EXAMPLE 2

$|x - 2| < 5$ if and only if $-5 < x - 2 < 5$.

> ✻ **Property 4:** If x is any real number and b is a positive real number, then $|x| > b$ if and only if $x > b$ or $x < -b$.

EXAMPLE 1

$|x| > 7$ if and only if $x > 7$ or $x < -7$.

EXAMPLE 2

$|x - 2| > 5$ if and only if $x - 2 > 5$ or $x - 2 < -5$.

We geometrically describe the definitions of absolute value, Property 3, and Property 4 in Figure 6.

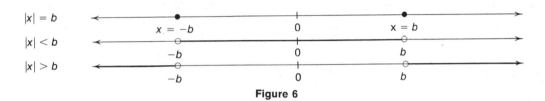

Figure 6

PROBLEM 1

Find all real numbers x for which (a) $|3x - 4| = 2$, (b) $|3x - 4| < 2$, and (c) $|3x - 4| > 2$.

Solution: (a) $|3x - 4| = 2$. Using the definition of absolute value we have two possible cases.

Case (i): $|3x - 4| = 2$ if $3x - 4 = 2$; that is, $x = 2$.

Case (ii): $|3x - 4| = 2$ if $-(3x - 4) = 2$; that is, $x = \dfrac{2}{3}$.

(b) $|3x - 4| < 2$ if and only if $-2 < 3x - 4 < 2$

$$2 < 3x < 6$$

$$\frac{2}{3} < x < 2$$

(c) $|3x - 4| > 2$ if and only if $3x - 4 > 2$ or $3x - 4 < -2$.

$3x - 4 > 2$	or	$3x - 4 < -2$
$3x > 6$	or	$3x < 2$
$x > 2$	or	$x < \dfrac{2}{3}$

1.4 EXERCISES

1. If $a = 2$ and $b = -3$, compute the following expressions:

   (a) $|a + b|$ (b) $|a - b|$ (c) $|b - a|$ (d) $|a| + |b|$

   (e) $|a| - |b|$ (f) $|ab|$ (g) $\left|\dfrac{a}{b}\right|$ (h) $2|a| - 3\big(\big||a| + |b|\big|\big)$

2. Compute the distance between the following pairs of real numbers:

   (a) 6 and 3   (b) 6 and -3 (c) -6 and -3 (d) -6 and 3

3. Graph the following:

   (a) $|x - 1| < 3$ (b) $|3x - 1| \le 4$ (c) $|x| > 5$

   (d) $|x - 2| > 8$ (e) $|2x + 1| = 3$ (f) $|x| \ge 0$

4. Find all real numbers x for which

   (a) $|2x| = 10$ (b) $|x + 4| = 7$ (c) $|2x + 1| < 3$

   (d) $|2 - 8x| \le 6$ (e) $|x + 2| > 5$ (f) $|2x - 1| > \dfrac{1}{10}$

5. Write each of the following inequalities as an absolute value inequality:

   (a) $x = 7$ or $x = -7$ (b) $-2 < x < 2$ (c) $x > 4$ or $x < -4$

   (d) $-8 < x - 3 < 8$ (e) $-18 < x < 20$ (f) $-4 < 3x < 0$

1.5 CARTESIAN COORDINATE SYSTEM

In Section 1.1 we saw that the real number line was a geometric representation of the real numbers. In this section we shall introduce the idea of "ordered pairs" as an algebraic representation of the plane. An application of this geometric representation will be to derive a formula which gives the distance between two points in a plane.

We start by taking two perpendicular real number lines intersecting at a point called the **origin.** The two number lines are called **axes.** The horizontal line, which extends indefinitely to the right and left, is called the **x-axis.** The vertical line, which extends indefinitely up and down, is called the **y-axis.** Figure 7 illustrates this coordinate system, called the *Cartesian coordinate system,* named after the famous French mathematician René Descartes (1596–1650).

To locate a point in the plane we select a pair of real numbers. The first real number will tell you how many units to the right or left of the origin. As an example the real number 3 indicates 3 units to the right of the origin. The second real number will tell you how many units up or down from the origin. As an example the real number 4 indicates 4 units up from the origin. The pair (3, 4) is shown in Figure 7. Note that the pair (3, 4) and (4, 3) are different points. Therefore, the order in which we select a pair of real numbers is very important in locating a point and thus we use the idea of *ordered pair.*

The real numbers that locate a point are called **coordinates** of the point. The first coordinate of the ordered pair (3, 4) is 3 and the second coordinate is

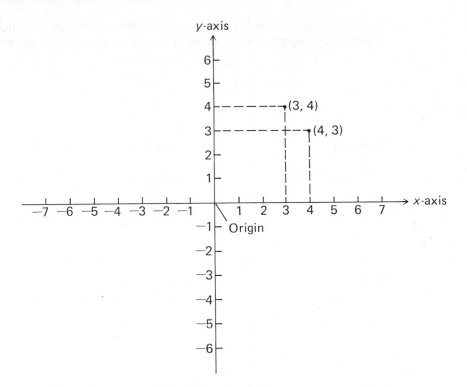

Figure 7

4. Sometimes the first coordinate is referred to as the **abscissa** of the point and the second coordinate is referred to as the **ordinate** of the point. The origin is represented by the ordered pair having zero as its abscissa and 0 as its ordinate—that is, the ordered pair (0, 0).

Every point P in the plane uniquely determines by an ordered pair of real numbers (x, y) and, conversely, every ordered pair of real numbers (x, y) is represented by a point P in the plane.

> **Definition:** Two ordered pairs (x_1, y_1) and (x_2, y_2) are considered to be **equal** if and only if $x_1 = x_2$ and $y_1 = y_2$.

EXAMPLES

$(3, 7) \neq (7, 3)$ and $(-2, 1) \neq (2, 1)$, whereas $(12, -6) = (12, -6)$.

PROBLEM

Find the values of x and y such that $(x, -4) = (8, y + 1)$.

Solution: $(x, -4) = (8, y + 1)$ if and only if $x = 8$ and $y + 1 = -4$; that is, $x = 8$ and $y = -5$.

Looking at the Cartesian coordinate system we see that it divides the plane into four regions called **quadrants** (see Fig. 8). In Quadrant I both coordinates

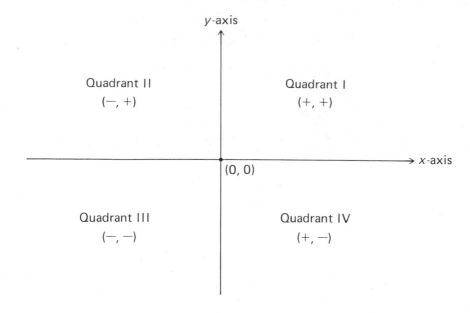

Figure 8

of a point are positive. Quadrant II has the first coordinate negative and the second coordinate positive. In Quadrant III we see that points have both coordinates negative. Finally, those points in Quadrant IV have a positive first coordinate and a negative second coordinate.

Note that points with one or both coordinates being zero will always lie on the axes. If the first coordinate is zero, $(0, y)$, then the point lies on the y-axis, while those with the second coordinate being zero, $(x, 0)$, lie on the x-axis. Of course, the point $(0, 0)$ is the representation for the origin.

The Distance Formula

Given any two points P_1 and P_2 with coordinates (x_1, y_1) and (x_2, y_2), respectively, we wish to derive a formula for the distance between those two points. We do so by considering three cases.

1. If the two points have the same first coordinates, that is, if $x_1 = x_2$, then the points P_1 and P_2 lie on a vertical line. The distance, d, between them is given by the absolute value of the difference of the y coordinates, $|y_2 - y_1|$ (see Fig. 9).
The distance between (x, y_1) and (x, y_2) is

$$|y_2 - y_1|$$

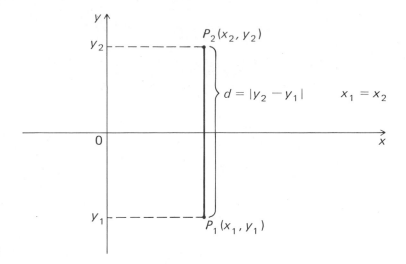

Figure 9

EXAMPLE 1

The distance between (4, 1) and (4, 6) is

$$|6 - 1| = |5| = 5$$

Recall that $|a - b| = |b - a|$, so it makes no difference which point is called P_1 or P_2.

PROBLEM

Find the distance between $(-2, -3)$ and $(-2, 4)$.

Solution: Let P_1 be $(-2, -3)$ and P_2 be $(-2, 4)$. The distance is

$$d = |4 - (-3)| = |4 + 3| = |7| = 7.$$

If we let P_1 be $(-2, 4)$ and P_2 be $(-2, -3)$, then

$$d = |-3 - 4| = |-7| = 7.$$

2. If the two points have the same second coordinates, that is, if $y_1 = y_2$, then the points P_1 and P_2 lie on a horizontal line. The distance, d, between them is given by the absolute value of the difference of the x coordinates, $|x_2 - x_1|$ (see Fig. 10).
The distance between (x_1, y) and (x_2, y) is

$$|x_2 - x_1|$$

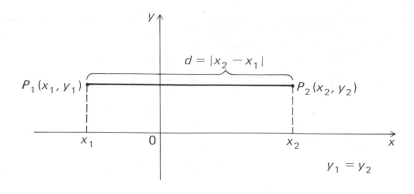

Figure 10

EXAMPLE

The distance between $(-2, 0)$ and $(-8, 0)$ is

$$|-2 - (-8)| = |-2 + 8| = |6| = 6.$$

PROBLEM

Find the distance between $\left(\frac{1}{8}, 7\right)$ and $(-2, 7)$.

Solution: $d = \left|-2 - \frac{1}{8}\right| = \left|-2\frac{1}{8}\right| = 2\frac{1}{8}$

We now consider the case in which the points P_1 and P_2 do not lie on a vertical or horizontal line.

3. If the two points P_1 and P_2 do not lie on a vertical line or a horizontal line, then the distance, d, between them is given by the following formula:

$$d = \sqrt{(x_2 - x_1)^2 + (y_2 - y_1)^2}$$

To see this, we note that if P_1 and P_2 are not on a vertical line or horizontal line, then the line segment joining P_1 and P_2 is the hypotenuse of a right triangle (see Fig. 11). The vertices of the right triangle are (x_1, y_1), (x_2, y_2) and (x_2, y_1). The lengths of the legs are $|x_2 - x_1|$ and $|y_2 - y_1|$. By the Pythagorean Theorem it follows that

$$d^2 = |x_2 - x_1|^2 + |y_2 - y_1|^2$$

and

$$d^2 = (x_2 - x_1)^2 + (y_2 - y_1)^2$$

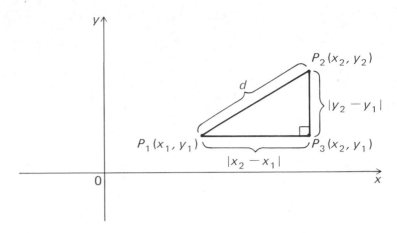

Figure 11

since for any real a, $|a|^2 = a^2$. Then

$$d = \sqrt{(x_2 - x_1)^2 + (y_2 - y_1)^2}$$

> Note the distance formulas for Cases 1 and 2 are special cases of the formula in Case 3. Hence, the formula
>
> $$d = \sqrt{(x_2 - x_1)^2 + (y_2 - y_1)^2}$$
>
> can be used for *any* two points.

PROBLEM 1

Find the distance between the points $(7, -2)$ and $(6, 3)$.

Solution: $d = \sqrt{(6 - 7)^2 + [3 - (-2)]^2}$

$d = \sqrt{(-1)^2 + (3 + 2)^2}$

$d = \sqrt{(-1)^2 + (5)^2}$

$d = \sqrt{1 + 25}$

$d = \sqrt{26}$

PROBLEM 2

Show that the triangle with vertices $(-1, 4)$, $(2, -1)$, and $(-3, -4)$ is a right triangle.

Solution: Let d_1 be the distance between $(-1, 4)$ and $(2, -1)$.
Let d_2 be the distance between $(2, -1)$ and $(-3, -4)$.
Let d_3 be the distance between $(-3, -4)$ and $(-1, 4)$.

Using the distance formula we compute d_1, d_2, and d_3.

$$d_1 = \sqrt{[2 - (-1)]^2 + (-1 - 4)^2} = \sqrt{3^2 + (-5)^2} = \sqrt{9 + 25}$$
$$= \sqrt{34}$$

$$d_2 = \sqrt{(-3 - 2)^2 + [-4 - (-1)]^2} = \sqrt{(-5)^2 + (-3)^2} = \sqrt{25 + 9}$$
$$= \sqrt{34}$$

$$d_3 = \sqrt{[-1 - (-3)]^2 + [4 - (-4)]^2} = \sqrt{(2)^2 + 8^2} = \sqrt{4 + 64}$$
$$= \sqrt{68} = \sqrt{4 \cdot 17} = 2\sqrt{17}$$

Note that $d_3^2 = d_1^2 + d_2^2$. Therefore, by the converse of the Pythagorean Theorem, the triangle with vertices $(-1, 4)$, $(2, -1)$, and $(-3, -4)$ is a right triangle.

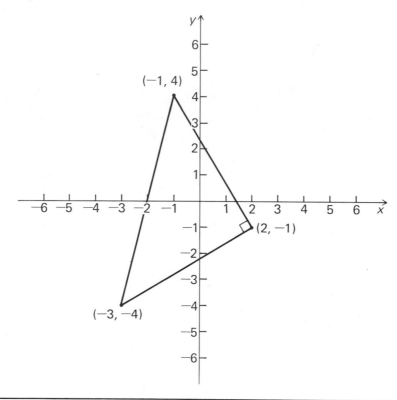

Figure 12

PROBLEM 3

Find the radius of a circle with center at $(2, 3)$ and which passes through the point $(-4, -2)$.

Solution: The radius of a circle is the distance from the center to any point on the circle. Thus, we must find the distance between $(2, 3)$ and $(-4, -2)$.

$$r = \sqrt{(-4 - 2)^2 + (-2 - 3)^2} = \sqrt{(-6)^2 + (-5)^2}$$
$$= \sqrt{36 + 25} = \sqrt{61}$$

1.5 EXERCISES

1. Determine in which quadrant each of the following points lies:
   (a) $(7, -6)$ (b) $(2, 3)$
   (c) $(-1, -3)$ (d) $(0, 7)$
   (e) $(0, 0)$ (f) $(-15, 2)$

2. Find the values of x and y such that $(3, y - 1) = (x - 2, 4)$.

3. Which of the following pairs of points lie on a vertical or horizontal line? Graph the line connecting each pair.
   (a) $(-1, 1)$; $(1, -1)$
   (b) $(7, 6)$; $(7, -2)$
   (c) $(-3, 4)$; $(2, 4)$
   (d) $(0, 1)$; $(5, 0)$

4. Find the distance between the following pairs of points:
   (a) $(4, 7)$ and $(0, 0)$
   (b) $(-5, -5)$ and $(5, 5)$
   (c) (a, b) and $(a + 2, b + 1)$
   (d) $(\sqrt{3}, 0)$ and $(0, \sqrt{6})$
   (e) (c, d) and $(3c, 3d)$
   (f) (c, d) and $(c + 3, d + 3)$

5. The equation of the circle with center at (a, b) and radius equal to r is given by $(x - a)^2 + (y - b)^2 = r^2$. Find the equation of the circle with center at $(1, -3)$ and radius equal to 5.

6. Prove that the points $(10, 5)$, $(15, 10)$, $(5, 10)$ and $(10, 15)$ are the vertices of a square.

7. Use the distance formula to show that the following points lie in a straight line: $(-2, 3)$, $(-6, 1)$, and $(-10, -1)$.

8. Show that the triangles whose vertices are given are isosceles:
   (a) $(2, 4)$, $(5, 1)$, and $(6, 5)$
   (b) $(2, 1)$, $(-1, 2)$ and $(2, 6)$
   (c) $(-2, 1)$, $(5, -2)$, and $(3, 3)$
   (d) $(a, 0)$, $(0, 0)$ and $(0, a)$

9. The ordinate of a point is 2 and its distance from $(7, 3)$ is $\sqrt{122}$. Find the two abscissa of the point.

10. Use the distance formula to show that the lengths of the diagonals of a square are equal. (Hint: Place one of the sides along the x-axis with the origin as an end point.)

SUMMARY

A. Real Number System

1. The set of **natural numbers,** $N = \{1, 2, 3, 4, \ldots\}$ used in counting is also called the set of **positive integers.**
2. The **negative integers,** $\{\ldots, -4, -3, -2, -1\}$, together with the positive integers and zero constitute the set of **integers.**
3. The set of **rational numbers,** Q, consists of all real numbers which can be written as the quotient of two integers, $\frac{p}{q}$, $q \neq 0$. The decimal representation of a rational number either terminates or is a repeating decimal.
4. **Irrational numbers** are numbers which are not rational, such as $\sqrt{2}$.
5. **Zero** is an unsigned number which allows us such operations as $7 - 7$. Thus, the **real number system** consists of the rational and irrational numbers.

B. The Real Number Line

The geometrical representation of the real numbers is achieved by representing real numbers as points on a line. First, we choose a point on the line to represent the number zero; this point is called the **origin.** The points to the *right* of the origin we associate with the positive real numbers while points to the *left* of the origin we associate with negative real numbers. This sets up a correspondence between each real number and point on the real number line.

The position of the real numbers on the real number line establishes an **order** to the real number system. The concepts of "greater than" and "less than" are used to express this ordering of the real numbers. If a point a lies to the right of another point b on the real number line we say that a is greater than b. The symbol for greater than is ">." If a point a lies to the left of another point b on the real number line we say that a is less than b. The symbol for less than is "<." In summary we have:

1. $a > b$ means "a is greater than b" or $a - b$ is positive.
2. $a < b$ means "a is less than b" or $a - b$ is negative.
3. $a \geq b$ means "a is greater than or equal to b."
4. $a \leq b$ means "a is less than or equal to b."
5. $a < x < b$ means "x is less than b but greater than a."

The above statements are called **inequalities.** Statement 5 is used to denote all real numbers that lie between a and b. A shorter notation for this statement is (a, b). Other **interval** notations are $[a, b]$, $[a, b)$, and $(a, b]$. **Infinite intervals** are of the form $(-\infty, +\infty)$, $(-\infty, b)$, $(-\infty, b]$, $(a, +\infty)$, $[a, +\infty)$.

C. Ordering Properties

1. If a and b are real numbers, then exactly one of the following is true: $a < b, a = b$, or $a > b$.
2. For a, b, and c real numbers:
   (i) If $a > b$ and $b > c$, then $a > c$.
   (ii) If $a < b$ and $b < c$, then $a < c$.
3. For a, b, and c real numbers:
   (i) If $a > b$, then $a + c > b + c$.
   (ii) If $a < b$, then $a + c < b + c$.
4. For a, b, and c real numbers:
   (i) If $a > b$ and $c > 0$, then $ac > bc$.
   (ii) If $a < b$ and $c > 0$, then $ac < bc$.
   (iii) If $a > b$ and $c < 0$, then $ac < bc$.
   (iv) If $a < b$ and $c < 0$, then $ac > bc$.

D. Absolute Value

$|x| = x$ if $x \geq 0$.

$|x| = -x$ if $x < 0$.

E. Cartesian Coordinate System

Two perpendicular number lines in a plane are called a **Cartesian coordinate system.** The number lines are called the **axes.** The intersection of these axes is called the **origin.** Thus, every point in the plane is represented by an ordered pair, called the **coordinates** of the point. If a point P has coordinates (x, y), then x, the first coordinate, is called the **abscissa** of P, and y, the second coordinate, is called the **ordinate** of P.

1. $(a, b) = (c, d)$ if and only if $a = c$ and $b = d$.
2. The distance between two points (x_1, y_1) and (x_2, y_2) is

$$d = \sqrt{(x_2 - x_1)^2 + (y_2 - y_1)^2}$$

CHAPTER 1 EXERCISES

1. Express the following fractions as non-terminating decimals:
   (a) $\dfrac{2}{7}$ (b) $\dfrac{3}{4}$

   (c) $\dfrac{12}{13}$ (d) $\dfrac{126}{110}$

2. Convert each of the following decimals to a fraction:
   (a) 0.7 (b) $.33\overline{33}$
   (c) $1.434\overline{343}$ (d) $7.263263\overline{263}$

3. Insert one of the symbols $<$, $\leq$, $>$, or $\geq$ into each square which makes the statement true.
   (a) $-2 \ \square \ -2.5$
   (b) $|-9| \ \square \ |3|$
   (c) $0 \ \square \ x^2$
   (d) $a < b$ if and only if $-b \ \square \ -a$

   (e) $\left(\dfrac{1}{2}\right)^2 \ \square \ \left(\dfrac{1}{2}\right)^3$

   (f) $-(3x^2 + 1) \ \square \ 0$

4. Solve and graph the following inequalities:
   (a) $2x - 5 > 7x + 1$

   (b) $\dfrac{2x}{5} + \dfrac{1}{5} < \dfrac{3}{10} + \dfrac{x}{10}$

   (c) $4x + 9 > 2x + 5$
   (d) $-4 < 5x - 6 \leq 14$

5. Solve for values of x and graph:
   (a) $x^2 + 4 < 4x$
   (b) $(x + 2)(x - 1)(x - 3) \geq 0$
   (c) $x^2 + 1 > 0$
   (d) $x^2 \geq 2x - 1$

6. Determine which of the following are true or false:
   (a) $|-7| = |9|$
   (b) $|-7 - 3| = |-7| + |-3|$
   (c) $|(3)(-2)| = |3| \cdot |-2|$
   (d) $|y - 4|^2 = (y - 4)^2$

7. Solve for y:
   (a) $|3y + 7| < 5$
   (b) $|2y - 9| > 7$

8. (a) $-9 < x < 9$ if and only if $|?| < ?$
   (b) $-7 \leq x \leq 4$ if and only if $|?| \leq ?$
   (c) $-10 < x < -8$ if and only if $|?| < ?$

9. Find the distance between the points
   (a) $(-6, 1)$ and $(3, 4)$
   (b) $(7, 5)$ and $(3, 2)$
   (c) $(-11, 3)$ and $(-8, -2)$
   (d) $(0, 0)$ and $(-1, -7)$
   (e) $(-2, 3)$ and $(1, 5)$
   (f) $(-3, 10)$ and $(1, 4)$

10. Show that the points $(5, 3)$, $(6, 2)$, and $(3, -1)$ lie on a circle whose center is $(4, 1)$.

CHAPTER 2

FUNCTIONS

INTRODUCTION

In this chapter we introduce the idea of a function and the properties of functions. To achieve this we introduce a fundamental concept in all branches of mathematics, the idea of a set. We define several operations on sets and show how we use this information to define a function. Once the set-definition of a function has been developed we shall give an equivalent definition of a function using what is called "functional notation."

We conclude the chapter describing operations on functions, types of functions, and inverse functions. Throughout we make extensive use of graphs and illustrations to give the student some geometric insight of functions and their properties.

2.1 SET AND SET OPERATIONS

The idea of a **set** is very basic in many branches of mathematics, and is itself a highly complex idea within mathematics. However, we shall not develop a rigorous definition of a set. Intuitively a set is any well-defined collection of objects. The objects characterize the set and these objects can be anything—students, teachers, cars, letters, or money—or more abstract objects such as points on the line, numbers, lines in the plane. These objects are called the **elements** or **members** of the set, and are denoted by {. . .}.

EXAMPLE 1

$A = \{1, 2, 3, 4, 5, 6, 7, 8, 9\}$

EXAMPLE 2

$B = \{$all the students reading this book$\}$

EXAMPLE 3

$C = \{$the positive integers$\}$

EXAMPLE 4

   $D = \{$the state capitals$\}$

EXAMPLE 5

   $E = \{$all the real numbers between 0 and 1$\}$

EXAMPLE 6

   $F = \{2, 4, 6, 8, 10, \ldots\}$

EXAMPLE 7

   $G = \{$all rational numbers$\}$

Sets will usually be denoted by capital letters $A, B, C, X, W, \ldots$. In Examples 1 and 6 we characterized the sets by listing its members while the other examples characterized the sets by listing those properties which describe the elements of the set. However, a shorter notation, called the **set builder notation,** can be very useful in characterizing sets. For example, the set A given above could be $A = \{x|x$ is an integer and $1 \le x \le 9\}$, which reads "A is the set of elements x such that x is an integer and x is greater than or equal to 1 and less than or equal to 9." Note that the vertical line "|" is read "such that." We now can rewrite the sets in Examples 2 through 7 using this notation.

EXAMPLE 2

   $B = \{x|x$ is a student and that student is reading this book$\}$

EXAMPLE 3

   $C = \{x|x$ is an integer and $x > 0\}$

EXAMPLE 4

   $D = \{x|x$ is a state capital$\}$

EXAMPLE 5

   $E = \{x|x$ is a real number and $0 < x < 1\}$

EXAMPLE 6

   $F = \{x|x$ is an even positive integer$\}$

EXAMPLE 7

   $G = \{x|x$ can be written as $\dfrac{p}{q}$ and p, q are integers, $q \ne 0\}$

Definition: If A is a given set, then "x is a member of A" or "x belongs to A" is written

$$x \in A$$

while the statement "x does not belong to A" is written

$$x \notin A$$

Graphically:

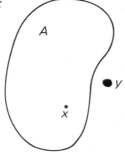

$x \in A$ while $y \notin A$

EXAMPLE 1

Let $F = \{x \mid x$ is an even positive integer$\}$. Then $4 \in F$, but $3 \in F$.

EXAMPLE 2

Let $D = \{x \mid x$ is a state capital$\}$. Then Chicago $\notin D$ and Albany $\in D$.

Definition: A set is said to be **finite** if it is possible to list all the elements of the set. A set which is neither empty nor finite is an **infinite** set.

EXAMPLE 1

The set $A = \{x \mid x$ is an integer and $1 < x < 9\}$ is finite.

EXAMPLE 2

The set $B = \{x \mid x$ is a student and that student is reading this book$\}$ is finite.

EXAMPLE 3

The set $E = \{x \mid x$ is a real number and $0 < x < 1\}$ is infinite.

EXAMPLE 4

The set $G = \{x \mid x$ is a rational number$\}$ is infinite.

EXAMPLE 5

The set $R = \{x|x \text{ is a real number}\}$ is infinite.

Definition: The sets A and B are **equal,** written

$$A = B$$

when they have exactly the same elements. If A and B are not equal, then we write

$$A \neq B$$

EXAMPLE

Let $A = \{1, 4, 3, 3, -5\}$, $B = \{1, 0, 4, -5\}$, and $C = \{3, 4, -5, 1\}$. Then we have $A = C$, but $A \neq B$. Note that a set does not change if its elements are repeated.

Definition: A special set that plays an important part in set theory is the **empty set,** sometimes called **null set,** which contains no elements. The notation used to denote the empty set is the symbol $\emptyset$.

EXAMPLE

Let $A = \{x|x \text{ is real and } x^2 = -10\}$. Then $A = \emptyset$.

Definition: If every element in a set A is also an element of set B, then A is called a **subset** of B. We denote this by $A \subseteq B$ and read "A is contained in B," or equivalently, "B contains A." Similarly, $A \not\subseteq B$ means A is not a subset of B. Using a shorter notation we have

$$A \subseteq B \text{ if } x \in A \text{ implies } x \in B$$

Graphically the idea of subset is illustrated in Figure 13.

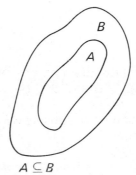

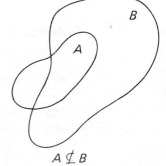

Figure 13 $A \subseteq B$ $A \not\subseteq B$

EXAMPLE 1

Let $A = \{1, 2, 3, 4, 5\}$, $B = \{4, 2, 2, 1\}$, and $C = \{5, 0, 2, 3\}$. Then $B \subseteq A$ and $C \not\subseteq A$.

EXAMPLE 2

Let *A, B,* and *C* be sets represented by **Venn diagrams** in Figure 14. Then $B \subseteq A$ and $C \subseteq A$, but $C \not\subseteq B$. $C \not\subseteq B$ can be seen because there is an $x \in C$ but $x \notin B$.

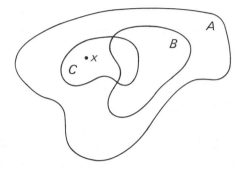

Figure 14

EXAMPLE 3

Let *A* be any set. Then $A \subseteq A$. This follows trivially from the definition of subset. For if $x \in A$, then this implies that $x \in A$ and thus $A \subseteq A$.

EXAMPLE 4

The null set $\varnothing$ is considered to be a subset of every set. That is, if *A* is any set, then $\varnothing \subseteq A$.

EXAMPLE 5

Let $A = \{x | x$ is an irrational number$\}$ and $R = \{x | x$ is a real number$\}$. Then $A \subseteq R$.

EXAMPLE 6

Let $B = \{x | x$ is a real number and $x^2 - x - 2 = 0\}$ and $C = \{-1, 2, 3\}$ and $D = \{-1, 2\}$. Then $C \not\subseteq B$ and $D \subseteq B$.

Using the definition and notation of subsets we can now give an equivalent definition for the equality of sets.

Definition: Two sets A and B are **equal,** written

$$A = B$$

if and only if $A \subseteq B$ and $B \subseteq A$.

EXAMPLE

Let $A = \{x | x$ is an even positive integer$\}$ and $B = \{2, 4, 6, 8, 10, \ldots\}$. Then $A \subseteq B$ and $B \subseteq A$ and thus $A = B$.

Definition: A set B is called a **proper subset** of a set A if B is a subset of A and B is not equal to A. When B is a proper subset of A, we shall write it $B \subset A$.

EXAMPLE 1

Let $A = \{a, b, c, d\}$. Then $B = \{a\}$, $C = \{c, d\}$, and $D = \{d, b, c\}$ are proper subsets of A. The set $\{b, a, d, c\}$ is a subset of A, but not a proper subset.

EXAMPLE 2

Let $I = \{x | x$ is a real number and $0 \leq x \leq 10\}$, $G = \{1, 2, 3, 4, 5, 6, 7, 8, 9, 10\}$, $K = \varnothing$, $L = \{-1, 7, 8, 2, 3\}$. Then $G \subset I$ and $K \subset I$, but L is neither a proper subset nor a subset of I. This can be seen by noticing that there is an element in L, the number -1, which does not belong to I.

PROBLEM

Does every set have a proper subset?

Solution: No. The null set $\varnothing$ does not have a proper subset. It does have itself as a subset; however, every other set has at least the null set $\varnothing$ as a proper subset.
(Some mathematicians do not consider the null set as a proper subset. If we adopted this convention, then sets which contain only one element would not contain a proper subset.)

2.1 EXERCISES (A)

1. List the elements of the following described sets:
   (a) {the natural numbers between -5 and 6}
   (b) {the real numbers satisfying $4x + 1 = -7$}
   (c) {the integers between 24 and 30}
   (d) {the odd positive integers}
   (e) {the real numbers satisfying $x^2 - x - 2 = 0$}

2. Using the **set builder notation,** represent the following sets:
   (a) {$-4, -3, -2, -1, 0, 1, 2, 3, 4, 5$}
   (b) {$-2, -4, -6, -8, -10, \ldots$}
   (c) {$1, \dfrac{1}{2}, \dfrac{1}{3}, \dfrac{1}{4}, \dfrac{1}{5}, \dfrac{1}{6}, \ldots$}
   (d) {set of all real numbers greater than 3 and less than 11}

3. Let $C = \{(x, y) | x^2 + y^2 \leq 25\}$
   $R = \{(x, y) | (x - 1)^2 + y^2 \leq 4\}$
   $T = \{(x, y) | (x - 2)^2 + (y + 3) \leq 9\}$
   Are R and T subsets of C?

4. List all the possible subsets of {1, 2, 3}.

5. What can be said about the following set?
   $A = \{(x, y) | x, y \text{ are real}$
   $\text{and } x^2 + y^2 = -5\}$

6. If B is a subset of the null set $\varnothing$, then what can be said about B?

7. Let $A = \{1\}$, $B = \{2, 1\}$, $C = \{3, 4, 2\}$, $D = \{3, 4\}$, and $E = \{3, 4, 1\}$. Determine if each of the following statements is true or false:
   (a) $B \subset D$ (b) $A \subset C$
   (c) $E \neq B$ (d) $A \subset E$
   (e) $A \not\subset B$ (f) $C = B$
   (g) $D \subset C$ (h) $C \not\subset E$

8. List all the proper subsets of {e, f, g}.

9. Let $B = \{1\}$. Which of the following relations are true?
   (a) $B \subset \{1\}$ (b) $1 \subset B$
   (c) $\{1\} \subset B$ (d) $\{1\} \in B$
   (e) $1 \in B$

10. Using the definitions of equality of sets, show that the following statements are true for sets A, B, and C:
    (a) $A = A$ (equality is reflexive).
    (b) If $A = B$, then $B = A$ (equality is symmetric).
    (c) If $A = B$ and $B = C$, then $A = C$ (equality is transitive).

Set Operations

Just as we had operations in arithmetic, we shall define operations on sets. These operations are the **union, intersection,** and **complement** of sets.

> **Definition:** The set consisting of the totality of elements under consideration in a particular discussion is called the **universal set** and is denoted by U.

EXAMPLE 1

When discussing sets of numbers, the universal set is taken to be all the real numbers. That is, U = {real numbers}.

EXAMPLE 2

When talking about students, we take

$$U = \{\text{all students}\}.$$

Definition: The set of all elements which belong to a given universal set, but which do not belong to a given subset A of U, is called the **complement** of A, written A'. That is,

$$A' = \{x | x \in U \text{ and } x \notin A\}$$

Graphically:

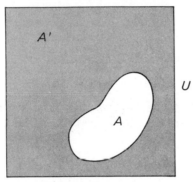

A' is shaded area
Figure 15

EXAMPLE 1

Let U = {1, 2, 3, 4, 5} and A = {1, 4, 5}. Then A' = {2, 3}.

EXAMPLE 2

Let U = {real numbers} and B = {$x | x$ is real and $x > 2$}. Then B' = {$x | x$ is real and $x \leq 2$}.

Definition: The collection of all elements which belong to A or B or to both sets is called the **union** of A and B, written $A \cup B$. That is,

$$A \cup B = \{x | x \in A \text{ or } x \in B \text{ or } x \in A \text{ and } B\}.$$

Graphically:

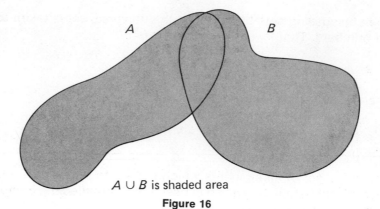

$A \cup B$ is shaded area

Figure 16

EXAMPLE 1

Let $A = \{1, 2, 5\}$ and $B = \{2, 5, 7, -3\}$. Then $A \cup B = \{1, 2, 5, 7, -3\}$.

EXAMPLE 2

Let $C = \{a, e, g\}$ and $D = \{b, c, d, f\}$. Then $C \cup B = \{a, b, c, d, e, f, g\}$.

Definition: The collection of all elements which belong to both A and B is called the **intersection** of the sets A and B, written $A \cap B$. That is,

$$A \cap B = \{x \mid x \in A \text{ and } x \in B\}.$$

Graphically:

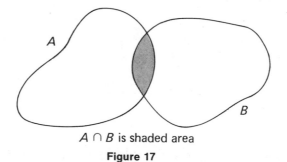

$A \cap B$ is shaded area

Figure 17

EXAMPLE 1

Let $A = \{1, 2, 5\}$ and $B = \{2, 7, 5, -3\}$. Then

$$A \cap B = \{2, 5\}.$$

EXAMPLE 2

Let $C = \{a, e, g\}$ and $D = \{b, c, d, f\}$. Then

$$C \cap D = \emptyset.$$

Definition: Two sets which have no common element are called **disjoint.** Using the idea of intersection, we can say that two sets A and B are disjoint if and only if $A \cap B = \emptyset$.

Graphically:

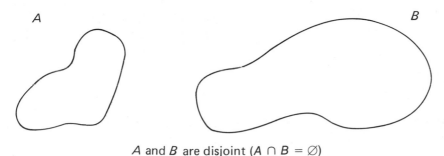

A and B are disjoint ($A \cap B = \emptyset$)

Figure 18

EXAMPLE 1

The sets $A = \{-1, 0, 1\}$ and $B = \{2, 3, -4\}$ are disjoint. This follows from the fact that $A \cap B = \emptyset$.

Properties of Set Operations

Let A, B, and C be any sets and U the universal set. Then

1. $A \cup A' = U$
2. $A \cap A' = \emptyset$
3. $U' = \emptyset$ and $\emptyset' = U$
4. $(A')' = A$
5. $(A \cap B) \subseteq A$ and $(A \cap B) \subseteq B$
6. $A \cap B = B \cap A$
7. $A \cup B = B \cup A$
8. $A \subseteq (A \cup B)$ and $B \subseteq (A \cup B)$

2.1 EXERCISES (B)

1. Let $U = \{a, b, c, d, e, f, g, h\}$, $A = \{a, b, c\}$, $B = \{d, c, g\}$, $C = \{a, e, f, h\}$, and $D = \{a, b, c, d, e, f, g, h\}$. Find A', B', C', and D'.

2. Using the sets in Exercise 1 find $(A \cap C)'$, $(A \cup C)'$, and $(D' \cap B')'$.

3. Prove set operation properties 1 through 8 using Venn diagrams.

4. Prove $A \cap A = A$ and $A \cup A = A$.

5. Prove $A \cap \emptyset = \emptyset$ and $A \cup \emptyset = A$.

6. Graph the following sets of real numbers on a number line:
   (a) $\{x|-2 < x < 1\} \cap \{x|0 < x < 3\}$
   (b) $\{x|4 \leq x \leq 10\} \cup \{x|-1 \leq x \leq 7\}$
   (c) $\{x|-7 \leq x \leq -2\} \cap \{x|-2 \leq x \leq 0\}$
   (d) $\{x|x \geq 5\} \cup \{x|x \leq 5\}$

7. If $C \subset D$, then what can be said about C' and D'? Use Venn diagrams.

8. Prove that if $A \subseteq B$ and $B \subseteq C$, then $A \subseteq C$ **(Transitive Law)**.

9. Let $U = \{1, 0, -3, 5\}$. If $C = \{1, 0\}$, $C \cap D = \{1\}$, and $C \cup D = U$, find the set D.

10. Let U be the universal set and A, B, C subsets of U. Use Venn diagrams to show that the following equations are true.
    (a) $A \cap (B \cup C) = (A \cap B) \cup (A \cap C)$
    (b) $(A \cap B)' = A' \cup B'$
    (c) $(A \cap B) \cup (A \cap B') = A$
    (d) $A \cup (B \cup C) = (A \cup B) \cup C$

2.2 RELATIONS

Recall that in Chapter 1 we showed that each point of the plane could be represented by an ordered pair (x, y). We can think of the plane in another way by the use of the cross product or Cartesian product. Let the real number line be represented by R. Then the plane is the set $\{(x, y)|x \in R \text{ and } y \in R\}$. This set is denoted by $R \times R$, which is read "R cross R."

Definition: The **cross product,** or **Cartesian product,** $A \times B$ (read "A cross B") of two sets A and B is the set of *all* ordered pairs (x, y) whose first components belong to A and whose second components belong to B. That is,

$$A \times B = \{(x, y)|x \in A \text{ and } y \in B\}$$

EXAMPLE 1

Let $A = \{1, 2, 3\}$ and $B = \{1, 2\}$. Then

$$A \times B = \{(1, 1), (1, 2), (2, 1), (2, 2), (3, 1), (3, 2)\}$$
$$B \times A = \{(1, 1), (1, 2), (1, 3), (2, 1), (2, 2), (2, 3)\}$$

Note that $A \times B \neq B \times A$.

EXAMPLE 2

Let $C = \{-2, 0, 3\}$. Then

$$C \times C = \{(-2, -2), (-2, 0), (-2, 3), (0, -2), (0, 0), (0, 3), (3, -2),$$
$$(3, 0), (3, 3)\}$$

EXAMPLE 3

Let $A = \{x|x$ is a real number and $0 \leq x \leq 1\}$ and $B = \{y|0 \leq y \leq 2\}$.

$$A \times B = \{(x, y)|0 \leq x \leq 1 \text{ and } 0 \leq y \leq 2\}$$

Graphically:

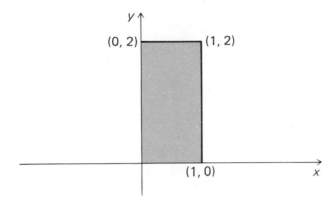

Definition: A **relation** is a set of ordered pairs. The set of first elements of the ordered pairs is called the **domain** of the relation, and the set of second elements is called the **range** of the relation.

EXAMPLE 1

Any subset of $R \times R$ is a relation on R.

EXAMPLE 2

The set of ordered pairs $\{(1, 2), (5, 5), (1, 3), (-3, 5), (0, 4), (-1, -1)\}$ defines a relation. The set $\{1, 5, -3, 0, -1\}$ is the domain and the set $\{2, 5, 3, 4, -1\}$ is the range. The graph of the relation is shown in Figure 19.

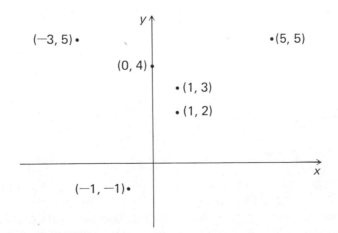

Figure 19

PROBLEM 1

Let $B = \{-1, 0, 1\}$. Does "equality" define a relation on B?

Solution: $B \times B = \{(-1, -1), (-1, 0), (-1, 1), (0, -1), (0, 0), (0, 1),$
$(1, -1), (1, 0), (1, 1)\}$

Choosing those ordered pairs whose components are equal, we have $C = \{(-1, -1), (0, 0), (1, 1)\}$. Also, $C \subseteq B \times B$, and therefore "equality" does define a relation on B.

PROBLEM 2

Plot the graph of the following relation on R: $\{(x, y)|x^2 + y^2 \leq 4\}$ (see Fig. 20). Note that the graph is a subset of $R \times R$.

Solution:

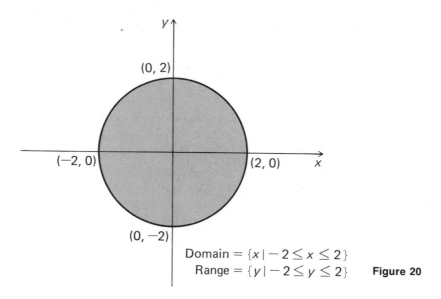

Domain $= \{x|-2 \leq x \leq 2\}$
Range $= \{y|-2 \leq y \leq 2\}$ **Figure 20**

PROBLEM 3

Graph $y = x^2$.

Solution: $y = x^2$ is a relation on R, R being the real numbers. The relation is denoted by the set $\{(x, y)|y = x^2\}$. The domain is R, and the range is the set of nonnegative real numbers. Therefore the graph is represented by a curve in the region above and on the x-axis (see Fig. 21).

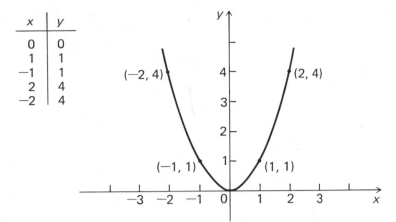

Figure 21

PROBLEM 4

Graph the relation "<" on R.

Solution: This relation is the set $L = \{(x, y)|x < y\}$. Geometrically the set L is that portion of the plane lying above the line $x = y$ (see Fig. 22). For example, the point $(3, 6)$ is in the set L, while $(5, 2)$ is not (see Fig. 22).

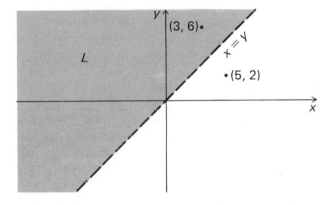

Figure 22

2.2 EXERCISES

Find the Cartesian product $A \times B$ and $B \times A$ in Exercises 1 to 5. List the domain and range of $A \times B$ and $B \times A$ in each case.

1. $A = \{1, 2, 3\}$; $B = \{0, 1\}$
2. $A = \{a, b, c\}$; $B = \{x, y\}$
3. $A = \{5\}$; $B = \{-1, -2\}$
4. $A = \left\{-\dfrac{2}{9}, 0, 4, 2\right\}$; $B = \{a, b, c, d, e\}$

5. $A = \{a, b, c, d, e\}$; $B = \{a, b, c, d, e\}$

6. Which of the following sets are relations?
   (a) $\{(1, 7), (-1, 6), (4, 3), (-7, -5)\}$
   (b) $\{1, 7, -5\}$
   (c) $\{(x, y), (y, x), (w, s), (t, v)\}$
   (d) $\{(9, -13)\}$

In Exercises 7 to 12, consider the set $A = \{-2, 1, 0, 4, 6\}$.

7. Find the set of ordered pairs determined by the relation $\geq$ (greater than or equal to).

8. Find the set of ordered pairs determined by the relation $\leq$ (less than or equal to).

9. Find the set of ordered pairs determined by the relation $=$.

10. Find the set of ordered pairs determined by the relation $>$ (greater than).

11. Find the set of ordered pairs determined by the relation $\neq$.

12. Find the set of ordered pairs determined by the relation $<$ (less than).

In Exercises 13 to 16 the relations are subsets of the Cartesian plane, $R \times R$. Graph these relations.

13. $\{(x, y) | x = -2\}$

14. $\{(x, y) | x^2 + y^2 > 1\}$.

15. $\{(x, y) | y = |x|\}$

16. $\{(x, y) | x \leq 0 \text{ and } y \geq 0\}$

17. State the domain of the following relations and graph.
    (a) $y = x + 1$ (b) $y = x^2$
    (c) $y > 12x - 2$ (d) $y = |x - 3|$
    (e) $y = 7$ (f) $x^2 + y^2 = 4$
    (g) $y = \sqrt{x - 3}$
    (h) $y = \dfrac{1}{x - 3}$

18. Give an example of a relation which illustrates that it is possible for two different ordered pairs of a relation to have the same first element.

19. If $A = \{4\}$ and $B = R$ (reals), describe the graph of $A \times B$.

20. Find the domain and range of (a) $y = \sqrt{x^2 - 9}$ and (b) $y = \sqrt{x^2 + 4}$.

2.3 FUNCTIONS

We are now interested in special types of relations called **functions**.

Definition: A **function** is a relation in which no two ordered pairs have the same first element and different second elements. The set of all first elements of the ordered pairs is called the **domain** of the function. The set of all second elements is called the **range** of the function.

EXAMPLE 1

Let $A = \{(1, 2), (-1, 3), (0, 0), (6, 3)\}$ and $B = \{(4, 6), (-3, 0), (4, -5), (2, -1)\}$. Then A is a function, while B is not a function. Note that the two ordered pairs (4, 6) and (4, −5) in *B* have the same first element but different second elements.

EXAMPLE 2

Consider the equation $y = x^2$. This represents a function because for any value chosen for *x*, there is one and only one corresponding value for *y*. Graphically the equation $y = x^2$ is a parabola (see Fig. 21).

EXAMPLE 3

The equation $y^2 = x$ *does not* represent a function. Choosing $x = 9$, we see that $y = 3$ or $y = -3$. Thus, the ordered pairs (9, −3) and (9, 3) satisfy the equation and have the same first element. Graphing $y^2 = x$ we obtain the following curve (Fig. 23):

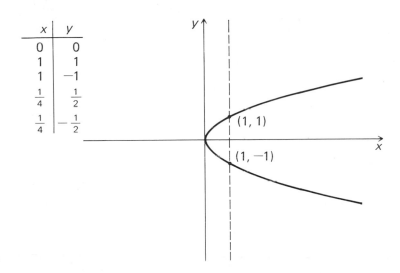

Figure 23

Graphically, how can one tell whether the graph of a relation represents a function? *If a vertical line intersects the graph at more than one point, then the graph does not represent a function.* The graph in Figure 21 represents a function while the graphs in Figures 22 and 23 do not represent functions.
The following graphs represent functions:

(i)

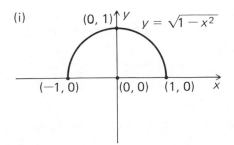

(ii)

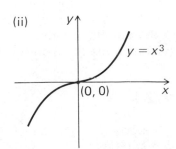

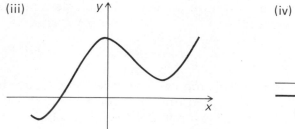 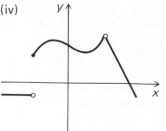

Functions

The following graphs do not represent functions:

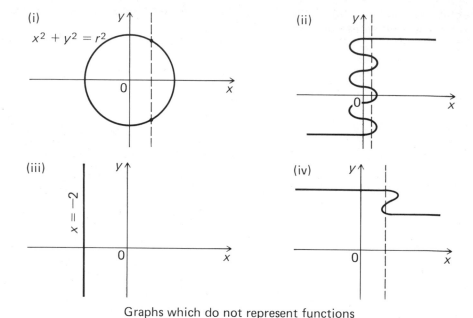

Graphs which do not represent functions

Functional Notation

Functions can be thought of as mappings. A function *maps* members of the domain to corresponding members of the range. In general, if (x, y) is a member of the function f, then f associates x in the domain with y in the range and we say that f maps x to y. The functional notation used is $y = f(x)$, which is read "y equals f of x." When using this notation y is called the **dependent variable** and x the **independent variable.** The graph of the function f is the set of points $(x, f(x))$ for which the function is defined.

EXAMPLE 1

Suppose $f = \{(1, 3), (-1, 1), (5, 6), (7, 1), (4, 4)\}$. Figure 24 illustrates f as a mapping:

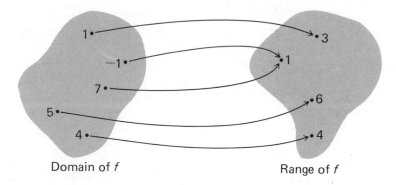

Figure 24   Domain of *f*   Range of *f*

For this function *f*, 1 is mapped into 3. We write this as $f(1) = 3$. We note that *f* maps -1 into 1 and also maps 7 into 1—that is, $f(-1) = 1$ and $f(7) = 1$. *Recall that the definition of a function allows more than one elements in the domain to be mapped to the same element in the range. It does not allow an element in the domain to have more than one image in the range* (see Fig. 25).

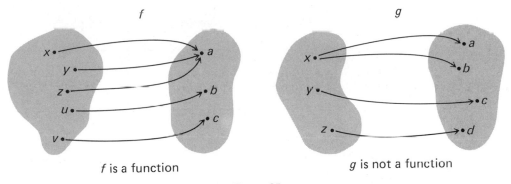

f is a function   *g* is not a function

Figure 25

EXAMPLE 2

Let $g(x) - 3x - 2$. Evaluating *g* at various values gives:

$$g(0) = 3(0) - 2 = -2$$

$$g(-1) = 3(-1) - 2 = -3 - 2 = -5$$

$$g(h) = 3(h) - 2 = 3h - 2$$

$$g\left(\frac{2}{7}\right) = 3\left(\frac{2}{7}\right) - 2 = \frac{6}{7} - 2 = -\frac{8}{7}$$

$$g(x + h) = 3(x + h) - 2 = 3x + 3h - 2$$

EXAMPLE 3

Consider the equation $x^2 + y^2 = 4$. Graphically this represents a circle of radius equal to 2 and center at the origin. Solving this equation for *y* we get

$$y = \pm\sqrt{4 - x^2}$$

For any value assigned to x, where $-2 < x < 2$, we get two values assigned to y. Thus, the equation of a circle does not represent a function of x. Note that this was also shown using the "vertical line" test. A vertical line can intersect the graph of the circle in more than one point.

EXAMPLE 4

Let $f(x) = x^2 + 3$. Then
(a) $f(0) = 0^2 + 3 = 3$
(b) $f(a) = a^2 + 3$
(c) $f(c - d) = (c - d)^2 + 3$
(d) $\dfrac{f(x + h) - f(x)}{h} = \dfrac{(x + h)^2 + 3 - (x^2 + 3)}{h}$

$$= \frac{x^2 + 2xh + h^2 - x^2 + 3 - 3}{h}$$

$$= \frac{2xh + h^2}{h}$$

$$= 2x + h \text{ provided } h \neq 0$$

Definition: The **domain,** D_f, of a function f is the set of values of x for which f is defined. The **range,** R_f, of f is the set of values $f(x)$.

PROBLEM 1

Find the domain and range of the function defined by $f(x) = 2x - 7$.

Solution: $D_f = R$(all the real numbers). Also, $R_f = R$.

PROBLEM 2

Find the domain and range of the function defined by

$$g(x) = \sqrt{x - 1}$$

Solution: A square root of a negative number is not a real number. This is true because the square of any real number is nonnegative. Thus, the domain of g consists of the set of all numbers such that $x - 1$ is nonnegative—that is,

$$x - 1 \geq 0$$

or
$$x \geq 1$$

The range is the set of nonnegative reals (see Fig. 26).

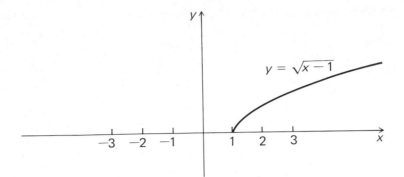

Figure 26

PROBLEM 3

Find the domain and range of the function defined by

$$h(x) = \frac{(x - 4)(x + 2)}{(x - 3)}$$

Solution: The function h is defined for all real numbers except for the value $x = 3$. At $x = 3$ the denominator becomes zero and h is undefined. It can be shown that the range is the set of real numbers.

PROBLEM 4

Find the domain of $f(x) = \dfrac{\sqrt{x^2 - 4}}{x^2 - 3x - 10}$.

Solution: The domain of f consists of all values for which the expression $(x^2 - 4)$ is nonnegative and the denominator is not zero. That is, $x^2 - 4 \geq 0$ and $x^2 - 3x - 10 \neq 0$. Using methods in Chapter 1, we find that $x^2 - 4 \geq 0$ if x belongs to the set $A = (-\infty, -2] \cup [2, +\infty)$. The expression $x^2 - 3x - 10$ is zero at the points 5 and -2. These points must be eliminated from the set A. So the domain of f is the set

$$D_f = \{x \mid x < -2 \text{ or } x \geq 2 \text{ and } x \neq 5\}$$

2.3 EXERCISES (A)

In Exercises 1 to 5 determine which relations are functions:

1. $f = \{(0, \sqrt{3}), (-1, 2), (6, \sqrt{3}), (7, 0)\}$

2. $g = \left\{ \left(\frac{1}{2}, 2\right), \left(-\frac{3}{5}, \frac{4}{3}\right), \left(\frac{1}{2}, 1\right), (7, -9) \right\}$

3. $h = \left\{ (4, 7), (6, 7) (-1, 7), \left(-\frac{1}{3}, 7\right), (7, 7) \right\}$

4. $p = \{(-1, 3), (2, 12), (0, 12), (-6, -4)\}$

5. $q = \{(\sqrt{3}, -7), (\sqrt{3}, -7), (1, 6), (5, -2)\}$

In Exercises 6 to 15, find the indicated values of the given functions:

6. $f(x) = 2x + 1$ $f(0);$ $f(-5);$ $f(a);$ $f(a + b)$

7. $g(x) = x^2$ $g(-2);$ $g(2);$ $g(b);$ $g(a - b)$

8. $h(x) = |x| - 1$ $h(0);$ $h(-4);$ $h(4);$ $h(a + b)$

9. $p(x) = \sqrt{x^2 - 9}$ $p(3);$ $p(-3);$ $p(a);$ $p(x + 1)$

10. $q(x) = 2x - \dfrac{1}{x}$ $q(1);$ $q\left(\dfrac{1}{2}\right);$ $q(y);$ $q(a + b)$

11. $s(x) = |x| - x$ $s(0);$ $s(1);$ $s(-1);$ $s\left(-\dfrac{1}{3}\right)$

12. $u(x) = \dfrac{2x - 1}{x}$ $u\left(\dfrac{1}{2}\right);$ $u(a - b);$ $u(y^2);$ $u(-1)$

13. $v(x) = \left|\dfrac{x - 7}{x + 1}\right|$ $v(0);$ $v(7);$ $v(1);$ $v(t)$

14. $w(x) = x^2 + 3x - 7$ $w(0);$ $w(2);$ $w(-4);$ $w(a)$

15. $y(x) = x^2(x - 2)$ $y(0);$ $y(1);$ $y(a);$ $y(a - b)$

In Exercises 16 to 20, form the difference quotient $\dfrac{f(x + h) - f(x)}{h}$, $h \neq 0$, and simplify the expression for the given functions:

16. $f(x) = x$ 17. $f(x) = x^2 - 1$ 18. $f(x) = 3$

19. $f(x) = \dfrac{1}{x}$ 20. $f(x) = 2x - .5$

Indicate which of the following graphs of relations represent functions of x:

21.

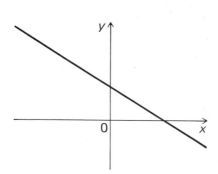

22.

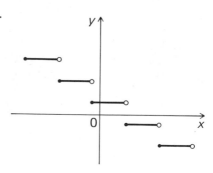

23.

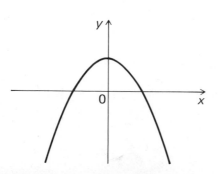

24.

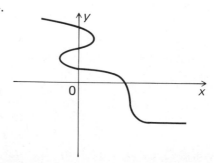

In Exercises 25 to 30, find the domain of the given function:

25. $f(x) = |x|$

26. $g(x) = \dfrac{x}{x - 2}$

27. $h(x) = 7$

28. $p(x) = 3x + \dfrac{4}{x}$

29. $q(x) = \sqrt{x^2 - 2x - 15}$

30. $f(x) = \dfrac{\sqrt{3x - 4}}{x^3 - 9x}$

31. Express the area of a circle of radius r as a function of the radius.

32. Express the Fahrenheit temperature as a function of the Celsius temperature.

Graphs of Functions

Often the graph of two functions have the same shape or orientation, the only difference between them being that one is a parallel displacement of the other. Any such parallel displacement of one graph to another is called a **transformation.** In this section we discuss how such transformations occur.

---

1. **Vertical Translations:** If c is any real number and f is any function, the sum $f + c$ is the function defined by $f(x) + c$. The graph of $f + c$ is the graph of f *shifted* $|c|$ units—upward if $c > 0$ and downward if $c < 0$.

EXAMPLE 1

Consider the function $y = x^2 + 2$. To graph this function we compare it to the function $y = x^2$. The graph of $y = x^2 + 2$ has the same shape as the graph of $y = x^2$, but it is shifted upward $|2|$ units (see Fig. 27).

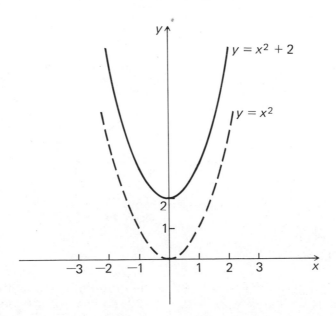

Figure 27

EXAMPLE 2

The graph of $y = |x| - 1$ is the graph of $y = |x|$ shifted downward $|1|$ units (see Fig. 28).

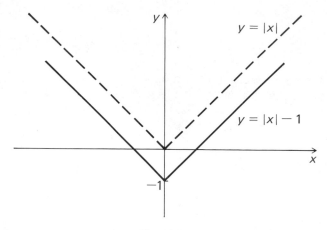

Figure 28

---

2. **Horizontal Translations:** If c is a real number and f is any function, then the function f_c defined by $f_c = f(x - c)$ represents a horizontal shift. The graph of f_c is the graph of f translated $|c|$ units—to the right if $c > 0$, and to the left if $c < 0$.

EXAMPLE 3

The graph of $y = |x - 3|$ is the graph of $y = |x|$ shifted to the right $|3|$ units (see Fig. 29).

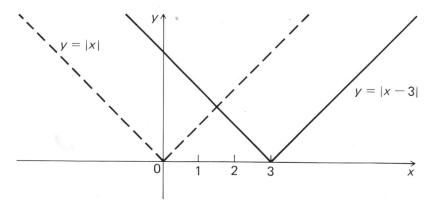

Figure 29

EXAMPLE 4

The graph of $y = (x + 2)^2$ is the graph of $y = x^2$ shifted to the left $|2|$ units. Note that $(x + 2)^2 = (x - (-2))^2$ so $c = -2 < 0$ (see Fig. 30).

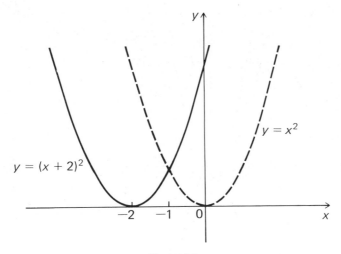

Figure 30

3. **Vertical Stretchings and Contractions:** If f is any function and c is a real number, then the graph of the function cf defined by $cf(x)$ is:
   (i) the graph of f stretched vertically if $|c| > 1$.
   (ii) the graph of f contracted vertically if $|c| < 1$.
   (iii) if c is negative, the graph is *also* reflected about the x-axis in addition to applying (i) or (ii).

EXAMPLE 5

The graph of $y = \frac{1}{2}|x - 3|$ is the graph of $y = |x - 3|$ contracted vertically (see Fig. 31).

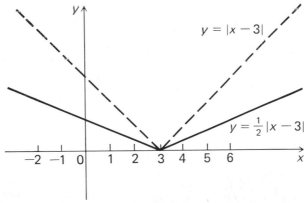

Figure 31

EXAMPLE 6

The graph of $y = -2x^2$ is the graph of $y = x^2$ stretched vertically and reflected about the x-axis (see Fig. 32).

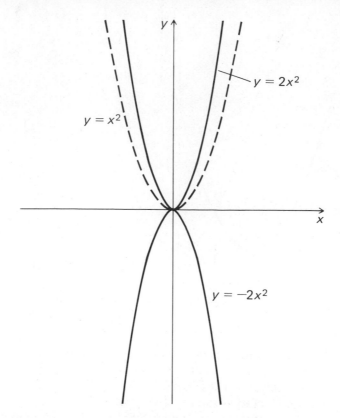

$y = 2x^2$

$y = x^2$

$y = -2x^2$

Figure 32

4. **Horizontal Stretchings and Contractions:** If f is any function and c is a real number, then the graph of the function defined by $f(cx)$ is
   (i) the graph of f contracted horizontally if $|c| > 1$.
   (ii) the graph of f stretched horizontally if $|c| < 1$.
   (iii) if c is negative, the graph is *also* reflected about the y-axis in addition to applying (i) or (ii).

EXAMPLE 7

If $f(x) = |x - 3|$, then $f(2x) = |2x - 3|$ is the graph of f contracted horizontally, and translated horizontally (see Fig. 33).

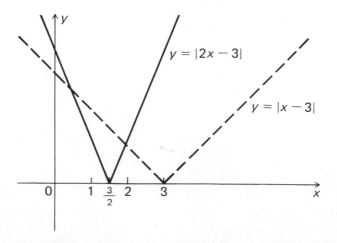

$y = |2x - 3|$

$y = |x - 3|$

Figure 33

EXAMPLE 8

If $g(x) = (x - 3)^2$, then the graph of $g(-x) = (-x - 3)^2 = (x + 3)^2$ is the graph of g reflected about the y-axis (see Fig. 34).

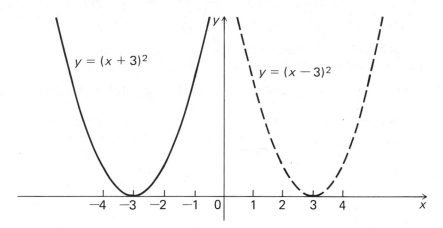

Figure 34

PROBLEM

Graph the equation $y = -2x^2 + 12x - 17$.

Solution: We "complete the square" on the right side of this equation:

$y = -2(x^2 - 6x) - 17$ make coefficient of x^2 equal to 1

$y = -2(x^2 - 6x + 9 - 9) - 17$ add and subtract $\left\{\dfrac{1}{2} \text{ coefficient of } x\right\}^2$

$y = -2(x^2 - 6x + 9) + 18 - 17$

$y = -2(x - 3)^2 + 1$

We now can see that the equation is a transformation of $y = x^2$. That is, the graph of $y = -2x^2 + 12x - 17 = -2(x - 3)^2 + 1$ is the same as the graph of $y = x^2$ but it is shifted horizontally to the right 3 units, vertically stretched 2 units, reflected about the x-axis and then vertically shifted 1 unit upward (see Fig. 35).

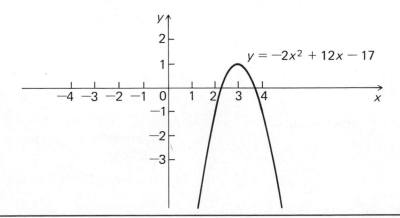

Figure 35

2.3 EXERCISES (B)

In Exercises 1 to 5, let $f(x) = x^3$ and graph the given function:

1. $g(x) = f(x) + 2$
2. $g(x) = 2f(x)$
3. $g(x) = -2f(x)$
4. $g(x) = f(x - 3)$
5. $g(x) = f(x + 2)$

Let h be the function defined by $h(x) = |x|$. Find the function whose graph is obtained from the graph of h by:

6. a vertical shift downward of 4 units.

7. a horizontal shift to the left of 3 units.

8. a vertical contraction of $\frac{1}{3}$ unit and also a reflection about the x-axis.

9. a horizontal stretch of 2 units and vertical shift of 1 unit.

10. a reflection about the y-axis.

Graph the functions in Exercises 11 to 15:

11. $f(x) = -x^2 + 2$
12. $g(x) = (x - 2)^2 - 1$
13. $h(x) = 2x^2 - 3$
14. $p(x) = 2(x + 1)^2 + 3$
15. $q(x) = -(x - 4)^2 - 2$

In Exercises 16 to 20, sketch the graph of g if the graph of f is

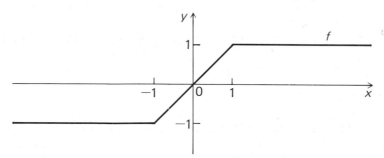

16. $g(x) = f(x - 3)$
17. $g(x) = f(x) - 2$
18. $g(x) = f(3x)$
19. $g(x) = 3f(x)$

20. $g(x) = -f\left(\dfrac{x}{2}\right) + 1$

21. Given $f(x) = \dfrac{1}{x}$, graph (a) $g(x) = -f(x)$; (b) $g(x) = f(-x)$.

22. Given the equation $y^2 = -x^2 + 16$, describe the graph of $(y - 2)^2 = -(x - 1)^2 + 16$.

23. Graph $y = x^2 - \dfrac{3}{2}x + \dfrac{9}{4}$.

24. Graph $y = 2x^2 - 20x + 50$.

25. Graph $y = -3x^2 - 18x - 29$.

2.4 ALGEBRA OF FUNCTIONS

In this section we develop the idea of combining functions to form new functions. We discuss how to add, subtract, multiply, and divide two functions. Finally, several properties concerning the algebra of functions are given.

Function Addition: Let f and g be two functions and suppose that D_f and D_g denote the domains of f and g, respectively. The function $f + g$ is defined by

$$(f + g)(x) = f(x) + g(x)$$

The domain of $f + g$ is $\{D_f \cap D_g\}$.

EXAMPLE 1

Let $f(x) = x^3 - 1$ and $g(x) = 3x$. Then $(f + g)(x) = f(x) + g(x) = x^3 - 1 + 3x = x^3 + 3x - 1$. The domain of $f + g$ is R, the real numbers.

EXAMPLE 2

Let $f(x) = x^3 - 1$ and $g(x) = 3x$. If $x = 2$, then $f(2) = (2)^3 - 1 = 8 - 1 = 7$ and $g(2) = 3 \cdot 2 = 6$. Thus, $(f + g)(2) = f(2) + g(2) = 7 + 6 = 13$.

Function Subtraction: Let f and g be two functions and suppose that D_f and D_g denote the domains of f and g, respectively. The function $f - g$ is defined by

$$(f - g)(x) = f(x) - g(x)$$

The domain of $f - g$ is $\{D_f \cap D_g\}$.

EXAMPLE 3

Let $f(x) = x^3 - 1$ and $g(x) = 3x$. Then $(f - g)(x) = f(x) - g(x) = x^3 - 1 - 3x = x^3 - 3x - 1$. The domain of $f - g$ is R, the real numbers.

Function Multiplication: Let f and g be two functions and let D_f and D_g be the domains of f and g, respectively. The function fg is defined by

$$(fg)(x) = f(x)g(x)$$

The domain of fg is $\{D_f \cap D_g\}$.

EXAMPLE 4

Let $f(x) = x + 3$ and $g(x) = x^2 - 1$. Then $(fg)(x) = f(x)g(x) = (x + 3)(x^2 - 1) = x^3 + 3x^2 - x - 3$.
The domain of fg is R, the real numbers.

EXAMPLE 5

If $f(x) = x^2$ and $g(x) = 4$, then $h(x) = (fg)(x) = 4x^2$ and $h(3) = f(3)g(3) = 9 \cdot 4 = 36$.

Function Division: Let f and g be two functions and let D_f and D_g be the domains of f and g, respectively. The function $\dfrac{f}{g}$ is defined by

$$\left(\frac{f}{g}\right)(x) = \frac{f(x)}{g(x)}, \quad g(x) \neq 0$$

The domain of $\dfrac{f}{g}$ is $\{D_f \cap D_g\}$ *excluding* those values of x for which $g(x) = 0$.

EXAMPLE 6

Let $f(x) = x + 3$ and $g(x) = x^2 - 1$. Then

$$\left(\frac{f}{g}\right)(x) = \frac{f(x)}{g(x)} = \frac{x + 3}{x^2 - 1}.$$

The domain of $\dfrac{f}{g}$ is $\{x \mid x$ is a real number and $x \neq 1$ nor $x \neq -1\}$, the real numbers excluding the numbers 1 and -1. Note that $g(1) = 0$ and $g(-1) = 0$.

PROBLEM

Let $f(x) = \sqrt{x - 4}$ and $g(x) = x^2 - 25$. Find $f + g$, $f - g$, fg, and $\dfrac{f}{g}$. Also, determine the domain of each.

Solution: (a) $(f + g)(x) = f(x) + g(x) = \sqrt{x - 4} + x^2 - 25$

(b) $(f - g)(x) = f(x) - g(x) = \sqrt{x - 4} - x^2 + 25$

(c) $(fg)(x) = f(x)g(x) = \sqrt{x - 4}(x^2 - 25)$

(d) $\left(\dfrac{f}{g}\right)(x) = \dfrac{f(x)}{g(x)} = \dfrac{\sqrt{x - 4}}{x^2 - 25}$

The domain $D_f = \{x \mid x \geq 4\}$, or equivalently, $[4, +\infty)$. The domain $D_g = \{x \mid x$ is a real number$\}$, or equivalently, $(-\infty, +\infty)$. Thus, the domain of $f + g, f - g$, and fg is

$$D_f \cap D_g = [4, +\infty).$$

The domain of $\dfrac{f}{g}$ is $D_f \cap D_g$ excluding those points in D_g which yield $g(x) = 0$. The values which yield $g(x) = 0$ are 5 and -5. Therefore, the domain of $\dfrac{f}{g}$ is $[4, +\infty)$ but $x \neq 5$.

We note that since addition and multiplication of functions are defined in terms of their values in their ranges, they inherit the corresponding properties of addition and multiplication of those values. For functions f, g, and h we have the following laws:

Commutative

$$f + g = g + f$$
$$fg = gf$$

Associative

$$(f + g) + h = f + (g + h)$$
$$(fg)h = f(gh)$$

Zero

$$f + f_0 - f, \text{ where } f_0(x) = 0 \text{ for all } x$$

Identity

$$ff_1 = f, \text{ where } f_1(x) = 1 \text{ for all } x$$

Distributive

$$f(g + h) = fg + fh$$

Theorem: Let f and g be real functions. Let c and d be real numbers. Then,

$$c(f + g) = cf + cg$$
$$(c + d)f = cf + df$$
$$c(fg) = (cf)g$$
$$(cd)f = c(df)$$
$$1f = f$$

EXAMPLE 7

Let $f(x) = 2x - 1$ and $g(x) = x^2 + 1$. Then
(a) $4(f + g)(x) = 4f(x) + 4g(x) = 4(2x - 1) + 4(x^2 + 1) = 4x^2 + 8x$
(b) $(3 + 5)f(x) = 3f(x) + 5f(x) = 3(2x - 1) + 5(2x - 1) = 6x - 3 + 10x - 5 = 16x - 8$
(c) $-2\left(\frac{1}{8} f(x)\right) = -2\left(\frac{x}{4} - \frac{1}{8}\right) = \frac{-x}{2} + \frac{1}{4}$
(d) $5(fg)(x) = 5[f(x)g(x)] = 5[2x^3 - x^2 + 2x - 1] = 10x^3 - 5x^2 + 10x - 5$

2.4 EXERCISES

1. Let $f(x) = 2x - 3$ and $g(x) = 3x^2 - 1$. Form each of the following functions:

(a) $f + g$ (b) $f - g$ (c) fg (d) $\dfrac{f}{g}$

In Exercises 2 to 10, find the domain of $f + g$, $f - g$, fg, and $\dfrac{f}{g}$. Assume that x is a real number.

2. $f(x) = x$ and $g(x) = \dfrac{1}{x}$

3. $f(x) = \sqrt{x}$ and $g(x) = 3x + 7$

4. $f(x) = x^2 - 4$ and $g(x) = x + 2$

5. $f(x) = \sqrt[3]{x + 1}$ and $g(x) = x^3 - 1$

6. $f(x) = |x|$ and $g(x) = \sqrt{1 - x^2}$

7. $f(x) = 4x$ and $g(x) = x^2 + 1$

8. $f(x) = \sqrt{9 - x^2}$ and $\sqrt{x^2 - 9}$

9. $f(x) = \begin{cases} 1 \text{ for } x \le 0 \\ -1 \text{ for } x > 0 \end{cases}$ and $g(x) = 1$

10. $f(x) = \begin{cases} 3 \text{ for } x \ge 0 \\ x \text{ for } x < 0 \end{cases}$ and $g(x) = \begin{cases} 3 \text{ for } x \ge 0 \\ -x \text{ for } x < 0 \end{cases}$

11. Use the graph of f and g to graph $h = f + g$ for the following pairs of functions:
(a) $f(x) = x^2$ $g(x) = 4$
(b) $f(x) = 2x^2$ $g(x) = 3x + 1$
(c) $f(x) = -4$ $g(x) = x^3 + 2$

In Exercises 12 to 25, determine whether the statement is true or false. If false, give an example to show that such is the case.

12. $(f + g)(x) = (g + f)(x)$

13. $f(x - 1) = f(x) - f(1)$

14. $(f - g)(x) = (g - f)(x)$

15. $f(3x) = 3f(x)$

16. $(3f)(x) = 3f(x)$

17. $(f - g)^2(x) = (f^2 - g^2)(x)$

18. $g(x) \cdot \dfrac{1}{g(x)} = 1$

19. $(|gf|)(x) = (|g||f|)(x)$

20. $(fg)(x) = \dfrac{1}{4}(f + g)^2(x) - \dfrac{1}{4}(f - g)^2(x)$

21. $[f + (-f)](x) = 0$

22. $g\left(\dfrac{1}{x}\right) = \dfrac{1}{g(x)}$

23. $[f(g - h)](x) = (fg)(x) - (fh)(x)$

24. $(g + 1)(x) = g(x) + 1$

2.5 TYPES OF FUNCTIONS

Special Functions

We begin this section by defining several functions which appear quite often in mathematics. Some have been described previously and some are new.

Identity Function. The *identity function I* is defined by the equation $y = x$. The domain of *I* is *R*, the set of real numbers, and the range is also *R*. The graph of *I* is given in Figure 36.

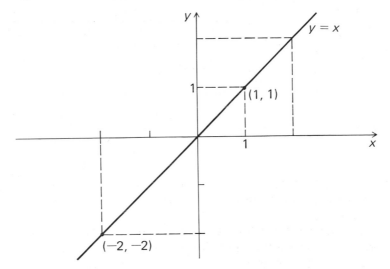

Figure 36

The Absolute Value Function. The *absolute value function* is defined by the equation $y = \begin{cases} x & \text{if } x \geq 0. \\ -x & \text{if } x < 0. \end{cases}$ The domain is *R* and the range is the set of all nonnegative real numbers (see Fig. 37).

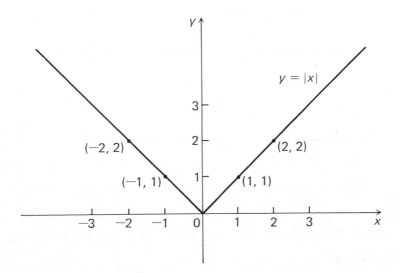

Figure 37

The Greatest Integer Function. The function that assigns to each number x the greatest integer that is $\leq x$ is called the **greatest integer function,** written $[x]$. That is, the greatest integer of a real number x is the integer n which satisfies $n \leq x < n + 1$. The domain of the greatest integer function is R and the range is the set of integers (see Fig. 38). The greatest integer function is an ex-

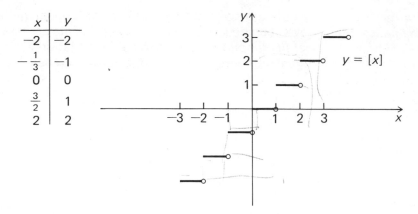

x	y
-2	-2
$-\frac{1}{3}$	-1
0	0
$\frac{3}{2}$	1
2	2

$y = [x]$

Figure 38

ample of a larger group of functions called **step functions.** Also, it is an example of what is called a **discontinuous** function because of the "jumps" at each of the integers.

The Constant Function. A function whose range consists of exactly one number is a constant function. For example, if that number is c, then the function is defined by the equation $f(x) = c$ (see Fig. 39).

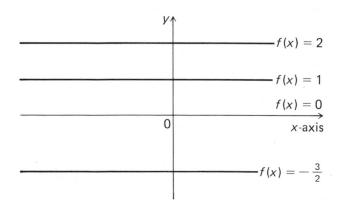

$f(x) = 2$

$f(x) = 1$

$f(x) = 0$

x-axis

$f(x) = -\frac{3}{2}$

Figure 39

A Reciprocal Function. The function $f(x) = \frac{1}{x}$, $x \neq 0$ (see Fig. 40) is a reciprocal function.

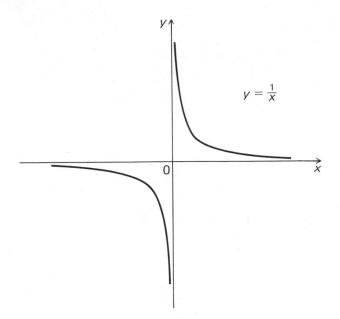

Figure 40

A Combination Function. Many times a function is the result of combining several other functions. Consider the function

$$f(x) = \begin{cases} -1 \text{ for } x < -1 \\ \;\;x \text{ for } -1 \le x \le 2 \\ \;\;2 \text{ for } x > 2 \end{cases}$$

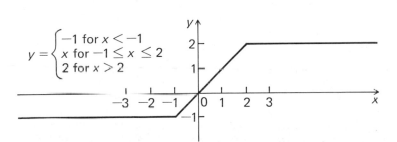

Figure 41

Symmetry

Two points P and Q are said to be **symmetric** with respect to a line l if the line l is the perpendicular bisector of the line segment $\overline{PQ}$. The line l is called a **line of symmetry** (see Fig. 42). Of particular interest to us is the case in which the x-axis or the y-axis is the line of symmetry. Also, we define symmetry with respect to the origin.

> **Definition:** A graph is **symmetric with respect to the x-axis** if, whenever the point (x, y) is on the graph, then the point $(x, -y)$ is also on the graph. That is, "flipping" the graph about the x-axis would not change the graph.

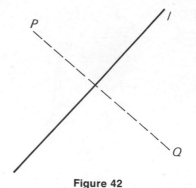

Figure 42

EXAMPLE 1

The graph of $x = |y|$ is symmetric with respect to the x-axis. Substituting y or $-y$ into the equation yields the same x; therefore (x, y) and $(x, -y)$ lie on the graph of $x = |y|$ (see Fig. 43).

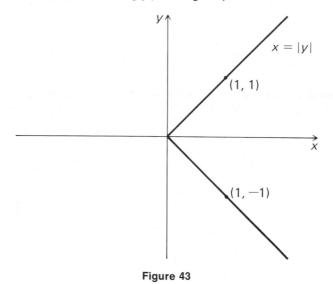

Figure 43

EXAMPLE 2

The graph of the function $y = \sqrt{x}$ is *not* symmetric with respect to the x-axis. This can be seen by noting that the range of this function is the set of nonnegative numbers and thus y cannot be negative. For example $(4, 2)$ belongs to this function, but $(4, -2)$ does not. Geometrically, if the graph of $y = \sqrt{x}$ were "flipped" about the x-axis, then the graph would be changed (see Fig. 44).

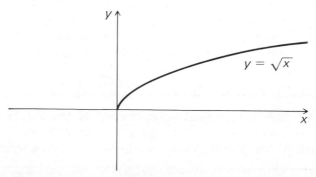

Figure 44

> **Definition:** A graph is **symmetric with respect to the y-axis** if, whenever the point (x, y) is on the graph, then the point $(-x, y)$ is also on the graph. That is, "flipping" the graph about the y-axis would not change the graph.

EXAMPLE 3

The graph of the function $y = -\sqrt{1 - x^2}$ is symmetric with respect to the y-axis (see Fig. 45).

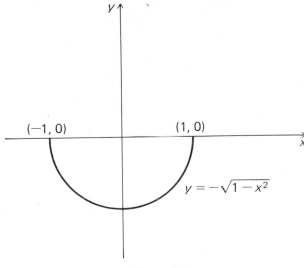

Figure 45

EXAMPLE 4

The graph of $y = x^3$ is *not* symmetric with respect to the y-axis. The point $(2, 8)$ lies on the graph, but the point $(-2, 8)$ does not (see Fig. 46).

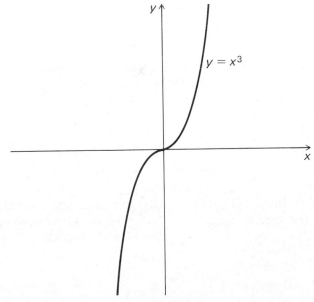

Figure 46

> **Definition:** A graph is **symmetric with respect to the origin** if, whenever the point (x, y) lies on the graph, then the point $(-x, -y)$ is also on the graph.

EXAMPLE 5

The graph $x^2 + y^2 = r^2$ is symmetric with respect to the origin (see Fig. 47).

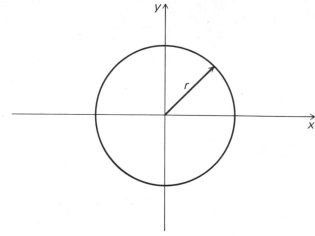

Figure 47

EXAMPLE 6

The graphs of $y = \dfrac{1}{x}$ (see Fig. 40) and $y = x^3$ (see Fig. 46) are symmetric with respect to the origin.

> **Fact:** If the graph of a relation satisfies any two of the above definitions, then it also satisfies the third.

EXAMPLE 7

The graph of $x^2 + y^2 = r^2$, a circle of radius equal to r, is symmetric with respect to the origin and symmetric with respect to the y-axis. Therefore, it is also symmetric with respect to the x-axis.

2.5 EXERCISES (A)

1. Examine the following relations for symmetry with respect to the x-axis, y-axis, and the origin:
   (a) $y = x$ (b) $y = x^2$
   (c) $4x^2 + 9y^2 = 36$ (d) $xy = 4$
   (e) $x = y^2 - 2$ (f) $y = 2x + 1$

2. Can the graph of a function, other than $f(x) = 0$, be symmetric with respect to the x-axis? Explain.

3. Let f and g be functions whose graphs are symmetric with respect to the y-axis. What can be said about symmetry with respect to the y-axis for:
   (a) $f + g$ (b) $f - g$
   (c) fg (d) $\dfrac{f}{g}$

4. Let f and g be functions whose graphs are symmetric with respect to the origin. What can be said about symmetry with respect to the origin for:
   (a) $f + g$ (b) $f - g$
   (c) fg (d) $\dfrac{f}{g}$

5. Consider the functions $f_n(x) = x^n$ for $n = 1, 2, 3, 4, \ldots$. If n is odd, what can be said about the symmetry of the graph of f_n? If n is even, what can be said about the symmetry?

Even and Odd Functions

Definition: A function f is called an **even function** if $f(-x) = f(x)$ for all values of x for which f is defined.

EXAMPLE 1

The function f, defined by $f(x) = x^2$, is an even function. Thus,

$$f(-x) = (-x)^2 = x^2 = f(x)$$

EXAMPLE 2

The function h, defined by $h(x) = |x|$, is an even function:

$$h(-x) = |-x| = |x| = h(x)$$

Note that the graph of an even function will be symmetric with respect to the y-axis.

Definition: A function f is called an **odd function** if $f(-x) = -f(x)$ for all values of x for which f is defined.

EXAMPLE 3

The function f, defined by $f(x) = x^3$, is an odd function:

$$f(-x) = (-x)^3 = -x^3 = -f(x)$$

EXAMPLE 4

If $h(x) = x^5 - 3x^3 + 4x$, then the function h is an odd function:

$$h(-x) = (-x)^5 - 3(-x^3) + 4(-x) = -x^5 + 3x^3 - 4x$$
$$= -(x^5 - 3x + 4x) = -h(x)$$

PROBLEM 1

Determine whether the function f defined by $f(x) = x^2 - x + 1$ is even, odd, or neither.

Solution: We examine $f(-x)$ and compare the results with $f(x)$.

$$f(-x) = (-x)^2 - (-x) + 1 = x^2 + x + 1$$

Note that $f(-x) \neq f(x)$, nor does $f(-x) = -f(x)$. Therefore, f is neither even nor odd.

PROBLEM 2

Let f and g be odd functions. Show that fg is an even function.

Solution: We must show that $(fg)(-x) = (fg)(x)$.

$(fg)(-x) = [f(-x)][g(-x)]$ definition of multiplication of functions

$\qquad = [-f(x)][-g(x)]$ because f and g are odd functions

$\qquad = f(x)g(x)$ (negative)(negative) = positive

$\qquad = (fg)(x)$ definition of multiplication of functions

Thus, $(fg)(-x) = (fg)(x)$ and fg is even.

2.5 EXERCISES (B)

In Exercises 1 to 10, determine whether each function is even, odd, or neither.

1. $f(x) = 4x^4 - 3x^2$

2. $g(x) = -3x^2 - 1$

3. $h(x) = x^3 - x$

4. $m(x) = \dfrac{x + 7}{x - 8}$

5. $p(x) = \dfrac{|x|}{x}$

6. $q(x) = \sqrt{x^2 - 1}$

7. $s(x) = [x]$

8. $u(x) = \dfrac{1}{x} - x$

9. $v(x) = \sqrt{x^3 - 3}$

10. $w(x) = 8$

11. Show that if f is odd, then $f(x) + f(-x) = 0$ for all x. Also, if f is even, then $f(x) - f(-x) = 0$.

12. Let f and g be even functions. Prove that $f + g$, $f - g$, fg and $\dfrac{f}{g}$ are even.

13. Which of the following graphs represent even functions, which represent odd functions, and which neither?

(a)

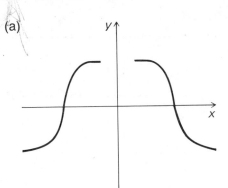

(b)

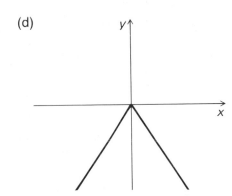

(c)

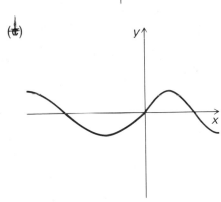

(d)

14. If f is an odd function, is cf odd? For what values of c would cf be an odd function?

15. Let $f(x) = ax + b$, where a and b are constants. For what values of a and b is f an even function? What values make f an odd function?

16. Let $f(x) = 3x - 2$. Show that $f(a + b) = f(a) + f(b) + 2$.

17. If f is an even function, what can be said about $\dfrac{1}{f}$?

18. Let f be an even function and g be an odd function. Prove that fg is an odd function.

19. What are the condition(s) on a, b, and c such that $g(x) = ax^2 + bx + c$ is even? Odd?

Increasing and Decreasing Functions

If the graph of a function f rises from left to right, then the function f is said to be strictly increasing. Similarly, if the graph of a function f falls from left to right, then the function f is said to be strictly decreasing. We now give algebraic definitions for strictly increasing and strictly decreasing functions.

Definition: A function *f* defined on an interval is said to be **strictly increasing** on that interval if

$$f(a) < f(b) \text{ whenever } a < b$$

where *a* and *b* are any numbers in the interval.

EXAMPLE 1

Let *f* be a function defined by $f(x) = x$; then *f* is a strictly increasing function on $(-\infty, -\infty)$ (see Fig. 48).

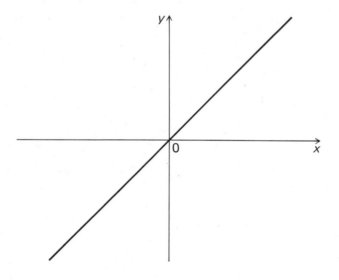

Figure 48

EXAMPLE 2

The following graph represents a strictly increasing function:

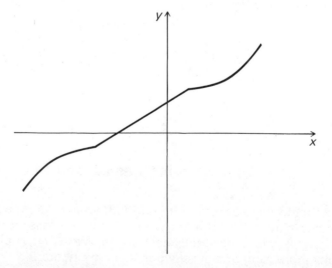

Figure 49

> **Definition:** A function f defined on an interval is said to be **strictly decreasing** on that interval if
>
> $$f(a) > f(b) \text{ whenever } a < b$$
>
> where a and b are any numbers in the interval.

EXAMPLE 3

The function defined by $f(x) = -x^3 + 1$ is a strictly decreasing function (see Fig. 50).

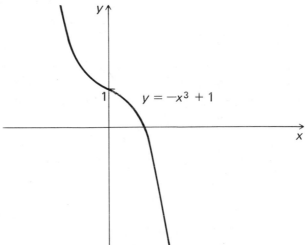

Figure 50

EXAMPLE 4

The linear equation $y = mx + b$ represents a strictly decreasing function provided that $m < 0$ (see Fig. 51).

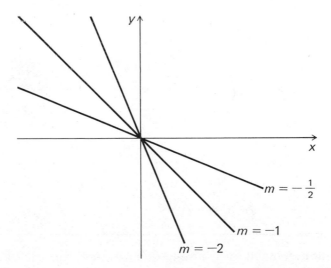

Figure 51

There are functions which are strictly increasing on one interval and strictly decreasing on another interval. Such an example is the function $f(x) = x^2$ (see Fig.

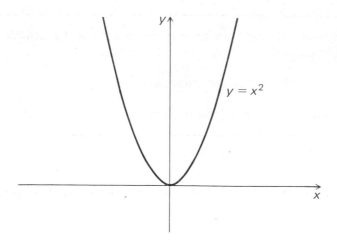

Figure 52

52). The domain of f is the whole real line and the function is strictly decreasing provided $x \le 0$. For $x \ge 0$, the function is strictly increasing. Also, there are functions which are neither increasing nor decreasing. The constant function is an example of such a function.

One-to-One Functions

Consider the functions $f = \{(2, 7), (-1, 7), (3, 6), (4, 1)\}$ and $g = \{(1, 3), (4, 2), (3, -6), (-2, 0), (-1, 5)\}$. Recall that the only restriction for a relation to be a function is that no element in the domain is mapped to more than one element in the range. However, functions such as f have several elements in the domain being mapped to the same element in the range. The function g has *no* two different elements in its domain being mapped to the same element in its range (see Fig. 53). Functions such as g form a special class of functions.

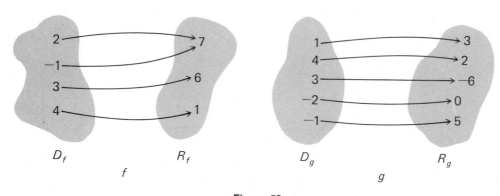

Figure 53

Definition: A function f is **one-to-one,** written f is 1-1, if for each a and b belonging to its domain $f(a) = f(b)$ implies $a = b$—that is, if no two different elements in the domain of f have the same image.

EXAMPLE 1

The function f, defined by $f(x) = x^3 + 1$, is 1-1. We need only show that $f(a) = f(b) \Rightarrow a = b$.

$$f(a) = f(b)$$
$$a^3 + 1 = b^3 + 1 \quad \text{subtract 1 from both sides}$$
$$a^3 = b^3 \quad \text{take cube root of both sides}$$
$$a = b$$

EXAMPLE 2

The function g, defined by $g(x) = x^2 - 4$, is *not* 1-1.

$$g(a) = g(b)$$
$$a^2 - 4 = b^2 - 4 \quad \text{add 4 to both sides}$$
$$a^2 = b^2 \quad \text{take square root of both sides}$$
$$a = \pm b$$

Thus $a = b$ or $a = -b$, and g is not 1-1. Rather than making use of the definition to determine whether or not a function is 1-1, we often use the graph of the function.

> If a function is one-to-one, then any horizontal line will intersect its graph at no more than one point.

EXAMPLE 3

Any linear function, defined by the equation $y = mx + b$ and $m \neq 0$, is a 1-1 function. This is easily seen from the graph of a linear function. A horizontal line intersects another line no more than once. The only possible line which would have more than one point of intersection would be another horizontal line, that is, $m = 0$ (see Fig. 54).

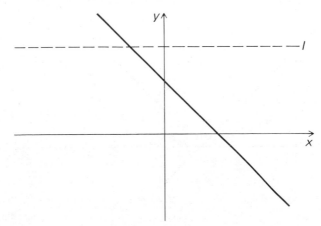

Figure 54

EXAMPLE 4

The function f, defined by $f(x) = x^2$, is *not* 1-1 (see Fig. 55). Note that (a, a^2) and $(-a, a^2)$ lie on the graph. Thus, two different values in the domain have the same image.

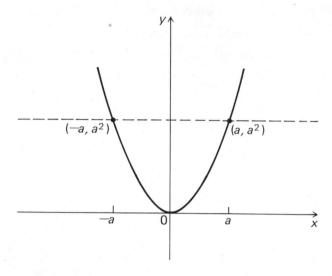

Figure 55

PROBLEM

Consider the function h defined by

$$h(x) = \begin{cases} -2x - 1 \text{ for } x \leq 0 \\ x + 1 \text{ for } x > 0 \end{cases}$$

Determine if h is 1-1.

Solution:

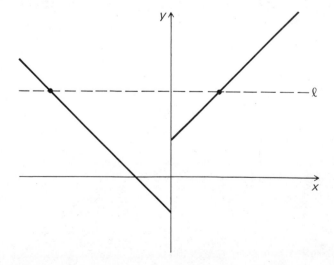

Figure 56

The horizontal line l intersects the graph of h twice (see Fig. 56). Also, if $x_1 = 2$ and $x_2 = -2$, then $x_1 \neq x_2$. But, since $2 > 0$, $h(2) = 2 + 1 = 3$. Also, since $-2 \leq 0$, $h(-2) = -2(-2) - 1 = 3$. Thus, $h(2) = h(-2)$, but $2 \neq -2$. Therefore, h is not 1-1.

Equivalent Definition. A function f is 1-1 if $a \neq b$ implies $f(a) \neq f(b)$.

Theorem 1: If f is a strictly increasing function or a strictly decreasing function, then f is 1-1.

2.5 EXERCISES (C)

1. Graph the equation $f(x) = x^3 + 2x - 1$ and determine whether it represents a strictly increasing or strictly decreasing function.

2. Do the same as in Exercise 1 for $g(x) = |x|$ where $x \leq 0$.

3. If f is a strictly increasing function, what can be said about $-f$?

4. Given that g is an even function, show that g can be neither strictly increasing nor strictly decreasing.

5. Determine whether the following represent strictly increasing functions or strictly decreasing functions:
   (a) $f(x) = \dfrac{1}{x}$

   (b) $g(x) = \sqrt{x}$
   (c) $h(x) = 4$
   (d) $p(x) = -x^2 + 2$, for $x \geq 0$

6. Prove that the following equations represent one-to-one functions:
   (a) $f(x) = 4x + 7$
   (b) $f(x) = x^3$
   (c) $f(x) = x^3 - 2x + 1$
   (d) $f(x) = |2x - 1|, \ x \leq \dfrac{1}{2}$
   (e) $f(x) = \sqrt{x - 1}$
   (f) $f(x) = x^5 - 4$

7. Can the constant function f, represented by $f(x) = c$, be one-to-one? Explain.

8. Determine the sets on which the identify function I, defined by $I(x) = x$, is 1-1.

9. Prove Theorem 1.

10. Which of the following statements are true for all functions?
    (a) If $x_1 = x_2$, then $f(x_1) = f(x_2)$.
    (b) If $f(x_1) = f(x_2)$, then $x_1 = x_2$.
    (c) If $x_1 \neq x_2$, then $f(x_1) \neq f(x_2)$.
    (d) If $f(x_1) \neq f(x_2)$, then $x_1 \neq x_2$.

11. Restrict the domains of the following functions such that they are one-to-one:

    (a) $f(x) = |x|$ (b) $g(x) = x^2$ (c) $h(x) = \dfrac{1}{x^2}$ (d) $p(x) = \sqrt{1 - x^2}$

2.6 COMPOSITE FUNCTIONS

Consider the function g, defined by $g(x) = 2x + 1$, and let the domain of g be the set $A = \{0, 1, -1\}$. Then the range of g is the set $B = \{1, 3, -1\}$. We now consider a second function f, defined by $f(x) = 3x$, whose domain we take to be the set $B = \{1, 3, -1\}$. That is, the range of g is the domain of f. Then, the range of f is the set $C = \{3, 9, -3\}$. Thus, the set $A = \{0, 1, -1\}$ has been mapped to the set $C = \{3, 9, -3\}$. This was achieved by applying the functions g and f in succession—first g, then f (see Fig. 57). This method of combining two functions is called the **composition** of two functions.

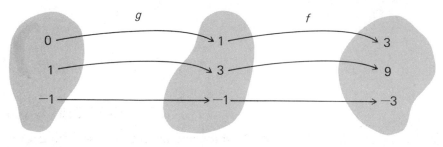

Domain of g Range of g = domain of f Range of f **Figure 57**

Definition: Let f and g be two functions. The **composite** of f and g, denoted by $f \circ g$, is the function defined for each x in D_g by

$$(f \circ g)(x) = f[g(x)]$$

provided that $g(x)$ is in the domain of f.

Geometrically, $f \circ g$ is shown in Figure 58. It is very important to note that in

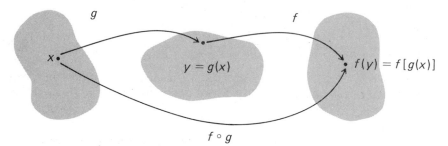

$f \circ g$ **Figure 58**

order to form the composition of two functions, we must have the range of the first function contained in the domain of the second function.

EXAMPLE 1

If $f(x) = x^2$ defines the function f and $g(x) = \sqrt{x + 1}$, then

$$(f \circ g)(x) = f[g(x)] = f[\sqrt{x + 1}] = (\sqrt{x + 1})^2 = x + 1$$

and

$$(g \circ f)(x) = g[f(x)] = g[x^2] = \sqrt{x^2 + 1}$$

EXAMPLE 2

Let $f(x) = x^2 + 1$ and $g(x) = 2x$. We now compute $(f \circ g)(3)$, $(g \circ f)(3)$, $(f \circ f)(3)$, and $(g \circ g)(3)$.
   (a) $(f \circ g)(3) = f[g(3)] = f[2 \cdot 3] = f[6] = (6)^2 + 1 = 37$
   (b) $(g \circ f)(3) = g[f(3)] = g[(3)^2 + 1] = g[10] = 2 \cdot 10 = 20$
   (c) $(f \circ f)(3) = f[f(3)] = f[(3)^2 + 1] = f[10] = (10)^2 + 1 = 101$
   (d) $(g \circ g)(3) = g[g(3)] = g[2 \cdot 3] = g[6] = 2 \cdot 6 = 12$

Note: Examples 1 and 2 show that in general composition does not commute. That is,

$$f \circ g \neq g \circ f$$

However, composition is associative. That is,

$$h \circ (g \circ f) = (h \circ g) \circ f$$

EXAMPLE 3

Let the three functions f, g, and h be defined by

$$f(x) = x - 1$$
$$g(x) = x^2$$
$$h(x) = 3x$$

Then $(h \circ g \circ f)(x) = h\{g[f(x)]\} = h\{g[x - 1]\}$

$$= h\{(x - 1)^2\} = 3(x - 1)^2$$

PROBLEM 1

Let f be defined by $f(x) = 3x - 1$ and g be defined by $g(x) = \dfrac{x + 1}{3}$.
Compute $f \circ g$ and $g \circ f$.

Solution: (a) $(f \circ g)(x) = f[g(x)] = f\left[\dfrac{x + 1}{3}\right]$

$$= 3\left(\dfrac{x + 1}{3}\right) - 1 = x$$

   (b) $(g \circ f)(x) = g[f(x)] = g[3x - 1]$

$$= \dfrac{3x - 1 + 1}{3} = x$$

In Problem 1 $f \circ g$ and $g \circ f$ act the same as the **identity function,** $I(x) = x$. The functions f and g are said to be **inverses** of one another.

PROBLEM 2

Let f be the function defined by $f(y) = 3y$ which converts y yards into feet. Let $g(m) = 1760\,m$ represent the function g which converts m miles into yards. Derive a function which converts miles to feet.

Solution:

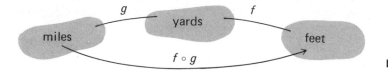

Figure 59

$$(f \circ g)(m) = f[g(m)] = f[1760\,m] = 3 \cdot 1760\,m = 5280\,m$$

PROBLEM 3

Let f and g be odd functions. What can be said about the function $f \circ g$?

Solution: The function f is odd if and only if

$$f(-y) = -f(y).$$

Similarly, the function g is odd if and only if

$$g(-x) = -g(x).$$

Thus,

$(f \circ g)(-x) = f[g(-x)]$; since $g(-x) = -g(x)$, we have
$\qquad = f[-g(x)]$; also, $f(-y) = -f(y)$
$\qquad = -f[g(x)]$; definition of composition
$\qquad = -(f \circ g)(x)$

Therefore, $(f \circ g)(x) = -(f \circ g)(x)$ and $f \circ g$ is odd.

2.6 EXERCISES

1. Let f and g be functions defined by $f(x) = 3x^2 - 1$ and $g(x) = \dfrac{4x + 1}{3}$. Find the following expressions:
   (a) $f \circ g$ (b) $g \circ f$
   (c) $f \circ f$ (d) $g \circ g$

2. Let f and g be functions defined by $f(x) = \dfrac{16}{4 - x^2}$ and $g(x) = \sqrt{x - 1}$. Compute, if they exist, the following:
   (a) $(g \circ f)(0)$ (b) $(f \circ g)(0)$
   (c) $(f \circ g)(1)$ (d) $(g \circ f)(1)$
   (e) $(f \circ g)(2)$ (f) $(g \circ f)(2)$

3. Let f be defined by $f(x) = x^3$ and g defined by $g(x) = x + 2$. Find $f \circ g$ and $g \circ f$. Also, state the domain and the range of $f \circ g$ and $g \circ f$.

4. Find the value of p such (g $\circ$ f)(x), where $f(x)$ $g(x) = 2x - p$.

5. Given that $h(x) = (f \circ g)(x)$, find $f(x)$ and $g(x)$ for each of the following:
   (a) $h(x) = (x + 1)^2 - 3$

   (b) $h(x) = \sqrt{(x - 1)^3} + \dfrac{1}{\sqrt[3]{x - 1}}$

   (c) $h(x) = |3x| - \dfrac{1}{|6x|^2}$

6. Let f and g be two functions. If $(f \circ g)(x) = 0$ for all real numbers x, does it follow that either $f = 0$ or $g = 0$? Explain your answer.

7. Let f be an odd function and g be an even function. Determine which of the following functions are even:
   (a) $g \circ f$ (b) $f \circ g$
   (c) $g \circ g$ (d) $f \circ f$

8. Let $f(x) = \dfrac{1}{x}$ and $g(x) = \dfrac{1}{x}$. Compute $(f \circ g)(x)$ and $(g \circ f)(x)$.

9. Derive a function f which converts gallons to quarts, and a function g which converts quarts to pints. Use composition to derive a function which converts gallons to pints.

In Exercises 10 to 13, find $f \circ g$ and $g \circ f$ and determine their domains.

10. $f(x) = x - \dfrac{1}{x}$; $g(x) = x - \dfrac{1}{x}$

11. $f(x) = x^3 + 1$; $g(x) = \sqrt[3]{x - 1}$

12. $f(x) = \dfrac{1}{x + 1}$; $g(x) = \dfrac{x + 1}{x}$

13. $f(x) = \sqrt{\dfrac{x + 1}{x - 1}}$; $g(x) = \sqrt{x}$

14. Given the linear functions defined by $f(x) = mx + b$ and $g(x) = px + c$, show that $f \circ g$ and $g \circ f$ are linear functions.

15. Prove that $h \circ (f \circ g) = (h \circ f) \circ g$.

2.7   INVERSE FUNCTIONS

In Problem 1 in Section 2.6 (p. 77) we saw that the composition of two functions can act the same as the identity function. That is, if one function converts x into $f(x)$, then the inverse function g converts $f(x)$ back into x (see Fig. 60).

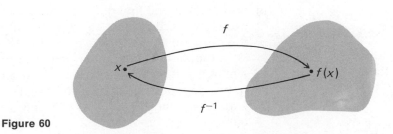

Figure 60

Definition: Let f and g be two functions. If $(f \circ g)(x) = f[g(x)] = x$ for each x in the domain of g, and if $(g \circ f)(x) = g[f(x)] = x$ for each x in the domain of f, then f and g are said to be **inverses** of one another. We write $f = g^{-1}$ and $g = f^{-1}$.

Note: f^{-1} is *not* to be taken as "f raised to the negative one power." We can give a simpler definition of inverse.

Definition: Let f be a given function. Then the **inverse function** of f is the function f^{-1} satisfying $y = f(x)$ if and only if $x = f^{-1}(y)$, provided such a function exists.

EXAMPLE

We show that the function g defined by $g(x) = \dfrac{2x - 7}{3}$ is the inverse of the function f defined by $f(x) = \dfrac{3x}{2} + \dfrac{7}{2}$. All we need to show is that $(f \circ g)(x) = x$ and $(g \circ f)(x) = x$.

(a) $(f \circ g)(x) = f[g(x)] = f\left[\dfrac{2x - 7}{3}\right] = \dfrac{3\left(\dfrac{2x - 7}{3}\right)}{2} + \dfrac{7}{2}$

$= \dfrac{2x - 7}{2} + \dfrac{7}{2} = x - \dfrac{7}{2} + \dfrac{7}{2} = x$

(b) $(g \circ f)(x) = g[f(x)] = g\left[\dfrac{3x}{2} + \dfrac{7}{2}\right] = \dfrac{2\left(\dfrac{3x}{2} + \dfrac{7}{2}\right) - 7}{3}$

$= \dfrac{3x + 7 - 7}{3} = \dfrac{3x}{3} = x$

We now discuss three important questions:

(a) When does the inverse of a function exist?
(b) How do we find the inverse of a function?
(c) If f^{-1} exists, what is the relation between the graph of f^{-1} and the graph of f?

(a) In answer to the first question, we recall that it is required that the domain of a function f be the range of its inverse function f^{-1} and the range of f^{-1} be the domain of f. Also, in order for a relation to be a function it must determine a unique y in the range for each element x in its domain. Likewise, in order for an inverse mapping to be a function, we must also have a unique x in its range for each y in its domain. Recall that this defines the class of functions we call 1-1.

> **Theorem:** A function f has an inverse if and only if f is 1-1.

(b) Finding f^{-1} for certain simple functions, given f, is a matter of using the definition of inverse. The following problems illustrate the method.

PROBLEM 1

Given $f(x) = x^3 - 1$, find f^{-1}, if it exists.

Solution: (a) $f(x) = x^3 - 1$ represents a 1-1 function. Thus, by the above theorem, f^{-1} exists. Recall, if f^{-1} exists, then $(f \circ f^{-1})(x) = x$. Thus,

$$(f \circ f^{-1})(x) = f[f^{-1}(x)] = x$$
$$\Rightarrow (f^{-1}(x))^3 - 1 = x$$
$$\Rightarrow (f^{-1}(x))^3 = x + 1$$
$$\Rightarrow f^{-1}(x) = \sqrt[3]{x + 1}$$

PROBLEM 2

Show that the function f defined by $f(x) = x^2$ does *not* have an inverse over the reals.

Solution: We recall that $f(x) = x^2$ does *not* define a 1-1 function. Therefore, by the theorem, f does not have an inverse.

PROBLEM 3

If f is defined by $f(x) = x^2$ for $x \geq 0$, then f^{-1} does exist. Find f^{-1}.

Solution: For $x \geq 0$, $f(x) = x^2$ defines a strictly increasing function, and is thus a 1-1 function. Therefore, by the theorem, f^{-1} exists.

$$(f \circ f^{-1})(x) = f[f^{-1}(x)] = x$$
$$\Rightarrow (f^{-1}(x))^2 = x$$
$$\Rightarrow f^{-1}(x) = \sqrt{x} \text{ which exists because } x \geq 0.$$

This last problem illustrates that by restricting the domain of certain functions, the function may have an inverse.

(c) Consider the graphs of f and f^{-1} in the following examples:

EXAMPLE 1

We have seen that if f is defined by $f(x) = x^2$ for $x \geq 0$, then f^{-1} was defined by $f^{-1}(x) = \sqrt{x}$. The graphs of f and f^{-1} are given in Figure 61.

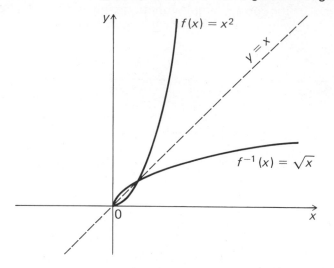

Figure 61

EXAMPLE 2

The function f defined by $f(x) = (x + 1)^3$ .has f^{-1} defined by $f^{-1}(x) = \sqrt[3]{x} - 1$ (see Fig. 62).

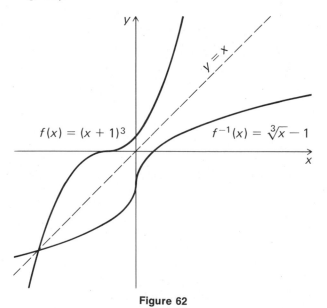

Figure 62

EXAMPLE 3

Recall that the function defined by $f(x) = \dfrac{1}{x}$ turned out to be its own inverse; that is, f^{-1} was also defined by $f^{-1}(x) = \dfrac{1}{x}$ (see Fig. 63).

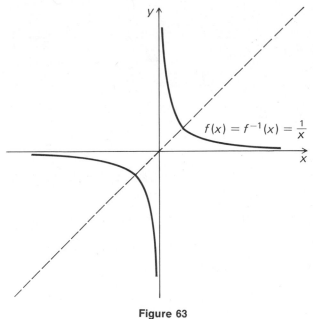

Figure 63

Note that, in all of the above examples, if the point (a, b) lies on the graph of f, then the point (b, a) lies on the graph of f^{-1}. To understand this, note that if (a, b) is on the graph of f, then $b = f(a)$, and since $f^{-1}[f(a)] = a$, we have $f^{-1}(b) = a$. Therefore, (b, a) is on the graph of f^{-1}.

> It follows that the graphs of f and f^{-1} are reflections of each other about the line $y = x$.

EXAMPLE 4

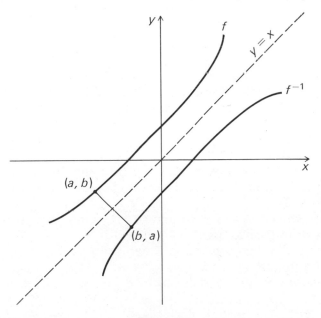

Figure 64

2.7 EXERCISES

In Exercises 1 to 10 determine if the inverse function exists and, if so, find the inverse function.

1. $f(x) = x$

2. $f(x) = \sqrt[3]{3x - 2}$

3. $f(x) = x^5$

4. $f(x) = 4x + 3$

5. $f(x) = 5$

6. $f(x) = |x - 3|$

7. $f(x) = \sqrt{\dfrac{x + 1}{x - 1}}$

8. $f(x) = x^3 - x$

9. $f(x) = \dfrac{1}{x + 1}$

10. $f(x) = x^2 + 3$ for $x > 0$

11. Let f be the linear function defined by $f(x) = mx + b$, $m \neq 0$. Find f^{-1}.

12. Prove that if f is a strictly increasing or strictly decreasing function, then f^{-1} exists.

13. Graph the function defined by $y = 3x - 5$ and then graph its inverse without solving for the inverse.

14. Given the function g defined by $g(x) = x^n$, n is an integer. What condition, if any, must be placed on n for g to have an inverse function?

15. Does the greatest integer function defined by $y = [x]$ have an inverse? Explain.

16. If f is a strictly increasing function, is f^{-1} strictly increasing?

17. Restrict the domain so that $y = |x|$ has an inverse.

18. Prove that if f has an inverse f^{-1} and if g has an inverse g^{-1}, then the inverse of $(g \circ f)$ is the function $(f^{-1} \circ g^{-1})$.

19. Let f be an even function. Does f^{-1} exist? Explain.

20. Prove that if g^{-1} exists, then g^{-1} is unique. (Hint: Assume that g has two inverses and show that this leads to a contradiction.)

SUMMARY

Set and Set Operations

1. A **set** is any well-defined collection of objects. These objects are called the **members** or **elements** of the set. The notation for a set is illustrated by the set $B = \{x \mid x$ is odd$\}$, which reads "B is the set of elements x such that x is odd."

2. If every element in a set A is also an element of a set B, then A is called a **subset** of B, written $A \subseteq B$.

3. Two sets A and B are **equal**, written $A = B$, if and only if $A \subseteq B$ and $B \subseteq A$.

4. A set C is called a **proper subset** of a set D, if C is a subset of D and also if C is not equal to D, i.e., $C \subset D$.

5. The **union** of two sets A and B, written $A \cup B$, is the set of all elements which belong to A or to B or to both.

6. The **intersection** of two sets A and B, written $A \cap B$, is the set of elements which belong to both A and B.

7. The **complement** of a set A, denoted by A', is the set of elements which do not belong to A.

Relations and Functions

1. The **Cartesian product (cross product)** of sets A and B, denoted by $A \times B$, is the set defined by

$$A \times B = \{(x, y)|x \in A \text{ and } y \in B\}.$$

2. Any subset of $A \times B$ is a **relation** on A and B.
3. A **function** is a relation in which no two ordered pairs have the same first component and different second components. The set of all first components is called the **domain** of the function, while the collection of all the second components is called the **range.**
4. A graph of a relation represents a function if any **vertical line** intersects the graph no more than once.

Algebra of Functions

1. $(f + g)(x) = f(x) + g(x)$
2. $(fg)(x) = f(x)g(x)$
3. $(f - g)(x) = f(x) - g(x)$
4. $\left(\dfrac{f}{g}\right)(x) = \dfrac{f(x)}{g(x)}$ when $g(x) \neq 0$
5. The **domain** of $f + g$, $f - g$, fg is the intersection of the domain of f with the domain of g. The same is true for $\dfrac{f}{g}$ except values of x for which $g(x) = 0$ must be excluded.

Types of Functions

1. A graph is said to be **symmetric with respect to the x-axis** if (x, y) lies on the graph implies that $(x, -y)$ also lies on the graph.
2. If (x, y) lies on the graph implies that $(-x, y)$ also lies on the graph, then the graph is said to be **symmetric with respect to the y-axis.**
3. **Symmetry with respect to the origin** means that if (x, y) lies on the graph, then $(-x, -y)$ also lies on the graph.
4. A function f is **odd** if and only if $f(-x) = -f(x)$. The function f is **even** if and only if $f(-x) = f(x)$.
5. A function f is **strictly increasing** if $a < b$ implies $f(a) < f(b)$. The function f is **strictly decreasing** if $a < b$ implies $f(a) > f(b)$.
6. A function f is **one-to-one** if $f(a) = f(b)$ implies $a = b$. Graphically, a function is 1-1 if any horizontal line intersects the graph no more than once.

Graphs

Suppose the graph of the function f is given and defined by $y = f(x)$. Also, suppose c and d are real numbers. Then the graphs of the following are transformations of the graph of f:

1. $y = cf(x)$ is a **vertical stretching** of f if $|c| > 1$ and a **vertical contraction** of f if $|c| < 1$. If $c < 0$, the graph of f is also reflected in the x-axis.
2. $y = f(dx)$ is a **horizontal stretching** of f if $|d| < 1$ and a **horizontal contraction** of f if $|d| > 1$. If $d < 0$, the graph of f is also reflected in the y-axis.

3. $y = f(x - c)$ is a **horizontal shift** of f. If $c > 0$, the graph is shifted to the right $|c|$ units. If $c < 0$, the graph is shifted to the left $|c|$ units.

4. $y = f(x) + c$ is a **vertical shift** of f. If $c > 0$, the graph is shifted upward $|c|$ units. If $c < 0$, the graph is shifted downward $|c|$ units.

Composite and Inverse Functions

1. The **composition** of two functions f and g, denoted by $f \circ g$ or $g \circ f$, is the function defined respectively by

$$(f \circ g)(x) = f[g(x)] \text{ or } (g \circ f)(x) = g[f(x)]$$

2. Let f be a given function. The **inverse** of f, denoted by f^{-1}, is the function satisfying

$$(f^{-1} \circ f)(x) = f^{-1}[f(x)] = x \text{ and } (f \circ f^{-1})(x) = f[f^{-1}(x)] = x$$

3. A function f has an inverse if and only if f is one-to-one.

CHAPTER 2 EXERCISES

1. Given sets $A = \{1, \sqrt{2}, -3\}$, $B = \{-1, -5\}$, and $C = \{\frac{1}{2}, 0, 4, -7\}$, compute the following:
   (a) $A \times B$ (b) $B \times A$
   (c) $B \times C$ (d) $C \times B$
   (e) $C \times A$ (f) $A \times A$
   (g) $B \times B$ (h) $C \times C$

2. Indicate which of the following relations is a function:
   (a) $\{(1, 7), (-3, 7), (0, -2), (-1, 6), (1, 9)\}$
   (b) $\{(x, y)|y = 16 - x^2\}$
   (c) $y = 2$
   (d) $x = 2$
   (e) $f(x) = 3x + 1$

3. Determine which of the following graphs represent functions:

(a)

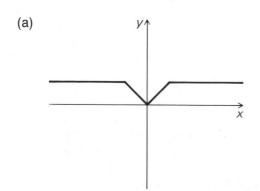

(b)

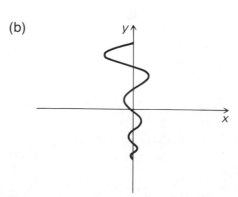

(c)

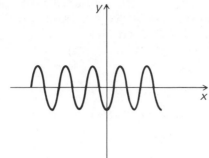

(d)

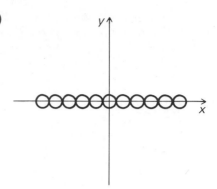

4. Find the domain and range of the following functions:

   (a) $f(x) = \dfrac{x^2 - 16}{x + 4}$

   (b) $g(x) = \begin{cases} x + 1, x \geq 0 \\ 1 - x, x < 0 \end{cases}$

   (c) $h(x) = \dfrac{1}{x^2}$

   (d) $f(x) = \dfrac{\sqrt{3x + 7}}{(x + 4)(x - 3)}$

5. Given $g(x) = 3x^2 + x - 2$, compute the following:

   (a) $g(0)$

   (b) $g(-3)$

   (c) $g(a + b)$

   (d) $\dfrac{g(x + h) - g(x)}{h}, h \neq 0$

6. Let $f(x) = x^3 - 1$ and $g(x) = \dfrac{1}{x}$. What is the domain of $f + g$, fg, $\dfrac{f}{g}$?

7. Determine the symmetry, if any, with respect to the x-axis, y-axis, and origin for the following relations:

   (a) $y = x^2 - 5$ (b) $4y^2 + 3x^2 = 12$

   (c) $xy = 4$ (d) $y = 2x + 7$

8. Graph the following functions:

   (a) $y = (x - 3)^2 - 2$

   (b) $y = -|x| + 3$

   (c) $y = \sqrt{16 - (x - 1)^2}$

   (d) $y = -x^3 + 7$

9. Determine whether the following functions are even, odd, or neither:

   (a) $f(x) = \left| \dfrac{1}{x} \right|$

   (b) $g(x) = 4 + \sqrt[3]{x}$

   (c) $h(x) = |x| + x$

   (d) $p(x) = \sqrt{x}(x^2 - 9)$

10. Determine the values of the constants $a, b, c,$ and d in order for the function defined by $f(x) = ax^3 + bx^2 + cx + d$ to be odd.

11. A ladder leaning against a 4-foot high wall is 3 feet away from the bottom of the wall. Use the Pythagorean Theorem to find the length of the ladder. If the perimeter of the right triangle formed by the ladder and the wall is held constant but the ladder is pulled down the wall 1 foot, then will the area of the triangle increase, decrease, or remain the same?

12. Given a function f mapping A into B, which is 1-1 and a function g mapping B into C, which is also 1-1. Prove that $g \circ f$ mapping A into C is 1-1.

13. Consider the functions f and g defined by $f(x) = 2x - 1$ and $g(x) = 4x^2 - 2x + 3$. Find:
    (a) $f \circ g$ (b) $g \circ f$
    (c) $f \circ f$ (d) $g \circ g$

14. Let f and g be defined by $f(x) = x^3 + 1$ and $g(x) = 3x - 1$. Compute:
    (a) $(g \circ f)(-1)$ (b) $(f \circ g)(0)$
    (c) $(f \circ f)(3)$ (d) $(g \circ g)(-5)$

15. Let $(-1, 1)$ be the domain of f, where f is defined by $f(x) = \dfrac{x}{1 + |x|}$. Prove that f is one-to-one.

16. Given $f(x) = \dfrac{4x - 1}{2x + 3}$, find $f^{-1}(x)$.

17. Given $g(x) = \sqrt[3]{5x + 2}$, find $g^{-1}(x)$.

18. Why is it that a polynomial of even degree cannot have an inverse?

19. Determine the value of the constant a so that the function defined by $f(x) = \dfrac{x + 5}{x + a}$ will be its own inverse.

20. Let f be the constant function defined by $f(x) = c$. Compute $\dfrac{f(x + h) - f(x)}{h}$, $h \neq 0$.

CHAPTER 3

POLYNOMIALS AND RATIONAL FUNCTIONS

INTRODUCTION

In this chapter we discuss certain types of functions called polynomial functions. We start by defining polynomials, polynomial functions, and the terminology associated with these concepts. In the following sections we analyze, in detail, particular polynomial functions: linear functions, systems of linear functions, quadratic functions, and some higher degree polynomials. In the last section we discuss a class of functions which are expressed as quotients of polynomials. Such a class of functions is called rational functions, and we shall investigate their properties.

3.1 POLYNOMIALS AND POLYNOMIAL FUNCTIONS

Recall that we have dealt with expressions such as $x^2 + 4x + 2$ and $x^3 - 2$. These expressions are examples of polynomials, and we now give a formal definition.

Definition: A **polynomial** is an expression of the form

$$a_n x^n + a_{n-1} x^{n-1} + \cdots + a_1 x + a_0,$$

where n is a nonnegative integer and $a_n, a_{n-1}, \ldots, a_1, a_0$ are called **coefficients.**

EXAMPLE 1

Consider the expression

$$-6x^2 + 4x + 7.$$

It is easy to see that this expression is a polynomial by putting $n = 2$, $a_2 = -6$, $a_1 = 4$, and $a_0 = 7$ in the definition.

EXAMPLE 2

The expression $\sqrt{2}\,x$ is a polynomial where $n = 1$, $a_1 = \sqrt{2}$, and $a_0 = 0$.

EXAMPLE 3

Consider the expression -5. If we put $n = 0$, $a_0 = -5$, then we see that the constant -5 is a polynomial.

EXAMPLE 4

The expression $x^{-1} + 3$ is *not* a polynomial because the exponent of x is -1 and the exponents of x must be nonnegative.

EXAMPLE 5

The expression $4x^3 - \sqrt{x} + 5$ is *not* a polynomial. Note that one of the terms, $\sqrt{x} \equiv x^{1/2}$, has a fraction as an exponent, and polynomials have nonnegative integers as exponents.

Terminology

Each $a_i x^i$, where $i = 0, 1, 2, \ldots, n$, in the definition of a polynomial is called a **term**. In particular, a_0 is called the **constant term** of the polynomial and a_n is the **leading coefficient**. If $a_n \neq 0$, we say that the polynomial has **degree** n. Therefore, the degree of a polynomial is the highest exponent of x that appears among the terms with a nonzero coefficient. The following examples illustrate these terms.

EXAMPLE	DEGREE	LEADING COEFFICIENT	CONSTANT TERM
1. $6x^3 - 2x^2 + 5$	3	6	5
2. $x^2 - x + \sqrt{2}$	2	1	$\sqrt{2}$
3. $2x - 3$	1	2	-3
4. 7	0	7	7
5. 0	no degree	0	0
6. $0x^3 + 0x^2 + 0x + 0$	no degree	0	0

Note that the constant polynomial 7 has a degree of zero; however, any polynomial having all its coefficients equal to zero has *no degree*. See Examples 5 and 6. Such a polynomial is called the **zero polynomial**. Other polynomials also have special names:

$$a_0 \quad \textbf{constant polynomial}$$
$$a_1 x + a_0 \quad \textbf{linear polynomial}$$
$$a_2 x^2 + a_1 x + a_0 \quad \textbf{quadratic polynomial}$$
$$a_3 x^3 + a_2 x^2 + a_1 x + a_0 \quad \textbf{cubic polynomial}$$
$$a_4 x^4 + a_3 x^3 + a_2 x^2 + a_1 x + a_0 \quad \textbf{quartic polynomial}$$
$$a_5 x^5 + a_4 x^4 + a_3 x^3 + a_2 x^2 + a_1 x + a_0 \quad \textbf{quintic polynomial}$$

Equality of Polynomials

Two polynomials

$$a_n x^n + a_{n-1} x^{n-1} + \cdots + a_1 x + a_0$$

and

$$b_m x^m + b_{m-1} x^{m-1} + \cdots + b_1 x + b_0$$

are said to be **equal** if and only if

$$m = n$$

and

$$a_i = b_i \text{ for } i = 0, 1, 2, \ldots, n.$$

That is, two polynomials are equal if they have the same degree and all their coefficients are the same.

EXAMPLE 6

The polynomials $4x^2 - 7x + 3$ and $4x^2 + 7x + 3$ are *not* equal because, while the degree of each is 2, the second terms have different coefficients, -7 and 7 respectively.

Polynomial Functions

If we substitute a number for x in a polynomial, then we get a unique number. Thus, a polynomial can be thought of as a single-valued mapping from a set of numbers to another set of numbers. For example, if we substitute a real number for x, then the polynomial yields a real number. Therefore, every polynomial defines a function.

Definition: A **polynomial function** p is a function defined by

$$p(x) = a_n x^n + a_{n-1} x^{n-1} + \cdots + a_1 x + a_0$$

EXAMPLE 7

Let p be the function defined by the quadratic polynomial

$$p(x) = x^2 - 2x + 5.$$

If x is a real number, then p maps the reals to the reals. In particular

$$p(0) = 0^2 - 2(0) + 5 = 5,$$
$$p(-1) = (-1)^2 - 2(-1) + 5 = 8,$$
$$p(1) = (1)^2 - 2(1) + 5 = 4.$$

EXAMPLE 8

The function p defined by $p(x) = c$, where c is a real number, is called the **constant function** (see Fig. 65).

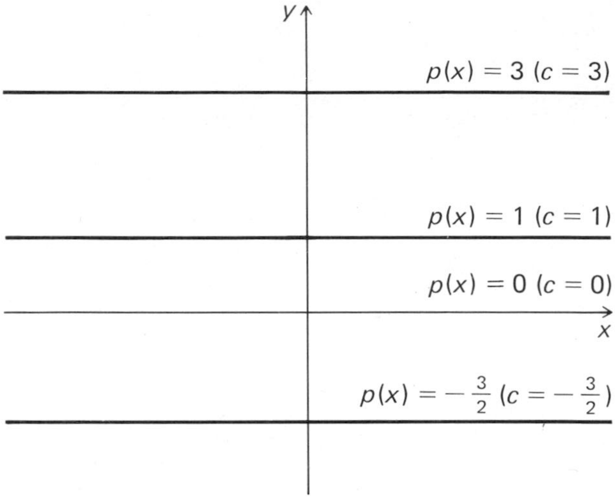

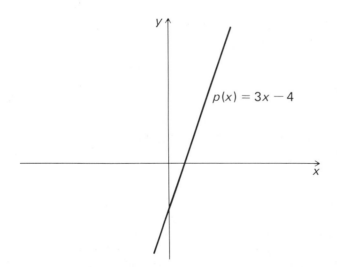

$p(x) = 3 \; (c = 3)$

$p(x) = 1 \; (c = 1)$

$p(x) = 0 \; (c = 0)$

$p(x) = -\frac{3}{2} \; (c = -\frac{3}{2})$

Figure 65

EXAMPLE 9

A function p defined by the expression $p(x) = mx + b$, where m and b are constants, is called a **linear function.** In particular, if we set $m = 3$ and $b = -4$, then we have the linear function defined by $p(x) = 3x - 4$ (see Fig. 66).

$p(x) = 3x - 4$

Figure 66

EXAMPLE 10

The function p defined by the expression $p(x) = ax^2 + bx + c$, where a, b, and c are constants, is called the **quadratic function.**

> **Definition:** If p is a polynomial function defined by $p(x)$, then any value of x which makes $p(x) = 0$ is called a **root** (or **zero**) of the polynomial function p.

EXAMPLES

If $p(x) = x^4 - 2x^2 - 3x - 2$, then

(1) $p(-1) = (-1)^4 - 2(-1)^2 - 3(-1) - 2 = 1 - 2 + 3 - 2 = 0$ and $x = -1$ is a root of $p(x)$.

(2) $p(0) = (0)^4 - 2(0)^2 - 3(0) - 2 = -2$ and $x = 0$ is *not* a root of $p(x)$.

(3) $p(2) = (2)^4 - 2(2)^2 - 3(2) - 2 = 16 - 8 - 6 - 2 = 0$ and $x = 2$ is a root of $p(x)$.

Graphically, roots are x-coordinates of points where the graph of the function crosses or touches the x-axis.

In later sections we shall discuss linear functions and quadratic functions in more detail. Of particular interest will be roots of linear and quadratic functions. Before doing so, we discuss the algebra of polynomials.

Addition of Polynomial Functions

The addition of polynomial functions is achieved by adding the coefficients of equal powers of the variable. We shall see that the sum of two polynomial functions is also a polynomial function.

EXAMPLE 11

Let p and q be polynominal functions defined by

$$p(x) = 4x^3 - 7x^2 + 5$$
$$q(x) = 2x^3 + 5x^2 + x - 2.$$

The sum $p + q$ is defined by

$$
\begin{aligned}
p(x) + q(x) &= (4x^3 - 7x^2 + 5) + (2x^3 + 5x^2 + x - 2) \\
&= (4x^3 + 2x^3) + (-7x^2 + 5x^2) + x + (5 - 2) \\
&= 6x^3 - 2x^2 + x + 3.
\end{aligned}
$$

Thus $p(x) + q(x) = 6x^3 - 2x^2 + x + 3$ is a polynomial function.

EXAMPLE 12

If q and r are polynomial functions defined by

$$q(x) = x^2 + 3x - 1$$
$$r(x) = -x^2 - 3x + 1$$

then the sum $q + r$ is defined by

$$
\begin{aligned}
q(x) + r(x) &= (x^2 + 3x - 1) + (-x^2 - 3x + 1) \\
&= (x^2 - x^2) + (3x - 3x) + (-1 + 1) \\
&= 0 + 0 + 0 \\
&= 0
\end{aligned}
$$

Thus the sum $q + r$ is the zero polynomial.

EXAMPLE 13

The sum of the polynomial functions p, q, and r which are defined by

$$p(x) = x + 10$$

$$q(x) = -x^2 + 7x - 3$$

$$r(x) = x^4 + 4x^3 - 6x^2 + 8x + 1$$

respectively, is given by

$$
\begin{aligned}
p(x) + q(x) + r(x) &= (x + 10) + (-x^2 + 7x - 3) + (x^4 + 4x^3 - 6x^2 + 8x + 1) \\
&= x^4 + 4x^3 + (-x^2 - 6x^2) + (x + 7x + 8x) + (10 - 3 + 1) \\
&= x^4 + 4x^3 - 7x^2 + 16x + 8.
\end{aligned}
$$

Thus, $p(x) + q(x) + r(x) = x^4 + 4x^3 - 7x^2 + 16x + 8$.

Note that to add polynomial functions we need only add the coefficients. Therefore, the laws which numbers obey should also apply to polynomial functions, and such is the case.

LAWS FOR THE ADDITION OF
POLYNOMIAL FUNCTIONS

If p, q, and r are polynomial functions defined by $p(x)$, $q(x)$, and $r(x)$ respectively, then

1. $p(x) + q(x) = q(x) + p(x)$	Commutative Law
2. $[p(x) + q(x)] + r(x) = p(x) + [q(x) + r(x)]$	Associative Law
3. $p(x) + 0 = p(x)$	Additive Identity
4. $p(x) + [-p(x)] = 0$	Additive Inverse

Multiplication of Polynomial Functions

In order to multiply polynomials we must recall some facts from algebra:

1. $x^m \cdot x^n = x^{m+n}$
2. When multiplying two polynomials, we must multiply each term in one polynomial by each term in the other polynomial.

EXAMPLE 14

Let p and q be polynomial functions defined by

$$p(x) = x^2 - 7x + 4$$

$$q(x) = 5x^3 + x^2 - 2x.$$

The product is given by the following scheme

$$
\begin{array}{lr}
x^2 - 7x + 4 & p(x) \\
5x^3 + x^2 \div 2x & q(x) \\
\hline
-2x^3 + 14x^2 - 8x & -2xp(x) \\
x^4 - 7x^3 + 4x^2 & x^2p(x) \\
5x^5 - 35x^4 + 20x^3 & 5x^3p(x) \\
\hline
5x^5 - 34x^4 + 11x^3 + 18x^2 - 8x & p(x) \cdot q(x)
\end{array}
$$

We now state a theorem which shortens the work involved in multiplying polynomial functions. What it states is that multiplication of polynomial functions distributes over addition.

Theorem 1: Let p, q, and r be polynomial functions defined by $p(x)$, $q(x)$, and $r(x)$. Then we have

$$p(x)[q(x) + r(x)] = p(x)q(x) + p(x)r(x).$$

EXAMPLE 15

Let p and q be polynomial functions defined by

$$p(x) = 2x - 1$$

$$q(x) - 7x^2 - 10x + 4$$

Using Theorem 1 we get

$$
\begin{aligned}
p(x)q(x) &= (2x - 1)(7x^2 - 10x + 4) \\
&= 7x^2(2x - 1) - 10x(2x - 1) + 4(2x - 1) \\
&= 14x^3 - 7x^2 - 20x^2 + 10x + 8x - 4 \\
&= 14x^3 - 27x^2 + 18x - 4
\end{aligned}
$$

We now list the laws for the multiplication of polynomial functions.

LAWS FOR THE MULTIPLICATION OF
POLYNOMIAL FUNCTIONS

If p, q, and r are polynomial functions defined by $p(x)$, $q(x)$, and $r(x)$, respectively, then

1. $p(x) \cdot q(x) = q(x) \cdot p(x)$ Commutative Law
2. $[p(x) \cdot q(x)]r(x) = p(x) \cdot [q(x) \cdot r(x)]$ Associative Law
3. $p(x) \cdot [q(x) + r(x)] = p(x) \cdot q(x) + p(x) \cdot r(x)$ Distributive Law
4. $p(x) \cdot 1 = p(x)$ Multiplicative Identity

Note that there is no law for the existence of a polynomial to serve as the multiplicative inverse for polynomial functions. If a polynomial function has an inverse, say $h(x)$, then $p(x) \cdot h(x) = 1$. The multiplicative identity for polynomial functions is the constant polynomial 1. If we let p be a polynomial function defined by $p(x) = x^4$, then its multiplicative inverse $h(x)$ is obtained using $p(x)h(x) = 1$. Thus $x^4 \cdot h(x) = 1$ and we have $h(x) = \dfrac{1}{x^4} = x^{-4}$. However, x^{-4} is not a polynomial and $p(x) = x^4$ does not have a multiplicative inverse which is a polynomial. Further investigation would show that the only polynomial functions which have multiplicative inverses which are polynomial functions are nonzero constant polynomial functions. For example, if p is a polynomial function defined by $p(x) = 7$, then the inverse polynomial function is defined by $h(x) = \dfrac{1}{7}$. This is shown to be the case because $p(x) \cdot h(x) = 7 \cdot \dfrac{1}{7} = 1$.

Division of Polynomial Functions

Recall from arithmetic that if we divide one positive integer P by another positive integer D, then there exist unique integers Q and R, where $0 \leq R < D$ such that $P = D \cdot Q + R$. For example if $P = 132$ and $D = 9$, we have

$$
\begin{array}{r}
14 \longleftarrow Q \\
D \longrightarrow 9)\overline{132} \longleftarrow P \\
9 \\
\overline{42} \\
36 \\
\overline{6} \longleftarrow R
\end{array}
$$

and $132 = 9 \cdot 14 + 6$. The number P is called the **dividend,** D the **divisor,** Q the **quotient,** and R the **remainder.** The process for determining Q and R is called **long division.** There is an analogous process for polynomial functions.

EXAMPLE 16

If p and d are polynomial functions defined by $p(x) = 2x^4 + 3x^3 - x^2 - 1$ and $d(x) = x - 2$, then we obtain the quotient and remainder in the following manner:

$$
\begin{array}{r}
2x^3 + 7x^2 + 13x + 26 \quad \longleftarrow \text{quotient} \\
x - 2)\overline{2x^4 + 3x^3 - x^2 - 1} \\
\underline{2x^4 - 4x^3} \\
7x^3 - x^2 - 1 \\
\underline{7x^3 - 14x^2} \\
13x^2 - 1 \\
\underline{13x^2 - 26x} \\
26x - 1 \\
\underline{26x - 52} \\
51 \longleftarrow \text{remainder}
\end{array}
$$

The $2x^3$ in the quotient is obtained by dividing x in the divisor into the first term of the dividend, and then proceeding as in grade school long division. The remainder will always be a polynomial of degree less than that of the divisor or the zero polynomial.

Thus the quotient is the polynomial q defined by the expression $q(x) = 2x^3 + 7x^2 + 13x + 26$ and the remainder is the constant polynomial r defined by $r(x) = 52$. Also, we have $p(x) = d(x) \cdot q(x) + r(x)$.

EXAMPLE 17

Let p and d be functions defined by

$$p(x) = x^3 - 8$$

$$d(x) = x - 2$$

Then

$$
\begin{array}{r}
x^2 + 2x\ \ + 4 \quad \longleftarrow \text{quotient} \\
x - 2 \overline{) x^3 \qquad\quad - 8\ \ } \\
\underline{x^3 - 2x^2} \\
2x^2 \qquad - 8 \\
\underline{2x^2 - 4x} \\
4x - 8 \\
\underline{4x - 8} \\
0 \longleftarrow \text{remainder}
\end{array}
$$

Thus q and r are defined by $q(x) = x^2 + 2x + 4$ and $r(x) = 0$, respectively. Note that the remainder polynomial is the zero polynomial; that is, $r(x) = 0$. In such cases we call $d(x)$ a **factor** of $p(x)$.

EXAMPLE 18

The expression $d(x) = x + 1$ is a factor of the expression $p(x) = x^3 + 1$. To see this we divide $p(x)$ by $d(x)$:

$$
\begin{array}{r}
x^2 - x\ \ + 1 \\
x + 1 \overline{) x^3 \qquad\quad + 1\ \ } \\
\underline{x^3 + x^2} \\
- x^2 \qquad + 1 \\
\underline{- x^2 - x} \\
x + 1 \\
\underline{x + 1} \\
0 \longleftarrow \text{remainder}
\end{array}
$$

Note that $x^3 + 1 = (x + 1)(x^2 - x + 1) + 0$, and $p(-1) = (-1 + 1)q(x) = 0$, which is the same as the remainder.

EXAMPLE 19

The expression $d(x) = x - 3$ is *not* a factor of $p(x) = x^3 + 2x^2 - x + 1$ because the remainder is not zero.

$$
\begin{array}{r}
x^2 + 5x\ + 14 \qquad \longleftarrow q(x) \\
x - 3\overline{)x^3 + 2x^2 -\quad x +\ \ 1} \\
\underline{x^3 - 3x^2} \\
5x^2 -\quad x +\ \ 1 \\
\underline{5x^2 - 15x} \\
14x +\ \ 1 \\
\underline{14x - 42} \\
43 \longleftarrow \text{remainder}
\end{array}
$$

Thus $p(x) = d(x) \cdot q(x) + r(x)$ or, equivalently,

$$x^3 + 2x^2 - x + 1 = (x - 3)(x^2 + 5x + 14) + 43$$

Also note that

$$p(3) = (3 - 3)q(x) + 43 = 43 = \text{remainder}$$

In general, if p and d are polynomial functions defined by $p(x)$ and $q(x)$, then there exist polynomial functions q and r, defined by $q(x)$ and $r(x)$, such that

$$p(x) = d(x) \cdot q(x) + r(x)$$

where $p(x)$ is the dividend, $d(x)$ is the divisor, $q(x)$ is the quotient, and $r(x)$ is the remainder. If $r(x) = 0$, then $d(x)$ is a factor of $p(x)$. However if $r(x) \neq 0$, then the degree of $r(x)$ is less than the degree of $d(x)$.

The Remainder and Factor Theorems

Examples 18 and 19 lead us to two very important theorems.

The Remainder Theorem: If p is a polynomial function defined by $p(x)$, and $p(x)$ is divided by $x - a$, then the remainder is given by $p(a)$. That is, the remainder is the number given by substituting the number a for x in the polynomial $p(x)$.

Proof. Recall that if we divide $p(x)$ by $x - a$, then there exist $q(x)$ and $r(x)$ such that

$$p(x) = (x - a)q(x) + r(x)$$

The remainder $r(x)$ must be 0 or have degree less than the degree of $x - a$. Since the degree of $x - a$ is 1, it follows that the remainder $r(x)$ must be a constant, say C. Thus we have

$$p(x) = (x - a)q(x) + C$$

Now substituting a for x in $p(x)$ we get

$$p(a) = (a - a)q(a) + C$$

$$p(a) = 0 \cdot q(a) + C$$

$$p(a) = C$$

Thus, $p(a) = C$, the remainder.

EXAMPLE 20

If p is a polynomial function defined by $p(x) = x^3 - 3x^2 + 2x + 7$, the remainder when $p(x)$ is divided by $x - 3$ is $p(3) = (3)^3 - 3(3)^2 + 2(3) + 7 = 27 - 27 + 6 + 7 = 13$. Note that since the remainder is not zero, $x - 3$ is not a factor of $p(x)$.

EXAMPLE 21

If p is defined by $p(y) = y^3 - 125$ and is divided by $y - 5$, then the remainder is $p(5) = (5)^3 - 125 = 125 - 125 = 0$. Thus $y - 5$ is a factor of $p(y)$.

EXAMPLE 22

Let p be a polynomial function defined by $p(x) = x^3 - 4x^2 + 5x + 1$. Divide $p(x)$ by $x + 2$. *Note that we must write $x + 2$ in the form $x - a$; that is, $x + 2 = x - (-2)$ and $a = -2$. Thus, the remainder is the value $p(-2) = (-2)^3 - 4(-2)^2 + 5(-2) + 1 = -8 - 4(4) - 10 + 1 = -8 - 16 - 10 + 1 = -33$.

EXAMPLE 23

If $p(x) = x^5 + x^2 + x + 1$ and we divide by $x + 1$, the remainder is $p(-1) = (-1)^5 + (-1)^2 + (-1) + 1 = -1 + 1 - 1 + 1 = 0$. Since the remainder equals zero, the polynomial $x + 1$ is a factor of $p(x)$.

A direct consequence of the Remainder Theorem is the following theorem.

The Factor Theorem: Let p be a polynomial function defined by $p(x)$. Then $x - a$ is a factor of $p(x)$ if and only if $p(a) = 0$.

Proof. By the Remainder Theorem

$$p(x) = (x - a)q(x) + r(x)$$

If $x - a$ is a factor of $p(x)$, then $r(x) = 0$. Therefore, $p(x) = (x - a)q(x)$, and $p(a) = (a - a)q(x) = 0 \cdot q(x) = 0$.
If $p(a) = 0$, then $p(a) = (a - a)q(x) + r(x) = 0 \cdot q(x) + r(x) = r(x) = 0$. Thus $r(x) = 0$ and $p(x) = (x - a)q(x)$, which says that $x - a$ is a factor of $p(x)$.

EXAMPLE 24

Let $p(x) = x^2 - 5x + 6$. To see whether $x - 2$ is a factor we evaluate $p(2)$:

$$p(2) = (2)^2 - 5(2) + 6 = 4 - 10 + 6 = 0$$

Therefore, the Factor Theorem says that $x - 2$ is a factor.

EXAMPLE 25

Let $p(x) = x^2 - x - 6$. If $x + 2$ is a factor of $p(x)$, then the Factor Theorem says that $p(-2) = 0$.

Synthetic Division

Let us consider a simplified method of dividing a polynomial $p(x)$ by a divisor of the form $x - a$. This method determines the values of the coefficients of the quotient and the value of the remainder can easily be determined. The method is called **synthetic division** and consists of working with only the coefficients of the variable.

Suppose we wish to divide $p(x) = 2x^4 + 5x^3 + 4x + 7$ by $d(x) = x + 3$. Note that $x + 3 = x - (-3)$ and therefore $a = -3$.

$$
\begin{array}{r|rrrrr}
-3 & 2 & 5 & 0 & 4 & 7 \\
   &   & -6 & 3 & -9 & 15 \\
\hline
   & 2 & -1 & 3 & -5 & 22 \\
\end{array}
$$

The top row gives the coefficients of the dividend $p(x)$. *Note that we insert zero as the coefficient for any missing power x;* in this case the x^2 term is missing. The -3 is the second term of the divisor $d(x)$ with the sign changed. We bring down the first coefficient (2), then multiply it by the divisor (-3), and write the result (-6) under the next coefficient (5). We then add to obtain (-1). Now we repeat the process. Multiplying (-1) by (-3) and writing the result under the next coefficient (0) we add to obtain (3). The procedure is continued until we get the last number (22). This bottom row gives the coefficients of the quotient with the last value (22) being the remainder. That is, the quotient is

$$q(x) = 2x^3 - x^2 + 3x - 5$$

with the remainder equal to 22. Note that since the dividend is of the fourth degree and the divisor is of the first degree, the quotient is of the third degree. To use synthetic division the coefficient of x in the divisor must be 1.

EXAMPLE 26

Let $p(x) = x^5 + 1$. We now use synthetic division to divide $p(x)$ by $x + 1$.

$$
\begin{array}{r|rrrrrr}
-1 & 1 & 0 & 0 & 0 & 0 & 1 \\
   &   & -1 & 1 & -1 & 1 & -1 \\
\hline
   & 1 & -1 & 1 & -1 & 1 & 0 \\
\end{array}
$$

Thus the quotient $q(x) = x^4 - x^3 + x^2 - x + 1$, with a remainder of 0. Since the remainder is equal to zero, the Remainder Theorem implies that $x + 1$ is a factor of $x^5 + 1$.

EXAMPLE 27

Let $p(x) = x^6 - 7x^5 + 2x^4 - 15x^3 + 48x + 5$. Suppose we wish to compute $p(7)$.

Of course, we could substitute 7 for x in $p(x)$. However, we immediately see that would involve computing a large number. A much shorter method would be to use synthetic division. Recall that the Remainder Theorem says that $p(7)$ is the remainder when $p(x)$ is divided by $x - 7$. Using synthetic division we get

$$
\begin{array}{r|rrrrrrr}
7 & 1 & -7 & 2 & -15 & 0 & 48 & 5 \\
  &   & 7 & 0 & 14 & -7 & -49 & -7 \\
\hline
  & 1 & 0 & 2 & -1 & -7 & -1 & -2 \leftarrow p(7)
\end{array}
$$

Thus, $p(7) = -2$.

Synthetic division can be used to determine if a number is a root of a polynomial.

EXAMPLE 28

If p is the polynomial function defined by $p(x) = 2x^3 + 4x^2 - 46x - 120$, then we use synthetic division to show that 5 is a root of $p(x)$; that is, $p(5) = 0$.

We compute $p(5)$ as the remainder of dividing $p(x)$ synthetically by 5.

$$
\begin{array}{r|rrrr}
5 & 2 & 4 & -46 & -120 \\
  &   & 10 & 70 & 120 \\
\hline
  & 2 & 14 & 24 & 0 \leftarrow p(5)
\end{array}
$$

Thus $p(5) = 0$ and 5 is a root of $p(x)$.

3.1 EXERCISES

1. List the coefficients $a_0, a_1, \ldots, a_n$ for the polynomial functions defined by the following expressions. State the degree.

   (a) $p(x) = 2x + 7$

   (b) $p(x) = -5$

   (c) $p(x) = -3x^2 + x$

   (d) $p(x) = 7x^3 + 4x^2 + \sqrt{2}$

   (e) $p(x) = mx + b$

   (f) $p(x) = 5 - x^3$

   (g) $p(x) = \sqrt{3}x^4 + 7x - 5$

   (h) $p(x) = x^{10}$

   (i) $p(x) = ax^2 + bx + c$

   (j) $p(x) = 17x^4 - 12x^3 + 5x^2 + 3x - 1$

2. Determine which of the following are expressions which define a polynomial function. State the degree of those which do represent polynomial functions.

(a) $p(x) = x + \sqrt{7}$

(b) $p(x) = 5\sqrt{x} + x - 2$

(c) $p(x) = \dfrac{2}{3}$

(d) $p(x) = \sqrt{3}x^2 - x + 5$

(e) $p(x) = \dfrac{7}{x} + x^2 - 1$

(f) $p(x) = x^{-2} + 3x^4 - 5x + 14$

(g) $p(x) = 0x^3 + 0x + 0$

(h) $p(x) = 0$

(i) $p(x) = 2^x$

(j) $p(x) = \dfrac{x^2 + 3x - 1}{x - 1}$

3. Let p be a polynomial function defined by $p(x)$ in each of the following. Evaluate each for the given values of x.

(a) $p(x) = x^3 - 2x^2 + x - 1; p(0), p(-1), p(1)$

(b) $p(x) = \sqrt{2}x - x; p(3), p(\sqrt{2}), p(a)$

(c) $p(x) = x^2 - x - 6; p(0), p(3), p(-2)$

(d) $p(x) = mx + b; \quad p(0), \quad p\left(-\dfrac{b}{m}\right), \quad p(x + h)$

(e) $p(x) = ax^2 + bx + c; \quad p(1), \quad p\left(\dfrac{1}{2}\right), \quad p(x + h)$

4. For each of the following pairs of polynomials $p(x)$ and $q(x)$ find $p(x) + q(x)$:

(a) $p(x) = 4x - 1, q(x) = -4x + 1$

(b) $p(x) = x^2 - 3x + 5, q(x) = x^3 - 5$

(c) $p(x) = 7, q(x) = 5 - x + 3x^3 + x^4$

(d) $p(x) = \dfrac{x^5}{5} - \dfrac{x^4}{4} + \dfrac{x^3}{3} - \dfrac{x^2}{2} + 1$,

$q(x) = -x^5 + \dfrac{x^4}{4} + \dfrac{x^2}{2} - 1$

5. Given $p(x)$ and $q(x)$, find a polynomial $r(x)$ such that $p(x) = q(x) + r(x)$ in each of the following pairs:

(a) $p(x) = x + 7, q(x) = x^2 - 7x - 5$

(b) $p(x) = 4x^3 - x^2 + x - 3, q(x) = 4x^3 - x^2 + x - 4$

(c) $p(x) = x^4 - 1, q(x) = 3x^3 - 7x^2 + 5x$

6. Perform the following multiplications:

(a) $(x + 4)(x - 4)$

(b) $(4x^2 + 2x + 1)(2x - 1)$

(c) $(3x - 4)(x^3 - 7x^2 + 5x - 2)$

(d) $(x^2 + 2)(2x^2 + 3x - 4)$

(e) $x(x^2 - ax + a^2)(x + a)$

7. Let $p(x)$ be a dividend and $d(x)$ a divisor. Find the quotient $q(x)$ and the remainder $r(x)$ if $p(x)$ is divided by $d(x)$.

(a) $p(x) = 4x^4 - 5x^3 + 7x^2 + x + 3;$ $d(x) = 4x^2$

(b) $p(x) = 4x^3 - 8x^2 + x - 2; d(x) = x - 1$

(c) $p(x) = 3x^3 - 10x^2 + 21x - 26;$ $d(x) = x - 2$

(d) $p(x) = x^3 + b^3; d(x) = x + b$

(e) $p(x) = 8x^8 - 1; d(x) = 2x^2 - 1$

8. Use the Remainder Theorem to find the remainder if we divide $p(x)$ by $d(x)$ in each of the following:

(a) $p(x) = 7x^{10} - 5x^5 + 1; d(x) = x + 1$

(b) $p(x) = 13x^{25} + 4x^{13} - 5;$ $d(x) = x - 1$

(c) $p(x) = x^3 - 3x^2 + 10; d(x) = x + 3$

9. If $p(x) = x^2 - 4x + 9$ is divided by $x + c$, the remainder is 14. Find c.

10. Use the Factor Theorem to determine if the first polynomial is a factor of the second polynomial in each of the following:
   (a) $x + 2$; $x^3 - 4x^2 + x - 5$
   (b) $x + 1$; $x^{13} + 1$
   (c) $x - \dfrac{1}{2}$; $2x^4 + 5x^3 - 3x^2 - 4x + 2$
   (d) $x - 3$; $x^4 - 4x^3 - 7x^2 + 22x + 24$
   (e) $x - a$; $x^n - a^n$

11. Find k such that $x + 1$ is a factor of $x^3 + 3x^2 + kx + 1$.

12. Which of the polynomials, if any, $x - 5$, $x + 1$, $x + 3$, are factors of $x^3 - 3x^2 - 13x + 15$?

13. Use synthetic division to find the quotient and remainder of:
   (a) $(x^4 - 5x^3 + 6x^2 + 4x - 8) \div (x - 2)$
   (b) $(6x^3 - 7x^2 + 1) \div \left(x + \dfrac{1}{3}\right)$
   (c) $(4x^4 - 17x^2 - 2) \div (x + 3)$
   (d) $(x^5 - 2) \div (x - \sqrt[5]{2})$
   (e) $(x^6 + a^6) \div (x + a)$

14. Use synthetic division to find the function values of the following:
   (a) $p(x) = x^4 - x^3 + x + 7$; find $p(-1)$, $p(1)$, $p(2)$
   (b) $p(x) = x^6 - 7x^5 + 2x^2 - 4$; find $p(1)$, $p(-2)$, $p(7)$
   (c) $p(x) = x^5 - 1$; find $p(-1)$, $p(1)$, $p(a)$

15. Determine whether the given numbers are roots of the given polynomial expression.
   (a) $2, 1$; $p(x) = 2x^3 - 2x - 12$
   (b) $\dfrac{7}{2}, -5$; $p(x) = x^3 - 2x^2 - 31x + 20$
   (c) $\sqrt{6}, 0$; $p(x) = 5x^3 - x^2 - 30x + 6$
   (d) $-1, -2$; $p(x) = 4x^6 + 5x^2 + 1$
   (e) $-2 + \sqrt{3}, -2 - \sqrt{3}$;
   $p(x) = x^2 + 4x + 1$

3.2   LINEAR FUNCTIONS

In this section we are interested in first degree polynomial functions. Such functions are also called **linear functions** because their graphs are straight lines. We shall investigate the properties of linear functions and special attention shall be given to their graphs.

Definition: A function f, defined by the equation $f(x) = mx + b$, where m and b are constants, is called a **linear function.** The graph of a linear function is a straight line.

EXAMPLE 1

The function f defined by $f(x) = 3x - 1$ is a linear function. The graph of $f(x) = 3x - 1$ (or $y = 3x - 1$) is obtained by plotting two points, since two distinct points uniquely determine a straight line (see Fig. 67). Note that m is positive ($m = 3$) and that the line slants upward from left to right.

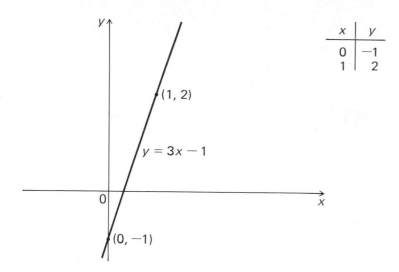

x	y
0	−1
1	2

Figure 67

EXAMPLE 2

The linear function defined by $y = -2x + 1$ is graphed in Figure 68. In this case m is negative ($m = -2$), and the line slants downward from left to right.

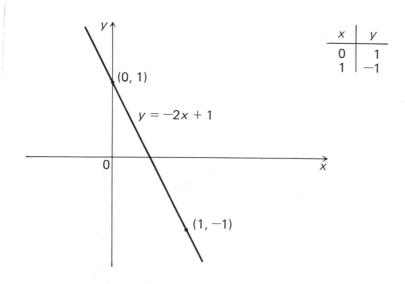

x	y
0	1
1	−1

Figure 68

EXAMPLE 3

Consider the linear function defined by $y = 2$. Here we have $m = 0$; Figure 69 shows that the line is horizontal.

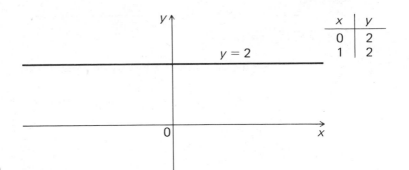

Figure 69

In Examples 1 through 3, we made a special point to note the constant m. The constant m in the expression $y = mx + b$ gives information about the graph of a line. In particular, it gives information about the steepness of a non-vertical line.

The Slope of a Line

Suppose P_1 and P_2 are points with coordinates (x_1, y_1) and (x_2, y_2) on a non-vertical line L (see Fig. 70).

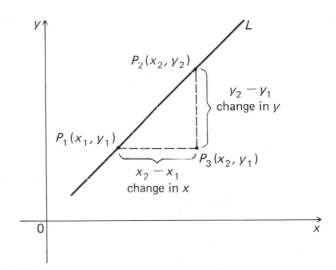

Figure 70

Definition: If (x_1, y_1) and (x_2, y_2) are two distinct points on a line not parallel to the y-axis, then the **slope** of the line is

$$m = \frac{y_2 - y_1}{x_2 - x_1} = \frac{\text{change in } y}{\text{change in } x}$$

PROBLEM 1

Find the slope of the line passing through the points (4, 3) and (2, 1).

Solution: Let $(x_1, y_1) = (4, 3)$ and $(x_2, y_2) = (2, 1)$. Then the slope $m = \dfrac{y_2 - y_1}{x_2 - x_1} = \dfrac{1 - 3}{2 - 4} = \dfrac{-2}{-2} = 1$. Note that if we interchange the points and let $(x_1, y_1) = (2, 1)$ and $(x_2, y_2) = (4, 3)$, then

$$m = \frac{y_2 - y_1}{x_2 - x_1} = \frac{3 - 1}{4 - 2} = \frac{2}{2} = 1.$$

Thus, the slope does not depend on order (see Fig. 71).

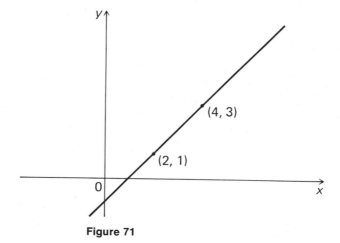

(4, 3)

(2, 1)

Figure 71

What if we choose other points which lie on the line? Will the slope change? To answer these questions we consider the graph in Figure 72. The tri-

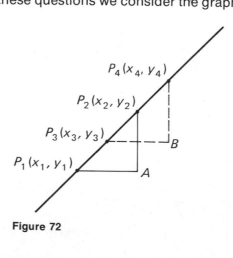

$P_4(x_4, y_4)$

$P_2(x_2, y_2)$

$P_3(x_3, y_3)$

B

$P_1(x_1, y_1)$

A

Figure 72

angles P_1AP_2 and P_3BP_4 are both right triangles and are similar triangles. Therefore, we have proportional sides. That is,

$$\frac{AP_2}{P_1A} = \frac{BP_4}{P_3B}$$

or, equivalently,

$$\frac{y_2 - y_1}{x_2 - x_1} = \frac{y_4 - y_3}{x_4 - x_3}$$

> Hence, the slope of a line does not depend on the particular points used.

PROBLEM 2

Find the slope of the line passing through the points (1, 3) and (3, 0).

Solution: The slope $m = \dfrac{\text{change in } y}{\text{change in } x} = \dfrac{0 - 3}{3 - 1} = \dfrac{-3}{2}$ (see Fig. 73).

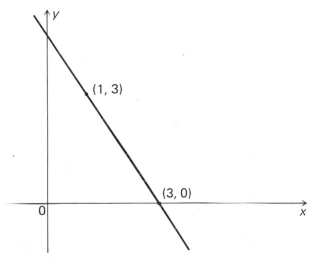

Figure 73

PROBLEM 3

Determine the slope of the line passing through (4, −1) and (−3, −1). Graph the line.

Solution: The slope $m = \dfrac{\text{change in } y}{\text{change in } x} = \dfrac{-1 - (-1)}{4 - (-3)} = \dfrac{0}{7} = 0$ (see Fig. 74).

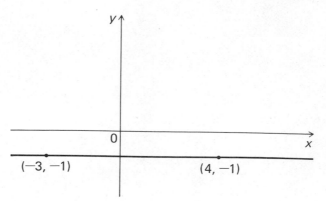

Figure 74

PROBLEM 4

Consider the line passing through (3, 2) and (3, −1). Find its slope, provided it exists. Graph the line.

Solution: The slope $m = \dfrac{\text{change in } y}{\text{change in } x} = \dfrac{-1 - 2}{3 - 3} = \dfrac{-3}{0}$, which is undefined

Thus, when the line contains points with the same first coordinates, the change in x will be zero. In such a case we have zero in the denominator and the slope is undefined. Lines which have the same first coordinates are parallel to the y-axis (see Fig. 75).

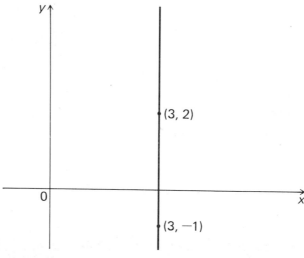

Figure 75

We now summarize information about the slope of a line.

1. If the slope of a line is positive, $m > 0$, then the line slants upward from left to right. Conversely, if the line slants upward from left to right, then the slope is positive.
2. If the slope of a line is negative, $m < 0$, then the line slants downward from left to right. Conversely, if the line slants downward from left to right, then the slope is negative.
3. If the slope of a line is zero, $m = 0$, then the line is parallel to the x-axis. That is, the line is horizontal. Conversely, if the line is horizontal, then the slope is zero.
4. If the slope of a line is undefined, then the line is parallel to the y-axis. That is, the line is vertical. Conversely, if the line is vertical, then the slope is undefined.

Equation of Lines

Up to now we have been given an equation which represents the line or given two points which graphically determine the unique line passing through the two points. We now reverse the situation and see what information is needed to determine an equation of a line.

THE TWO-POINT EQUATION OF A LINE

Suppose we wish to find an equation of a line which passes through two points $P_1(x_1, y_1)$ and $P_2(x_2, y_2)$. Let (x, y) be any other point on the line passing through (x_1, y_1) and (x_2, y_2) (see Fig. 76). Recall that for lines the slope does not

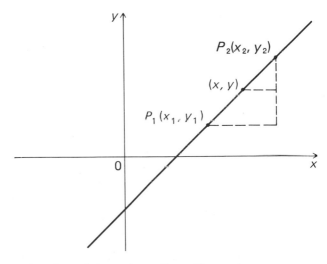

depend on the points used: slope through (x, y) and (x_1, y_1) = slope through (x_1, y_1) and (x_2, y_2). Thus,

$$\frac{y - y_1}{x - x_1} = \frac{y_2 - y_1}{x_2 - x_1}$$

or, equivalently,

$$y - y_1 = \frac{y_2 - y_1}{x_2 - x_1}(x - x_1)$$

$y - y_1 = m(x - x_1)$

Theorem: Two-point Equation
An equation of a nonvertical line passing through points P_1 (x_1, y_1) and P_2 (x_2, y_2) is given by

$$y - y_1 = \frac{y_2 - y_1}{x_2 - x_1}(x - x_1)$$

PROBLEM 5

Find an equation of the line containing the points (4, 3) and (1, -2).

Solution: If we let P_1 be (4, 3) and P_2 be (1, -2), we have

$$y - 3 = \frac{-2 - 3}{1 - 4}(x - 4)$$

which simplifies to $y - 3 = \frac{5}{3}(x - 4)$ or $y = \frac{1}{3}(5x - 11)$. The graph of the line is given in Figure 77.

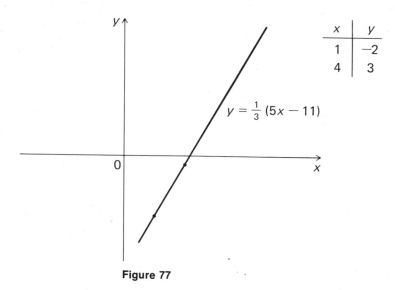

x	y
1	-2
4	3

$y = \frac{1}{3}(5x - 11)$

Figure 77

PROBLEM 6

Find an equation of the line passing through the points (2, 1) and $(-4, 1)$.

Solution: Using the two-point formula gives

$$y - 1 = \frac{1 - 1}{-4 - 2}(x - 2), \text{ or } y - 1 = 0(x - 2)$$

Simplifying, we get $y = 1$ (see Fig. 78).

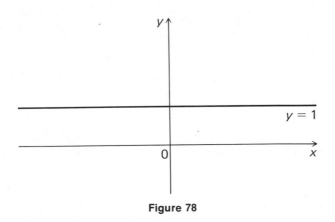

Figure 78

THE POINT-SLOPE EQUATION OF A LINE

Suppose that we are given the slope m of a line and a point $P(x_1, y_1)$ through which the line passes. The following theorem gives an equation of the line satisfying these conditions.

Theorem: Point-slope Equation
An equation of a nonvertical line passing through the point $P_1(x_1, y_1)$ and having slope m is given by

$$y - y_1 = m(x - x_1)$$

Proof. The two-point theorem gives

$$y - y_1 = \frac{y_2 - y_1}{x_2 - x_1}(x - x_1)$$

We know that the slope $m = \dfrac{y_2 - y_1}{x_2 - x_1} = \dfrac{\text{change in } y}{\text{change in } x}$ and we substitute this into the two-point equation. Thus,

$$y - y_1 = m(x - x_1)$$

PROBLEM 7

Find an equation of the line having slope $m = -8$ and passing through the point $(2, -3)$.

Solution: We have $m = -8$ and $(x_1, y_1) = (2, -3)$. Using the point-slope equation we get

$$y - (-3) = -8(x - 2)$$

or, equivalently,

$$y + 3 = -8x + 16$$

which simplifies to

$$y = -8x + 13$$

PROBLEM 8

Find an equation of the horizontal line passing through the point $(5, -4)$.

Solution: Recall that a horizontal line has slope $m = 0$. The point-slope equation gives

$$y - (-4) = 0(x - 5), \text{ or } y + 4 = 0$$

Simplifying, we have

$$y = -4$$

THE SLOPE-INTERCEPT EQUATION OF A LINE

Before we derive the slope-intercept equation we shall discuss the y-intercept of a line. Consider the point at which the line in Figure 79 crosses the y-axis.

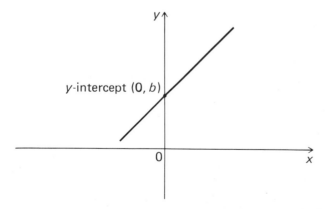

Figure 79

Definition: If a line intersects the y-axis at a point with coordinates $(0, b)$, then b is called the **y-intercept** of the line. Note that the y-intercept is the value we get if we set $x = 0$.

EXAMPLE 4

Given the linear equation $y = 3x + 7$, the y-intercept is then $y = 3(0) + 7 = 7$. We now state a theorem which gives an equation of a line given the slope and y-intercept.

Theorem: Slope-intercept Equation
    An equation of a nonvertical line with slope m and y-intercept b is given by

$$y = mx + b$$

Proof. The point-slope theorem says that given the slope m and point (x_1, y_1), an equation of the line satisfying these conditions is

$$y - y_1 = m(x - x_1)$$

If we are given y-intercept b, then the point which this represents is $(0, b)$. Substituting $(0, b)$ for (x_1, y_1) into the point-slope formula we have

$$y - b = m(x - 0)$$

or, equivalently,

$$y - b = mx$$

Thus,

$$y = mx + b$$

PROBLEM 9

Find an equation of the lIne with slope 4 and y-intercept 5.

Solution: We have $m = -4$ and $b = 5$; therefore, an equation is $y = -4x + 5$.

PROBLEM 10

Find the slope-intercept equation of the line whose equation is $3x - 2y - 4 = 0$.

Solution: The slope-intercept form can be obtained by solving the linear equation $3x - 2y - 4 = 0$ for the variable y. Solving for y we get

$$y = \frac{3}{2}x - 2$$

Thus, we have slope $\frac{3}{2}$ and y-intercept -2.

Note that there is no slope-intercept equation for vertical lines. Such lines have an undefined slope. In general, vertical lines have the equation $x = c$, where c is a constant.

EXAMPLE 5

An equation of the vertical line passing through the point $(5, -2)$ is $x = 5$. It is also interesting to note that, as mentioned in Chapter 2, a vertical line does not represent a function of x.

Parallel Lines and Perpendicular Lines

EXAMPLE 6

Consider the graphs of the two linear equations $-x + y - 2 = 0$ and $-x + y = 0$ (Fig. 80). If we write both equations in slope-intercept form

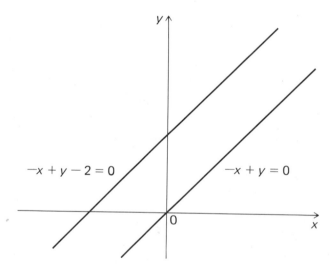

Figure 80

we have $y = x + 2$ and $y = x$, respectively. These equations show that both lines have the same slope, 1, but different y-intercepts. Figure 80 shows that the lines are parallel.

EXAMPLE 7

If we graph the equations $2y = 6x + 4$ and $-3y = -9x - 6$, we see that each equation represents the same line (see Fig. 81). Putting the equations in slope-intercept form, we get $y = 3x + 2$ for both equations. Thus, the lines have the same slope, 3, and the same y-intercept, 2.

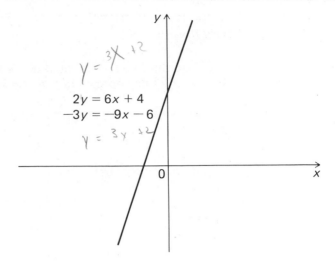

Figure 81

We now state a theorem which describes parallel lines. The proof will be omitted.

Theorem: Parallel Lines
    Let L_1 and L_2 be two nonvertical lines with slopes m_1 and m_2, respectively. Then L_1 and L_2 are parallel if and only if $m_1 = m_2$.

PROBLEM 11

    Find an equation of the line passing through the point (1, 1) and parallel to the line $4x - 8y = 20$.

    Solution: The line we are looking for is parallel to $4x - 8y = 20$. Therefore, the two lines have the same slope. Putting $4x - 8y = 20$ into slope-intercept form we get $y = \frac{1}{2}x - \frac{5}{2}$ and the slope is $\frac{1}{2}$. Thus the line we seek has slope $\frac{1}{2}$ and contains the point (1, 1). Using the point-slope formula we have the equation

$$y - 1 = \frac{1}{2}(x - 1)$$

or, equivalently,

$$y = \frac{1}{2}x + \frac{1}{2}$$

PROBLEM 12

Find the value of k such that the line represented by $7x + ky = 5$ is parallel to the line represented by the equation $-4x + 6y = 20$.

Solution: We put both equations in slope-intercept form to get $y = \dfrac{-7}{k}x + \dfrac{5}{k}$ and $y = \dfrac{2}{3}x + \dfrac{10}{3}$. If the lines are to be parallel, then the slopes must be equal. That is, $\dfrac{-7}{k} = \dfrac{2}{3}$, so $k = \dfrac{-21}{2}$.

PROBLEM 13

The general equation of a line is $Ax + By + C = 0$. If lines $Ax + By + C = 0$ and $Dx + Ey + F = 0$, then what condition must be imposed on the coefficients if the lines are parallel?

Solution: If the lines are parallel and nonvertical, then their respective slopes must be equal. Writing the equations in slope-intercept form we get $y = -\dfrac{A}{B}x - \dfrac{C}{B}$ and $y = \dfrac{-D}{E}x - \dfrac{F}{E}$. Thus equal slopes imply that

$$\frac{-A}{B} = \frac{-D}{E}$$

or, equivalently,

$$BD - AE = 0$$

EXAMPLE 8

Consider the graphs of $y = -2x + 1$ and $y = \dfrac{1}{2}x + 2$ (see Fig. 82).

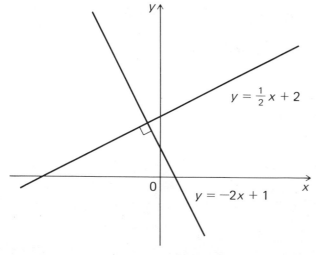

Figure 82

We see that the lines are perpendicular and that their slopes are negative reciprocals. That is, the product of the two slopes is -1.

We state this result as a theorem.

Theorem: Perpendicular Lines

Let L_1 and L_2 be two nonvertical lines with slopes m_1 and m_2, respectively. Then L_1 and L_2 are perpendicular if and only if $m_1 m_2 = -1$. That is,

$$m_1 = \frac{-1}{m_2}.$$

PROBLEM 14

Write an equation of the line passing through $\left(\frac{1}{3}, -2\right)$ and perpendicular to $3x - 4y - 6 = 0$.

Solution: Writing the given line in slope-intercept form we obtain

$$y = \frac{3}{4}x - \frac{3}{2}$$

and we see that it has slope $\frac{3}{4}$. Thus, the slope of any line perpendicular to it is $\frac{-1}{\frac{3}{4}} = \frac{-4}{3}$. We now use the point-slope formula to obtain the desired equation

$$y - (-2) = \frac{-4}{3}\left(x - \frac{1}{3}\right)$$

Simplifying, we get $y + 2 = \frac{-4x}{3} + \frac{4}{9}$ or

$$9y + 12x + 14 = 0$$

PROBLEM 15

Find the conditions on the coefficients such that the lines $Ax + By + C = 0$ and $Dx + Ey + F = 0$ are perpendicular.

Solution: In Problem 13 we saw that the slopes of the lines were $\frac{-A}{B}$ and $\frac{-D}{E}$, respectively. If the lines are to be perpendicular, then $\frac{-A}{B} = \frac{-1}{\frac{-D}{E}}$, or $\dfrac{-A}{B} = \dfrac{E}{D}$. Simplifying, we have the condition that in order for the lines to be perpendicular $BE + AD = 0$.

3.2 EXERCISES

1. Graph the following linear equations:
   (a) $y = x + 3$
   (b) $f(x) = -x$
   (c) $y = 7$
   (d) $3y = 2x + 4$
   (e) $4x - y + 5 = 0$
   (f) $\dfrac{2}{3}x + \dfrac{4}{9}y = \dfrac{1}{9}$

2. Find the slope of the line passing through each of the following pairs of points:
   (a) $(0, 4)$ and $(5, -1)$
   (b) $(-2, 5)$ and $(2, -3)$
   (c) $(0, 0)$ and $(-2, 1)$
   (d) $\left(\dfrac{1}{2}, -2\right)$ and $\left(3, \dfrac{1}{5}\right)$
   (e) $(-1, 6)$ and $(4, 6)$
   (f) $\left(-\dfrac{1}{3}, \dfrac{1}{4}\right)$ and $\left(\dfrac{1}{2}, -1\right)$

3. Discuss whether each line in Exercise 2 slants upward from left to right, slants downward from left to right, or is horizontal.

4. Find an equation of the line passing through each of the following pairs of points:
   (a) $(1, 0)$ and $(0, 1)$
   (b) $(3, 4)$ and $(-2, 5)$
   (c) $\left(\dfrac{1}{2}, -2\right)$ and $\left(3, \dfrac{1}{5}\right)$
   (d) $(0, 0)$ and $(2, -1)$
   (e) $(-7, 4)$ and $(1, 4)$
   (f) $\left(\dfrac{1}{3}, \dfrac{1}{4}\right)$ and $\left(-\dfrac{1}{2}, 1\right)$
   (g) $(0, a)$ and $(b, 0)$
   (h) $(\sqrt{3}, 2)$ and $(1, \sqrt{5})$
   (i) (π, e) and $(3, 0)$
   (j) (A, A) and $(-A, -A)$

5. Write an equation of the vertical line passing through each of the following points:
   (a) $(0, 0)$ (b) $(-1, -1)$
   (c) $(4, 6)$ (d) $(\pi, 5)$
   (e) $(a, -7)$ (f) $\left(\dfrac{2}{3}, b\right)$

6. In each of the following, write an equation of the line which passes through the given point and has slope m:
   (a) $(0, 3)$; $m = -1$
   (b) $(1, -2)$; $m = 3$
   (c) $(-4, -1)$; $m = -2$
   (d) $(5, 3)$; $m = \dfrac{1}{3}$
   (e) $(6, 6)$; $m = 0$
   (f) $\left(\dfrac{1}{3}, -\dfrac{1}{2}\right)$; $m = -\dfrac{2}{5}$

7. Determine the slope and y-intercept of each of the following lines:
   (a) $y = 4x - 5$ (b) $3y = 9x - 6$
   (c) $y = -x$ (d) $2x - y + 5 = 0$
   (e) $\frac{2}{3}x + \frac{5}{4}y = 1$ (f) $y - 3 = 0$
   (g) $ax + by = 0$ (h) $Px + Qy + R = 0$

8. In each of the following, write an equation of the line which has the given slope m and given y-intercept b:
   (a) $m = 1; b = -3$
   (b) $m = -\frac{1}{2}; b = 2$
   (c) $m = 0; b = 5$
   (d) $m = 9; b = -9$
   (e) $m = \frac{1}{3}; b = \frac{1}{2}$
   (f) $m = 1; b = 0$
   (g) $m = -5; b = \pi$
   (h) $m = \sqrt{3}; b = e$
   (i) $m = p; b = q$
   (j) $m = \frac{1}{\sqrt{2}}; b = \frac{1}{\sqrt{5}}$

9. Find an equation of the line with y-intercept 4 and x-intercept -3. Also, determine an equation of a line with y-intercept $-\frac{c}{b}$ and x-intercept $-\frac{c}{a}$, where $a \neq 0$ and $b \neq 0$.

10. Determine which of the following pairs of linear equations represent parallel or perpendicular lines. Graph each pair.
    (a) $4x + 2y = 7; y = \frac{1}{2}x - 5$
    (b) $x + y - 7 = 0; -x - y + 4 = 0$
    (c) $3x - 2y = 5; 6x - 4y = 10$

11. Find an equation of the line which passes through $(1, -3)$ and which is perpendicular to the line $4x - 3y - 10 = 0$.

12. Find an equation of the line which is parallel to the line $7x + 6y - 1 = 0$ and which passes through $(-2, 5)$.

13. Determine the value of k in the equation $4x - 3ky + 1 = 0$ such that this line passes through $(1, 5)$.

14. Find the value of k such that the line $kx + (3 - k)y + 7 = 0$ has slope equal to 7.

15. Three points are said to be **collinear** if they all lie on the same line. Determine, using slope, whether the following points are collinear:
    (a) $(-3, -2)$, $(5, 2)$, and $(9, 4)$
    (b) $(-1, 2)$, $(1, 3)$, and $(-3, 0)$

16. Show that the points $(7, 5)$, $(2, 3)$, and $(6, -7)$ are vertices of a right triangle.

17. Show that if f is a linear function and $f(0) = 0$, then $f(x + y) = f(x) + f(y)$ for all real x and y.

18. Let f be the linear function defined by $f(x) = mx + b$. Compute
    $$\frac{f(x + h) - f(x)}{h},$$
    where $h \neq 0$. Graphically, what does this "difference quotient" represent?

19. The perpendicular distance from a point (x_0, y_0) to the line $Ax + By + C = 0$ is given by the formula

$$d = \frac{|Ax_0 + By_0 + C|}{\sqrt{A^2 + B^2}}$$

Use the above formula to find the distance from the given point to the given line:
(a) $(-2, -3)$; $-8x - 15y + 24 = 0$
(b) $(-1, 7)$; $6x - 8y = -5$

20. Graph the following pairs of equations and find the point of intersection:
(a) $2x + y - 3 = 0$; $y - x + 3 = 0$
(b) $3x - y + 5 = 0$; $3y - 5x - 11 = 0$
(c) $5x + 3y + 26 = 0$; $4x - 9y - 2 = 0$

3.3 SYSTEMS OF LINEAR EQUATIONS

Systems of Equations in Two Variables

We begin our discussion of systems of linear equations by investigating how to determine the solution(s), if they exist, graphically. Then we describe an algebraic method for solving systems of equations.

THE GRAPHICAL METHOD

For graphs of two linear equations in two variables there are three possible cases:

1. The lines intersect in exactly one point.
2. The lines are parallel and have no points in common.
3. The lines coincide and have an infinite number of points in common.

Geometrically, the solution of two linear equations in two variables is a set of ordered pairs representing the points of intersection of these lines, if such points of intersection exist.

EXAMPLE 1

Consider the system

$$\begin{cases} x + y = 1 \\ x - y = 2 \end{cases}$$

The graph of each line in the system is given in Figure 83. The point of

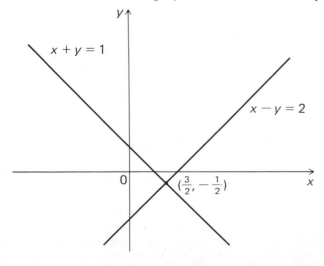

Figure 83

intersection of the two lines has coordinates $x = \dfrac{3}{2}$ and $y = -\dfrac{1}{2}$. If we substitute these values into the equations, we see that they satisfy both equations.

Definition: A system of linear equations in which the lines intersect in exactly one point is called a **consistent** system and the lines are said to be **independent.** Such a system has a unique solution.

EXAMPLE 2

Consider the graphs of the linear equations in the following system.

$$\begin{cases} x + y = 2 \\ 2x + 2y = 2 \end{cases}$$

To graph the equations we put each equation in slope-intercept form (see Fig. 84).

$$\begin{cases} y = -x + 2 \\ y = -x + 1 \end{cases}$$

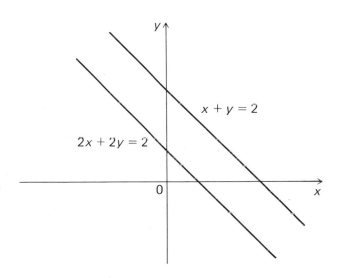

$x + y = 2$

$2x + 2y = 2$

Figure 84

Note that the lines have the same slope and different y-intercepts, and thus are parallel lines as illustrated in Figure 84. Also, there are no points of intersection and therefore no solution to the system.

Definition: A system of linear equations in which the lines are parallel is called **inconsistent** and the lines are **independent.** Such a system has no solution.

EXAMPLE 3

Suppose we have the following system of linear equations:

$$\begin{cases} x - y = -2 \\ -2x + 2y = 4 \end{cases}$$

Putting each of the equations in slope-intercept form, we have

$$\begin{cases} y = x + 2 \\ y = x + 2 \end{cases}$$

Both equations have the same slope, 1, and the same y-intercept, 2. Thus, the graphs of both equations are represented by the same line (see Fig. 85). Since the lines coincide, there are an infinite number of points

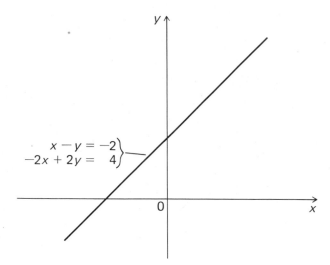

Figure 85

which satisfy both equations—for example, $(0, 2)$, $(-2, 0)$, $(1, 3)$, $(-1, 1)$.

Definition: A system of linear equations in which the lines coincide is called **consistent** and the lines are **dependent.** Such a system has an infinite number of solutions.

In general, a system of two linear equations in two variables

$$\begin{cases} a_1x + b_1y = c_1 \\ a_2x + b_2y = c_2 \end{cases}$$

results in one of three possible systems.

> (i) If the lines intersect in exactly one point, then the system is con-
> sistent and the equations are independent. This system has a
> unique solution given by the point of intersection.
> (ii) If the lines are parallel, the system is inconsistent and the equa-
> tions are independent. Since there are no points of intersection,
> there is no solution to this system.
> (iii) If the lines coincide, the system is consistent and the equations
> are dependent. The solutions are all the points on the line, which
> represents both equations. Therefore, there are an infinite
> number of solutions.

Although the graphical method gives useful information, it is not always fea-
sible. For example, the point of intersection may be too far from the origin and
thus difficult to graph. If the coefficients are very small fractional numbers, the
graphical method may not yield the desired accuracy. Therefore, we now introduce a method which eliminates these limitations.

THE ELIMINATION METHOD

We now introduce an algebraic procedure for solving systems of linear
equations. This method involves replacing a system of equations with a simpler
equivalent system. **Equivalent systems** are systems which have the same solu-
tion(s). What algebraic operations can be performed on a system of linear equations to produce an equivalent system? The following theorem indicates which
algebraic operations are used in replacing a given system with an equivalent
system.

Theorem: Given a system of linear equations, each of the following operations can be performed on the equations of the system to produce an
equivalent system.

(i) Interchange any two equations.
(ii) Multiply any equation by a nonzero real number.
(iii) Multiply any equation by a nonzero real number and add the resulting
equation to another equation.

EXAMPLE 4

Consider the system

$$\begin{cases} x + 4y = 1 \\ 3x - 2y = 5 \end{cases}$$ **(I)**

(i) Interchanging the equations gives the equivalent system

$$\begin{cases} 3x - 2y = 5 \\ x + 4y = 1 \end{cases}$$

(ii) Multiplying the first equation in system (I) by -3 yields the equivalent
system

$$\begin{cases} -3x - 12y = -3 \\ 3x - 2y = 5 \end{cases}$$

(iii) Multiplying the first equation in system (I) by -3 and adding the result to the second equation gives the equivalent system

$$\begin{cases} x + 4y = 1 \\ -14y = 2 \end{cases} \tag{II}$$

To complete this example, we apply several more operations.

(iv) We multiply the second equation in system (II) by $-\dfrac{1}{14}$ and obtain the equivalent system

$$\begin{cases} x + 4y = 1 \\ y = -\dfrac{1}{7} \end{cases} \tag{III}$$

(v) Finally, we multiply the second equation in system (III) by -4 and add the result to the first equation, thus giving the equivalent system

$$\begin{cases} x = \dfrac{11}{7} \\ y = -\dfrac{1}{7} \end{cases}$$

Note that in this example we used the theorem to obtain an equivalent system of the form

$$\begin{cases} x = p \\ y = q \end{cases}$$

The solution (p, q) to system (I) is easily obtained. This method for solving a system of linear equations by "eliminating" variables is aptly called the **elimination method.**

Although elimination was used to find the solution in Example 4, the question arises as to whether the method gives information about systems which have a unique solution and systems which have an infinite number of solutions or no solutions. Fortunately, the elimination method does yield such information. The following problems illustrate how the elimination method achieves this.

PROBLEM 1

Solve the system

$$\begin{cases} -3x + 4y = 5 \\ 6x - 8y = -2 \end{cases}$$

Solution: We try to eliminate x from the second equation. Multiplying the first equation by 2 and adding the result to the second equation we have

$$\begin{cases} -3x + 4y = 5 \\ \qquad\quad 0 = 8 \end{cases}$$

The resulting equivalent system has a contradiction—that is, $0 = 8$. Since this cannot occur, we have no solution. This can be seen by graphing each of the equations in the given system. The graph consists of two parallel lines (see Fig. 86).

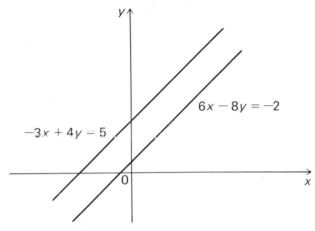

Figure 86

PROBLEM 2

Solve the system

$$\begin{cases} 1.5x - .3y - 3 \\ .5x - .1y = 1 \end{cases} \qquad \textbf{(I)}$$

Solution: We interchange the two equations to get the equivalent system

$$\begin{cases} .5x - .1y = 1 \\ 1.5x - .3y = 3 \end{cases} \qquad \textbf{(II)}$$

Next, we multiply the first equation in system (II) by -3 and add the result to the second equation, so that system (II) becomes

$$\begin{cases} .5x - .1y = 1 \\ \qquad\quad 0 = 0 \end{cases} \qquad \textbf{(III)}$$

Thus, system (I) is reduced to an equivalent system with only the equation $.5x - .1y = 1$. Any solution of this equation satisfies both equations in system (I). We see that there are an infinite number of solutions to $.5x - .1y = 1$ and hence to the system (I). These results are reflected in the fact that the graphs of the equations in system (I) are represented by the same line (see Fig. 87).

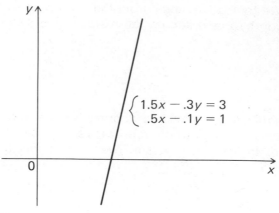

Figure 87

PROBLEM 3

The elimination method can be used to solve certain systems of **non-linear** equations. Consider the following system of nonlinear equations.

$$\frac{1}{x} - \frac{1}{y} = 1 \tag{I}$$

$$\frac{2}{x} + \frac{3}{y} = 7$$

Solution: If we let $u = \frac{1}{x}$ and $v = \frac{1}{y}$, then the system of nonlinear equations in x and y becomes a system of linear equations in u and v. That is,

$$\begin{cases} u - v = 1 \\ 2u + 3v = 7 \end{cases} \tag{II}$$

We now apply the elimination method to system (II). First, we multiply the first equation by -2 and add the result to the second equation.

$$\begin{cases} u - v = 1 \\ \quad\ 5v = 5 \end{cases} \tag{III}$$

Multiplying the second equation by $\frac{1}{5}$ we get $v = 1$. Substituting the value $v = 1$ into the first equation we have $u = 2$. Setting $u = \frac{1}{x} = 2$, we get $x = \frac{1}{2}$. Similarly, $v = \frac{1}{y} = 1$ gives $y = 1$. Thus the solution to system (I) is the ordered pair $\left(\frac{1}{2}, 1\right)$.

We now illustrate the many applications of the elimination method in solving problems with two equations and two unknowns.

PROBLEM 4:   ABSTRACT NUMBER PROBLEM

The difference of two numbers is 7, and twice the smaller number is 3 less than the larger number. Find the numbers.

Solution: Let x be the larger number.
Let y be the smaller number.
Then $x - y = 7$ and $x - 2y = 3$. The system representing the conditions is

$$\begin{cases} x - y = 7 \\ x - 2y = 3 \end{cases}$$

To eliminate x, multiply the first equation by -1 and add the result to the second equation.

$$\begin{cases} x - y = 7 \\ - y = -4 \end{cases}$$

Thus, $y = 4$ and $x = 11$

PROBLEM 5:   AGE PROBLEM

Five years ago a woman was four times as old as her daughter. In seven years she will be twice as old as her daughter. Find their present ages.

Solution: Let $x =$ the woman's present age and $y =$ her daughter's present age.
Five years ago: $x - 5 = 4(y - 5)$
Seven years from now: $x + 7 = 2(y + 7)$
Thus we have the system

$$\begin{cases} x - 4y = -15 \\ x - 2y = 7 \end{cases}$$

Using the elimination method, we get

$$\begin{cases} x - 4y = -15 \\ y = 11 \end{cases}$$

Therefore, the daughter is 11 years old and the woman is 29 years old.

PROBLEM 6:   MIXTURE PROBLEM

A nutritionist wishes to prepare a food mixture that contains 40 grams of vitamin A and 50 grams of vitamin B. The two mixtures that are available contain the following percentages of vitamin A and vitamin B:

	Vitamin A	Vitamin B
Mixture I	10%	4%
Mixture II	5%	12%

How many grams of each mixture should be used to obtain the desired diet?

Solution: Let x be the grams of mixture I to be used. Let y be the grams of mixture II to be used. The equation describing the amount of vitamin A we must have from the two mixtures is

$$.1x + .05y = 40$$

Similarly, for vitamin B we have

$$.04x + .12y = 50$$

We have the system

$$\begin{cases} .1x + .05y = 40 \\ .04x + .12y = 50 \end{cases}$$

Solving by the elimination method we get the equivalent system

$$\begin{cases} .1x + .05y = 40 \\ \qquad .1y = 34 \end{cases}$$

Thus, $y = 340$ and $x = 230$. Thus, the nutritionist needs 230 grams of mixture I and 340 grams of mixture II.

Systems of Equations in Three or More Variables

The elimination method can be applied to systems of linear equations in three or more variables. We shall now study in detail systems of three equations in three unknowns. The method is equally applicable to higher order systems.

Just as the solution(s) of systems in two variables were described by ordered pairs, solutions of systems in three variables can be represented by ordered triples. For example the ordered triple $\left(\frac{1}{3}, \frac{2}{3}, 5\right)$ is a solution to the system

$$\begin{cases} 3y - z = -3 \\ -6x \qquad + 2z = 8 \\ x + y \qquad = 1 \end{cases}$$

However, the ordered triple $(-1, 2, -2)$ is *not* a solution to the system

$$\begin{cases} x + 2y - 3z = 9 \\ 3x - y - 4z = 3 \\ 2x - y - 2z = 7 \end{cases}$$

While the ordered triple $(-1, 2, -2)$ satisfies the first and second equations, it does not satisfy the third equation.

We will now illustrate how the elimination method is used to solve higher-order systems.

PROBLEM 7

Solve the system

$$\begin{cases} x - 2y + z = -1 \\ 3x + y - 2z = 4 \\ y - z = 1 \end{cases}$$

Solution: First, we multiply the first equation by -3 and add the result to the second equation to obtain

$$\begin{cases} x - 2y + z = -1 \\ 7y - 5z = 7 \\ y - z = 1 \end{cases}$$

Now, multiply the second equation by $-\dfrac{1}{7}$ and add the result to the third equation to get the equivalent system

$$\begin{cases} x - 2y + z = -1 \\ 7y - 5z = 7 \\ -\dfrac{2}{7}z = 0 \end{cases}$$

The third equation gives $z = 0$, and substituting this value into the second equation gives $y = 1$. Finally, the first equation yields $x = 1$. Thus, the solution is the ordered triple $(1, 1, 0)$.

PROBLEM 8

Consider the system

$$\begin{cases} x - 2y + 3z = 2 \\ 2x + 3y - 2z = 5 \\ 4x - y + 4z = 1 \end{cases}$$

We apply the elimination method to find the solution of the system, provided it exists. First, multiply the first equation by -2 and add the results to the second equation. Also, multiply the first equation by -4 and add the resulting equation to the third equation. This gives the following equivalent system:

$$\begin{cases} x - 2y + 3z = 2 \\ \quad\quad 7y - 8z = 1 \\ \quad\quad 7y - 8z = -7 \end{cases}$$

Next, multiply the second equation by -1 and add the results to the third equation to obtain

$$\begin{cases} x - 2y + 3z = 2 \\ \quad\quad 7y - 8z = 1 \\ \quad\quad\quad\quad 0 = -8 \end{cases}$$

Thus, the given system is equivalent to a system which contains a contradiction—that is, $0 = -8$. Therefore, the system is inconsistent and has *no solution*.

PROBLEM 9

Solve the system

$$\begin{cases} x + 2y - 3z = 1 \\ x - y + z = 5 \\ -2x - 7y + 10z = 2 \end{cases}$$

Solution: Adding -1 times the first equation to the second equation and adding 2 times the first equation to the third equation gives the equivalent system

$$\begin{cases} x + 2y - 3z = 1 \\ \quad -3y + 4z = 4 \\ \quad -3y + 4z = 4 \end{cases}$$

Next, we multiply the second equation by -1 and add the result to the third equation to obtain

$$\begin{cases} x + 2y - 3z = 1 \\ \quad -3y + 4z = 4 \\ \quad\quad\quad 0 = 0 \end{cases} \qquad \textbf{(D)}$$

Thus, the given system is equivalent to the reduced system (D), which is a dependent system. If we eliminate y from the first equation by adding $\frac{2}{3}$ times the second equation to the first equation, we get

$$\begin{cases} x \quad\quad - \dfrac{1}{3}z = \dfrac{11}{3} \\ \quad -3y + 4z = 4 \end{cases}$$

or, equivalently,

$$\begin{cases} x = \dfrac{11}{3} + \dfrac{1}{3}z \\ y = \dfrac{4}{3}z - \dfrac{4}{3} \end{cases}$$

If z is chosen arbitrarily, say $z = t$, then

$$\begin{cases} x = \dfrac{1}{3}(11 + t) \\ y = \dfrac{1}{3}(4t - 4) \end{cases}$$

and the solution is given by the ordered triple

$$\left(\frac{11 + t}{3}, \frac{4t - 4}{3}, t\right)$$

Since t is arbitrary and ranges over all the real numbers, the system has an infinite number of solutions. For example: if $t = 0$, then we have the solution $\left(\frac{11}{3}, \frac{-4}{3}, 0\right)$; if $t = 1$, $(4, 0, 1)$; and if $t = -1$, we have the solution $\left(\frac{10}{3}, -\frac{8}{3}, -1\right)$.

PROBLEM 10

Solve the system

$$\begin{cases} x + y + z = 0 \\ 2x - y + z = 0 \\ x + y - z = 0 \end{cases}$$

Solution: Multiply the first equation by -2 and add the result to the second equation. Also, multiply the first by -1 and add the result to the third equation. These operations yield the following equivalent system:

$$\begin{cases} x + y + z = 0 \\ -3y - z = 0 \\ -2z = 0 \end{cases}$$

We see that the system has the trivial solution of the ordered triple $(0, 0, 0)$.

Such systems as those in Problem 10 are of great interest and have a special name. Any system of the form

$$\begin{cases} a_1x + b_1y + c_1z = 0 \\ a_2x + b_2y + c_2z = 0 \\ a_3x + b_3y + c_3z = 0 \end{cases}$$

where all values on one side are zero, is called a **homogeneous system.** Note that a homogeneous system always has the trivial solution (0, 0, 0). However, a homogeneous system may have an infinite number of solutions.

PROBLEM 11

Solve the system

$$\begin{cases} x + 2y - z = 0 \\ 2x + 5y + 2z = 0 \\ x + 4y + 7z = 0 \end{cases}$$

Solution: Multiply the first equation by -2 and add the results to the second equation. Next, multiply the first equation by -1 and add the results to the third equation. We get

$$\begin{cases} x + 2y - z = 0 \\ y + 4z = 0 \\ 2y + 8z = 0 \end{cases}$$

Eliminating y from the third equation gives

$$\begin{cases} x + 2y - z = 0 \\ y + 4z = 0 \\ 0 = 0 \end{cases}$$

Finally, add -2 times the second equation to the first to eliminate y

$$\begin{cases} x - 9z = 0 \\ y + 4z = 0 \end{cases}$$

or, equivalently,

$$\begin{cases} x = 9z \\ y = -4z \end{cases}$$

If we let $z = t$, then the solutions take the form $(9t, -4t, t)$ where t is any real number. Thus, the given system has infinitely many solutions. Some of the solutions are obtained by assigning values to t. For example: if $t = 0$, the trivial solution is (0, 0, 0); if $t = 1$, the solution is $(9, -4, 1)$; and if $t = -2$, it is $(-18, 8, -2)$.

Matrices and Systems of Linear Equations

A closer look at the elimination method shows that we worked mostly with the coefficients and not variables. The elimination procedure can be speeded up by listing the constants of a system in a rectangular array called a **matrix.**

Definition: A **matrix** is a rectangular array of numbers arranged in rows and columns. Brackets are used to indicate a matrix.

EXAMPLES

The following are matrices:

(a) $\begin{bmatrix} 1 & 3 \\ -4 & 0 \end{bmatrix}$ (b) $\begin{bmatrix} 1 \\ 5 \end{bmatrix}$ (c) $[-6]$ (d) $[7 \quad -3 \quad 9]$

(e) $\begin{bmatrix} 7 & 1 & -4 \\ -2 & 0 & 16 \\ -1 & 3 & 9 \end{bmatrix}$ (f) $\begin{bmatrix} 1 & 0 \\ 0 & 1 \end{bmatrix}$ (g) $\begin{bmatrix} a_1 & b_1 & c_1 \\ a_2 & b_2 & c_2 \end{bmatrix}$

We consider the system

$$\begin{cases} a_1x + b_1y = c_1 \\ a_2x + b_2y = c_2 \end{cases}$$

Associated with this linear system is a **coefficient matrix**

$$\begin{bmatrix} a_1 & b_1 \\ a_2 & b_2 \end{bmatrix}$$

and an **augmented matrix**

$$\begin{bmatrix} a_1 & b_1 & | & c_1 \\ a_2 & b_2 & | & c_2 \end{bmatrix}$$

EXAMPLE 1

The system

$$\begin{cases} 3x - 4y = 1 \\ 2x + y = 5 \end{cases}$$

has the matrix $\begin{bmatrix} 3 & -4 \\ 2 & 1 \end{bmatrix}$ as a coefficient matrix, while the augmented ma-

trix is $\begin{bmatrix} 3 & -4 & | & 1 \\ 2 & 1 & | & 5 \end{bmatrix}$

EXAMPLE 2

Associated with the system

$$\begin{cases} a_1x + b_1y + c_1z = d_1 \\ a_2x + b_2y + c_2z = d_2 \\ a_3x + b_3y + c_3z = d_3 \end{cases}$$

is the coefficient matrix

$$\begin{bmatrix} a_1 & b_1 & c_1 \\ a_2 & b_2 & c_2 \\ a_3 & a_3 & c_3 \end{bmatrix}$$

and the augmented matrix

$$\begin{bmatrix} a_1 & b_1 & c_1 & \bigm| & d_1 \\ a_2 & b_2 & c_2 & \bigm| & d_2 \\ a_3 & b_3 & c_3 & \bigm| & d_3 \end{bmatrix}$$

EXAMPLE 3

Consider the system

$$\begin{cases} x + 2y - 3z = 9 \\ 2x - y + 2z = -8 \\ 3x - y - 4z = 3 \end{cases}$$

The augmented matrix of the system is

$$\begin{bmatrix} 1 & 2 & -3 & \bigm| & 9 \\ 2 & -1 & 2 & \bigm| & -8 \\ 3 & -1 & -4 & \bigm| & 3 \end{bmatrix}$$

We now use matrices to solve systems of linear equations. The procedure is illustrated in the following example.

EXAMPLE 4

Consider the system

$$\begin{cases} 3x - 4y = 1 \\ 12x + y = 5 \end{cases}$$

Linear System	Matrix Operations
$\begin{cases} 3x - 4y = 1 \\ 12x + y = 5 \end{cases}$	$\begin{bmatrix} 3 & -4 & \vert & 1 \\ 12 & 1 & \vert & 5 \end{bmatrix}$

Multiply the first equation by $\frac{1}{3}$:
$\begin{cases} x - \frac{4}{3}y = \frac{1}{3} \\ 12x + y = 5 \end{cases}$

Multiply the first row by $\frac{1}{3}$:
$\begin{bmatrix} 1 & -\frac{4}{3} & \vert & \frac{1}{3} \\ 12 & 1 & \vert & 5 \end{bmatrix}$

Multiply the first equation by -12 and add to the second equation:
$\begin{cases} x - \frac{4}{3}y = \frac{1}{3} \\ 17y = 1 \end{cases}$

Multiply the first row by -12 and add the result to the second row:
$\begin{bmatrix} 1 & -\frac{4}{3} & \vert & \frac{1}{3} \\ 0 & 17 & \vert & 1 \end{bmatrix}$

Multiply the second equation by $\frac{1}{17}$:
$\begin{cases} x - \frac{4}{3}y = \frac{1}{3} \\ y = \frac{1}{17} \end{cases}$

Multiply the second row by $\frac{1}{17}$:
$\begin{bmatrix} 1 & -\frac{4}{3} & \vert & \frac{1}{3} \\ 0 & 1 & \vert & \frac{1}{17} \end{bmatrix}$

Add $\frac{4}{3}$ times the second equation to the first equation:
$\begin{cases} x = \frac{21}{51} \\ y = \frac{1}{17} \end{cases}$

Add $\frac{4}{3}$ times the second row to the first row:
$\begin{bmatrix} 1 & 0 & \vert & \frac{21}{51} \\ 0 & 1 & \vert & \frac{1}{17} \end{bmatrix}$

Thus, the solution to the set is $\left(\frac{21}{51}, \frac{1}{17}\right)$.

Note that we reduced the augmented matrix of the system to the form

$$\begin{bmatrix} 1 & 0 & \vert & a \\ 0 & 1 & \vert & b \end{bmatrix}$$

Such a matrix is said to be in **reduced-echelon form.** The matrix in each step of Example 4 is the augmented matrix of a system which is equivalent to the original system. We will call two augmented matrices **row-equivalent** if they are augmented matrices of equivalent systems of equations. The following theorem gives the operations which transform any matrix into a row-equivalent matrix.

Theorem: Row-equivalent matrices are obtained by the following row operations:

  (i) Two rows are interchanged.
 (ii) Any row is multiplied by a nonzero number.
(iii) Multiply any row by a nonzero number and add the resulting row to another row.

EXAMPLE 5

We solve the system given in Example 3:

$$\begin{cases} x + 2y - 3z = 9 \\ 2x - y + 2z = -8 \\ 3x - y - 4z = 3 \end{cases}$$

The augmented matrix of this system is

$$\left[\begin{array}{ccc|c} 1 & 2 & -3 & 9 \\ 2 & -1 & 2 & -8 \\ 3 & -1 & -4 & 3 \end{array}\right]$$

Multiply the first row by -2 and add the results to the second row. Also, multiply the first row by -3 and add the results to the third row:

$$\left[\begin{array}{ccc|c} 1 & 2 & -3 & 9 \\ 0 & -5 & 8 & -26 \\ 0 & -7 & 5 & -24 \end{array}\right]$$

Multiply the second row by $-\dfrac{1}{5}$:

$$\left[\begin{array}{ccc|c} 1 & 2 & -3 & 9 \\ 0 & 1 & -\dfrac{8}{5} & \dfrac{26}{5} \\ 0 & -7 & 5 & -24 \end{array}\right]$$

Multiply the second row by -2 and add the results to the first row. Also, add 7 times the second row to the third row:

$$\left[\begin{array}{ccc|c} 1 & 0 & \dfrac{1}{5} & -\dfrac{7}{5} \\ 0 & 1 & -\dfrac{8}{5} & \dfrac{26}{5} \\ 0 & 0 & -\dfrac{31}{5} & \dfrac{62}{5} \end{array}\right]$$

Multiply the third row by $-\dfrac{5}{31}$:

$$\begin{bmatrix} 1 & 0 & \frac{1}{5} & \bigg| & -\frac{7}{5} \\ 0 & 1 & -\frac{8}{5} & \bigg| & \frac{26}{5} \\ 0 & 0 & 1 & \bigg| & -2 \end{bmatrix}$$

Add $\dfrac{8}{5}$ times the third row to the second row. Also, multiply the third row by $-\dfrac{1}{5}$ and add the results to the first row:

$$\begin{bmatrix} 1 & 0 & 0 & \big| & -1 \\ 0 & 1 & 0 & \big| & 2 \\ 0 & 0 & 1 & \big| & -2 \end{bmatrix}$$

The solution is $x = -1$, $y = 2$, and $z = -2$. The solution can be represented by the ordered triple $(-1, 2, -2)$.

Often it is not necessary to reduce the augmented matrix to the form

$$\begin{bmatrix} 1 & 0 & 0 & \big| & p \\ 0 & 1 & 0 & \big| & q \\ 0 & 0 & 1 & \big| & r \end{bmatrix}$$

Sometimes just reducing the matrix to a triangular form is sufficient. The following problems illustrate this method.

PROBLEM 1

Solve the system

$$\begin{cases} x + 2y + 3z = 5 \\ 2x + 9y + 3z = -1 \\ x \quad\quad + 4z = 9 \end{cases}$$

Solution: The augmented matrix is

$$\begin{bmatrix} 1 & 2 & 3 & \big| & 5 \\ 2 & 9 & 3 & \big| & -1 \\ 1 & 0 & 4 & \big| & 9 \end{bmatrix}$$

Add -2 times row one to row two:

$$\begin{bmatrix} 1 & 2 & 3 & \big| & 5 \\ 0 & 5 & -3 & \big| & -11 \\ 1 & 0 & 4 & \big| & 9 \end{bmatrix}$$

Add -1 times row one to row three:

$$\begin{bmatrix} 1 & 2 & 3 & | & 5 \\ 0 & 5 & -3 & | & -11 \\ 0 & -2 & 1 & | & 4 \end{bmatrix}$$

Multiply row two by $\frac{1}{5}$:

$$\begin{bmatrix} 1 & 2 & 3 & | & 5 \\ 0 & 1 & -\dfrac{3}{5} & | & -\dfrac{11}{5} \\ 0 & -2 & 1 & | & 4 \end{bmatrix}$$

Add 2 times row two to row three:

$$\begin{bmatrix} 1 & 2 & 3 & | & 5 \\ 0 & 1 & -\dfrac{3}{5} & | & -\dfrac{11}{5} \\ 0 & 0 & -\dfrac{1}{5} & | & -\dfrac{2}{5} \end{bmatrix}$$

So we have an augmented matrix of the triangular system

$$\begin{cases} x + 2y + 3z = 5 \\ \quad y - \dfrac{3}{5}z = -\dfrac{11}{5} \\ \quad\quad -\dfrac{1}{5}z = -\dfrac{2}{5} \end{cases}$$

From the last equation we have $z = 2$. From the second equation we obtain $y = -1$. Finally, the first equation yields $x = 1$. Thus the solution is $(1, -1, 2)$.

What about systems which have infinitely many solutions or no solution? This information can also be obtained using matrices.

PROBLEM 2

Solve the homogeneous system

$$\begin{cases} x + y + 13z = 0 \\ x - y - 6z = 0 \end{cases}$$

Solution:

$$\begin{bmatrix} 1 & 1 & 13 & | & 0 \\ 1 & -1 & -6 & | & 0 \end{bmatrix}$$

Add -1 times row one to row two:

$$\begin{bmatrix} 1 & 1 & 13 & | & 0 \\ 0 & -2 & -19 & | & 0 \end{bmatrix}$$

Multiply row two by $-\dfrac{1}{2}$:

$$\begin{bmatrix} 1 & 1 & 13 & | & 0 \\ 0 & 1 & \dfrac{19}{2} & | & 0 \end{bmatrix}$$

Add -1 times row two to row one:

$$\begin{bmatrix} 1 & 0 & \dfrac{7}{2} & | & 0 \\ 0 & 1 & \dfrac{19}{2} & | & 0 \end{bmatrix}$$

We cannot reduce the matrix any further. The augmented matrix represents the system

$$\begin{cases} x + \dfrac{7}{2}z = 0 \\ y + \dfrac{19}{2}z = 0 \end{cases}$$

or, equivalently,

$$\begin{cases} x = -\dfrac{7}{2}z \\ y = -\dfrac{19}{2}z \end{cases}$$

If we choose $z = t$, then the solutions take the form $\left(-\dfrac{7}{2}t, \ -\dfrac{19}{2}t, \ t \right)$.

Since t is any real number, the system has an infinite number of solutions. If $t \neq 0$, then the solution is nontrivial.

PROBLEM 3

Solve the system

$$\begin{cases} x + y - z = 1 \\ -x + y + 2z = 3 \\ -x - y + z = -2 \end{cases}$$

Solution: Perform the usual row operations on the augmented matrix

$$\left[\begin{array}{ccc|c} 1 & 1 & -1 & 1 \\ -1 & 1 & 2 & 3 \\ -1 & -1 & 1 & -2 \end{array}\right]$$

Add row one to row two and row three:

$$\left[\begin{array}{ccc|c} 1 & 1 & -1 & 1 \\ 0 & 2 & 1 & 4 \\ 0 & 0 & 0 & -1 \end{array}\right]$$

The system corresponding to this augmented matrix is

$$\begin{aligned} x + y - z &= 1 \\ 2y + z &= 4 \\ 0 &= -1 \end{aligned}$$

This is an inconsistent system because of the contradiction $0 = -1$. Therefore, the system has no solution.

3.3 EXERCISES

1. Determine which of the following systems are consistent and independent, inconsistent and independent, or consistent and dependent. Graph each pair of equations.

(a) $\begin{cases} 3x - 2y = 2 \\ 2x + y = 8 \end{cases}$

(b) $\begin{cases} 4x - 2y = 7 \\ -2x + y = 3 \end{cases}$

(c) $\begin{cases} -x + y = 1 \\ x - y = -1 \end{cases}$

(d) $\begin{cases} 4x + 2y = 0 \\ 2x + 6y = 0 \end{cases}$

(e) $\begin{cases} x + y = 5 \\ -x + y = -1 \end{cases}$

(f) $\begin{cases} x - 3y = 2 \\ -2x + 6y = -8 \end{cases}$

2. Determine which of the following pairs of systems are equivalent:

(a) $\begin{cases} 3x - 2y = 4 \\ x + y = 6 \end{cases}$ and $\begin{cases} 3x - 2y = 4 \\ 4x - y = 10 \end{cases}$

(b) $\begin{cases} 2x + 2y = 2 \\ x - y = 1 \end{cases}$ and $\begin{cases} x + y = 1 \\ x - y = 1 \end{cases}$

(c) $\begin{cases} 2x + 3y = 1 \\ 4x + 2y = 10 \end{cases}$ and $\begin{cases} 2x + 3y = 1 \\ 5x + 7y = 3 \end{cases}$

(d) $\begin{cases} 5x - 3y = 1 \\ 8x - 2y = 0 \end{cases}$ and $\begin{cases} 9x - 4y = 1 \\ 8x - 2y = 0 \end{cases}$

(e) $\begin{cases} x - y = 2 \\ x + y = 0 \end{cases}$ and $\begin{cases} x - y = 2 \\ 2x = 2 \end{cases}$

3. Solve the following systems for x and y:

(a) $\begin{cases} cx + dy = c \\ c^2x + d^2y = c^2 \end{cases}$ Assume $c \neq 0$, $d \neq 0$ and $c \neq d$.

(b) $\begin{cases} ax + by = c \\ bx + ay = d \end{cases}$ Assume $a^2 - b^2 \neq 0$

(c) $\begin{cases} a_1x + b_1y = c_1 \\ a_2x + b_2y = c_2 \end{cases}$ Assume $a_1b_2 - a_2b_1 \neq 0$

4. Solve the following systems by means of the augmented matrix:

(a) $\begin{cases} x + y + z = 3 \\ -2x + y - z = 4 \\ x + 2y + z = 5 \end{cases}$

(b) $\begin{cases} x + 2y + 3z = 3 \\ 2x + 3y + 8z = 4 \\ 3x + 2y + 17z = 1 \end{cases}$

(c) $\begin{cases} x + y + 2z = 0 \\ y + z = 0 \\ 2x - 3y - z = 0 \end{cases}$

(d) $\begin{cases} 2x + y - z = -3 \\ x + 3y + 5z = -1 \\ -4x - 2y + 2z = 7 \end{cases}$

5. Given $\begin{bmatrix} 1 & -4 & 1 & -1 \\ 0 & 1 & 1 & -1 \\ 0 & 0 & -2 & 7 \\ 0 & 0 & 0 & 1 \end{bmatrix}$, explain why the following are row-equivalent to the given

matrix:

(a) $\begin{bmatrix} 1 & -4 & 1 & 0 \\ 0 & 1 & 1 & 0 \\ 0 & 0 & -2 & 0 \\ 0 & 0 & 0 & 1 \end{bmatrix}$

(b) $\begin{bmatrix} 1 & -4 & 1 & 0 \\ 0 & 1 & 1 & 0 \\ 0 & 0 & 1 & 0 \\ 0 & 0 & 0 & 1 \end{bmatrix}$

(c) $\begin{bmatrix} 1 & -4 & 0 & 0 \\ 0 & 1 & 0 & 0 \\ 0 & 0 & 1 & 0 \\ 0 & 0 & 0 & 1 \end{bmatrix}$

6. The general equation of a circle is given by

$$x^2 + y^2 + Dx + Ey + F = 0.$$

Find the equation of the circle passing through the three points $(5, 3)$, $(6, 2)$, and $(3, -1)$ by determining the constants D, E, and F.

7. Solve $\begin{cases} \dfrac{1}{x} + \dfrac{1}{y} + \dfrac{1}{z} = 5 \\ -\dfrac{2}{x} + \dfrac{3}{y} + \dfrac{4}{z} = 11 \\ \dfrac{3}{x} + \dfrac{2}{y} - \dfrac{1}{z} = -6 \end{cases}$

8. A psychologist is conducting an experiment using rats and she must administer to them a blend which contains 46 ounces of carbohydrates, 32 ounces of protein, and 12.4 ounces of cholesterol. The following compositions are available:

	Carbohydrate	Protein	Cholesterol
Blend A	40%	30%	4%
Blend B	20%	20%	12%
Blend C	30%	10%	10%

How many ounces of each blend should be used to meet the required amounts of carbohydrates, protein, and cholesterol?

9. The sum of the second and third digits of a three-digit number is equal to the first digit. The sum of the first digit and the second digit is 2 more than the third digit. If the second and third digits are interchanged, the new number will be 54 more than the original number. Find the number.

10. Solve the following system using matrices:

$$\begin{cases} x + y + z - w = 0 \\ 3x + 3y - z + 2w = 7 \\ -x + y + 2z - 2w = -1 \\ x + 3w = 9 \end{cases}$$

3.4 DETERMINANTS

In the previous section we discussed a matrix method for solving a system of linear equations. We shall now consider the special case of linear systems in which the number of equations is the same as the number of unknowns. Such systems have a coefficient matrix which is a **square matrix,** a square matrix being one with the same number of rows as columns. We introduce a special function called the **determinant** of a square matrix. The determinant function maps the square matrices to the real numbers (Fig. 88). Also, the determinant

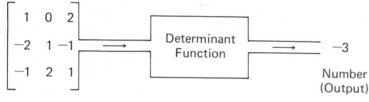

Square Matrix
(Input)

Number
(Output)

Figure 88

plays a role in the study of whether or not a system of equations has a solution. It was used extensively in solving systems of equations by **Cramer's Rule,** a method which was widely used until just a few years ago. In computer operations, this method is very inefficient.

Second-order Determinants

Consider the solutions of simultaneous equations in two unknowns. The problem is to find all ordered pairs (x, y) of real numbers for which

$$\begin{cases} a_1 x + b_1 y = c_1 \\ a_2 x + b_2 y = c_2 \end{cases} \tag{I}$$

As in the previous section, we apply the elimination process to the augmented matrix of system (I).

$$\begin{bmatrix} a_1 & b_1 & | & c_1 \\ a_2 & b_2 & | & c_2 \end{bmatrix}$$

Adding $-\dfrac{a_2}{a_1}$ times row one to row two and then multiplying row two by a_1, we see that we have the row-equivalent matrix

$$\begin{bmatrix} a_1 & b_1 & \vdots & c_1 \\ 0 & a_1b_2 - a_2b_1 & \vdots & a_1c_2 - a_2c_1 \end{bmatrix}$$

The equivalent system is

$$\begin{cases} a_1x + b_1y = c_1 \\ (a_1b_2 - a_2b_1)y = a_1c_2 - a_2c_1 \end{cases}$$

Therefore the system has a solution if and only if $a_1b_2 - a_2b_1 \neq 0$. The solution of system (I) is

$$x = \frac{b_2c_1 - b_1c_2}{a_1b_2 - a_2b_1}, \quad y = \frac{a_1c_2 - a_2c_1}{a_1b_2 - a_2b_1} \tag{II}$$

The number $a_1b_2 - a_2b_1$ is called the **determinant** of the square matrix

$$A = \begin{bmatrix} a_1 & b_1 \\ a_2 & b_2 \end{bmatrix}$$

and is denoted by

$$\begin{vmatrix} a_1 & b_1 \\ a_2 & b_2 \end{vmatrix} \quad \text{or} \quad |A|$$

Closer investigation shows that the determinant of A arises from **cross-multiplying** the elements of A. That is,

$$|A| = \begin{matrix} a_1 & b_1 \\ & \diagdown \\ a_2 & b_2 \end{matrix} = a_1b_2 - a_2b_1$$

EXAMPLE 1

$$\begin{vmatrix} 1 & 3 \\ -2 & 4 \end{vmatrix} = (1)(4) - (-2)(3) = 4 + 6 = 10$$

EXAMPLE 2

$$\begin{vmatrix} 1 & 1 \\ -2 & 0 \end{vmatrix} = (1)(0) - (-2)(1) = 0 + 2 = 2$$

EXAMPLE 3

$$\begin{vmatrix} 1 & 3 \\ 0 & 4 \end{vmatrix} = (1)(4) - (0)(3) = 4 - 0 = 4$$

We can write the solution (II) of system (I) using determinants. Given the system

$$\begin{cases} a_1x + b_1y = c_1 \\ a_2x + b_2y = c_2 \end{cases} \qquad \text{(I)}$$

the solution is

$$\text{(S): } x = \frac{D_x}{D} = \frac{\begin{vmatrix} c_1 & b_1 \\ c_2 & b_2 \end{vmatrix}}{\begin{vmatrix} a_1 & b_1 \\ a_2 & b_2 \end{vmatrix}} \quad \text{and} \quad y = \frac{D_y}{D} = \frac{\begin{vmatrix} a_1 & c_1 \\ a_2 & c_2 \end{vmatrix}}{\begin{vmatrix} a_1 & b_1 \\ a_2 & b_2 \end{vmatrix}}$$

The equations (S): $x = \frac{D_x}{D}$ and $y = \frac{D_y}{D}$ are referred to as **Cramer's Rule** for two linear equations in two unknowns. Note that D is the determinant of the coefficient matrix: D_x is the determinant of the coefficient matrix, except that a_1 and a_2 are replaced by the numbers c_1 and c_2, and D_y is the determinant of the coefficient matrix except that b_1 and b_2 are replaced by the numbers c_1 and c_2.

There are three possible cases for such a system, as we described in Section 3.3.

Case 1: If $D = \begin{vmatrix} a_1 & b_1 \\ a_2 & b_2 \end{vmatrix} \neq 0$, then system (I) is **consistent** and **independent**. Thus, there is exactly one solution given by (S).

Case 2: If $D = \begin{vmatrix} a_1 & b_1 \\ a_2 & b_2 \end{vmatrix} = 0$ and *both* $D_x = \begin{vmatrix} c_1 & b_1 \\ c_2 & b_2 \end{vmatrix} = 0$ and $D_y = \begin{vmatrix} a_1 & c_1 \\ a_2 & c_2 \end{vmatrix} = 0$, then system (I) is **consistent** and **dependent**. Therefore, the system has infinitely many solutions.

Case 3: If $D = \begin{vmatrix} a_1 & b_1 \\ a_2 & b_2 \end{vmatrix} = 0$ and *either* $D_x = \begin{vmatrix} c_1 & b_1 \\ c_2 & b_2 \end{vmatrix} \neq 0$ or $D_y = \begin{vmatrix} a_1 & c_1 \\ a_2 & c_2 \end{vmatrix} \neq 0$, then system (I) is **inconsistent** and **independent**. Hence, the system has no solution.

PROBLEM 1

Solve the following system using Cramer's Rule:

$$\begin{cases} x + 3y = 1 \\ -2x + 4y = 0 \end{cases}$$

Solution: From Examples 1 to 3 we have $D = \begin{vmatrix} 1 & 3 \\ -2 & 4 \end{vmatrix} = 10$, $D_x = \begin{vmatrix} 1 & 3 \\ 0 & 4 \end{vmatrix} = 4$, and $D_y = \begin{vmatrix} 1 & 1 \\ -2 & 0 \end{vmatrix} = 2$.

Therefore $x = \frac{D_x}{D} = \frac{4}{10} = \frac{2}{5}$, and $y = \frac{D_y}{D} = \frac{2}{10} = \frac{1}{5}$.

PROBLEM 2

Solve $\begin{cases} 2x + 3y = 2 \\ x - 2y = -6 \end{cases}$

Solution: First we compute D, D_x, and D_y.

$$D = \begin{vmatrix} 2 & 3 \\ 1 & -2 \end{vmatrix} = (2)(-2) - (1)(3) = -4 - 3 = -7.$$

$$D_x = \begin{vmatrix} 2 & 3 \\ -6 & -2 \end{vmatrix} = (2)(-2) - (-6)(3) = -4 + 18 = 14.$$

$$D_y = \begin{vmatrix} 2 & 2 \\ 1 & -6 \end{vmatrix} = (2)(-6) - (1)(2) = -12 - 2 = -14.$$

Thus, $x = \dfrac{D_x}{D} = \dfrac{14}{-7} = -2$ and $y = \dfrac{D_y}{D} = \dfrac{-14}{-7} = 2$.

PROBLEM 3

Solve $\begin{cases} 7x + 3y = 4 \\ 14x + 6y = 8 \end{cases}$

Solution: $D = \begin{vmatrix} 7 & 3 \\ 14 & 6 \end{vmatrix} = 0$, $D_x = \begin{vmatrix} 4 & 3 \\ 8 & 6 \end{vmatrix} = 0$, and $D_y = \begin{vmatrix} 7 & 4 \\ 14 & 8 \end{vmatrix} = 0$.

Therefore, by Case 2, the system is consistent and the equations are dependent, and there are infinitely many solutions.

PROBLEM 4

Solve $\begin{cases} 4x + 6y = 12 \\ 2x + 3y = -6 \end{cases}$

Solution:

$$D = \begin{vmatrix} 4 & 6 \\ 2 & 3 \end{vmatrix} = 0, \quad D_x = \begin{vmatrix} 12 & 6 \\ -6 & 3 \end{vmatrix} = 72, \quad \text{and } D_y = \begin{vmatrix} 4 & 12 \\ 2 & -6 \end{vmatrix} = -48.$$

Case 3 says such a system is inconsistent and the equations are independent. There are no solutions.

Higher-order Determinants

The general determinant is not simply the extension of the cross-product method used for a 2 by 2 matrix. Therefore, we introduce a process to calculate the determinant for higher-order matrices.

A determinant of a square matrix of order n, denoted by

$$\det \begin{bmatrix} a_{11} & a_{12} & \cdots & a_{1n} \\ a_{21} & a_{22} & \cdots & a_{2n} \\ & \cdots\cdots\cdots & \\ a_{n1} & a_{n2} & \cdots & a_{nn} \end{bmatrix} = \begin{vmatrix} a_{11} & a_{12} & \cdots & a_{1n} \\ a_{21} & a_{22} & \cdots & a_{2n} \\ & \cdots\cdots\cdots & \\ a_{n1} & a_{n2} & \cdots & a_{nn} \end{vmatrix}$$

is called a **determinant of order n**.

PROPERTIES OF DETERMINANTS

There are certain properties of determinants which simplify the job of evaluating a higher-order determinant.

I. If each element in a row (or column) is zero, then the value of the determinant is zero.

Example:

$$\begin{vmatrix} 0 & 0 & 0 \\ a_2 & b_2 & c_2 \\ a_3 & b_3 & c_3 \end{vmatrix} = 0 \quad \text{and} \quad \begin{vmatrix} a_1 & b_1 & 0 \\ a_2 & b_2 & 0 \\ a_3 & b_3 & 0 \end{vmatrix} = 0$$

II. If each of the elements in a single row (or column) of a determinant is multiplied by a real number k, then the value of the determinant is multiplied by k.

Example:

$$\begin{vmatrix} a_1 & kb_1 & c_1 \\ a_2 & kb_2 & c_2 \\ a_3 & kb_3 & c_3 \end{vmatrix} = k \begin{vmatrix} a_1 & b_1 & c_1 \\ a_2 & b_2 & c_2 \\ a_3 & b_3 & c_3 \end{vmatrix}$$

III. If two rows (or columns) of a determinant are identical, then the value of the determinant is zero.

Example:

$$\begin{vmatrix} a_1 & b_1 & c_1 \\ a_2 & b_2 & c_2 \\ a_1 & b_1 & c_1 \end{vmatrix} = 0 \quad \text{and} \quad \begin{vmatrix} a_1 & b_1 & b_1 \\ a_2 & b_2 & b_2 \\ a_3 & b_3 & b_3 \end{vmatrix} = 0$$

IV. Interchanging any two rows (or columns) reverses the sign of the determinant.

Example:

$$\begin{vmatrix} c_1 & b_1 & a_1 \\ c_2 & b_2 & a_2 \\ c_3 & b_3 & a_3 \end{vmatrix} = - \begin{vmatrix} a_1 & b_1 & c_1 \\ a_2 & b_2 & c_2 \\ a_3 & b_3 & c_3 \end{vmatrix} \quad \begin{array}{l} \text{(interchanging column} \\ \text{one and column three)} \end{array}$$

EVALUATING HIGHER-ORDER DETERMINANTS

We now define two terms that are used to evaluate higher-order determinants—the minor and the cofactor of an element.

> **Definition:** The **minor** of an element of a determinant of order n is the determinant of order $n - 1$ obtained by deleting the row and column which contain the given element.

EXAMPLE 4

In the third-order determinant $\begin{vmatrix} 4 & -1 & 7 \\ 3 & 6 & 8 \\ 0 & 5 & -2 \end{vmatrix}$, the minor of 6 is obtained by crossing out the row and column containing 6,

$$---\begin{vmatrix} 4 & -1 & 7 \\ 3 & 6 & 8 \\ 0 & 5 & -2 \end{vmatrix}---$$

and writing the resulting second-order determinant

$$\begin{vmatrix} 4 & 7 \\ 0 & -2 \end{vmatrix}$$

EXAMPLE 5

If we consider the third-order determinant $\begin{vmatrix} a_{11} & a_{12} & a_{13} \\ a_{21} & a_{22} & a_{23} \\ a_{31} & a_{32} & a_{33} \end{vmatrix}$, then the minor of a_{23} is the second-order determinant $\begin{vmatrix} a_{11} & a_{12} \\ a_{31} & a_{32} \end{vmatrix}$.

> **Definition:** The **cofactor** of an element a_{ij}, ith row and jth column, is the product of the minor of a_{ij} and $(-1)^{i+j}$. Note that $(-1)^{i+j}$ is equal to 1 if $i + j$ is even, and equal to -1 if $i + j$ is odd.

EXAMPLE 6

The cofactor of 6 in Example 4 is

$$(-1)^{2+2}\begin{vmatrix} 4 & 7 \\ 0 & -2 \end{vmatrix} = (-1)^4\begin{vmatrix} 4 & 7 \\ 0 & -2 \end{vmatrix} = (1)(-8 - 0) = -8.$$

Note that $i = 2$ and $j = 2$ because 6 is in the second row and second column.

EXAMPLE 7

The cofactor of a_{23} in Example 5 is

$$(-1)^{2+3}\begin{vmatrix} a_{11} & a_{12} \\ a_{31} & a_{32} \end{vmatrix} = -\begin{vmatrix} a_{11} & a_{12} \\ a_{31} & a_{32} \end{vmatrix}$$

PROBLEM 5

Find the cofactor of -3 in the determinant

$$\begin{vmatrix} 1 & 6 & 8 & 0 \\ -3 & 9 & -5 & -4 \\ 8 & 0 & 2 & 0 \\ -1 & 11 & 1 & -7 \end{vmatrix}$$

Solution:

The cofactor of $(-3) = (-1)^{2+1} \begin{vmatrix} 6 & 8 & 0 \\ 0 & 2 & 0 \\ 11 & 1 & -7 \end{vmatrix}$

PROBLEM 6

Find the cofactor of each element of row one in the determinant

$$\begin{vmatrix} 1 & -3 & 2 \\ 0 & 6 & 4 \\ -1 & 0 & 5 \end{vmatrix}$$

Solution:

The cofactor of $1 = (-1)^{1+1} \begin{vmatrix} 6 & 4 \\ 0 & 5 \end{vmatrix} = (1)(30) = 30.$

The cofactor of $-3 = (-1)^{1+2} \begin{vmatrix} 0 & 4 \\ -1 & 5 \end{vmatrix} = (-1)(4) = -4.$

The cofactor of $2 = (-1)^{1+3} \begin{vmatrix} 0 & 6 \\ -1 & 0 \end{vmatrix} = (1)(6) = 6.$

> ### DETERMINANTS BY COFACTOR EXPANSIONS
>
> The value of any determinant can be obtained as follows:
>
> 1. Choose any row (or column).
> 2. Multiply each element in the row (or column) by its corresponding cofactor.
> 3. Add the products obtained in step 2.

PROBLEM 7

Evaluate the determinant in Problem 6.

Solution: We expand using the elements in the first row,

$$
\begin{vmatrix} 1 & -3 & 2 \\ 0 & 6 & 4 \\ -1 & 0 & 5 \end{vmatrix} = (1)(\text{cofactor of } 1) + (-3)(\text{cofactor of } -3) \\ \qquad\quad + (2)(\text{cofactor of } 2)
$$
$$
= (1)(30) + (-3)(-4) + (2)(6)
$$
$$
= 30 + 12 + 12
$$
$$
= 54
$$

Suppose we had chosen column two. Then

$$
\begin{vmatrix} 1 & -3 & 2 \\ 0 & 6 & 4 \\ -1 & 0 & 5 \end{vmatrix} = (-3)(\text{cofactor of } -3) + (6)(\text{cofactor of } 6) \\ \qquad\quad + (0)(\text{cofactor of } 0)
$$
$$
= (-3)\left\{(-1)^{1+2}\begin{vmatrix} 0 & 4 \\ -1 & 5 \end{vmatrix}\right\} + (6)\left\{(-1)^{2+2}\begin{vmatrix} 1 & 2 \\ -1 & 5 \end{vmatrix}\right\} + 0
$$
$$
= (-3)(-1)(4) + (6)(1)(7)
$$
$$
= 12 + 42
$$
$$
= 54
$$

It can be shown that the value of the determinant is the same no matter which row or column we choose.

PROBLEM 8

Evaluate $\begin{vmatrix} 4 & 3 \\ -8 & -6 \end{vmatrix}$.

Solution: Since the second row is -2 times the first, Property II and Property III say the determinant equals zero.

HIGHER-ORDER SYSTEMS OF EQUATIONS

Now that we know how to evaluate higher-order determinants, we can state Cramer's Rule for finding the solutions of a system of n linear equations in n unknowns.

Theorem: Cramer's Rule

Given the system $\begin{cases} a_{11}x_1 + a_{12}x_2 + \cdots + a_{1n}x_n = b_1 \\ a_{21}x_1 + a_{22}x_2 + \cdots + a_{2n}x_n = b_2 \\ \quad\cdots\cdots\cdots\cdots\cdots\cdots\cdots\cdots \\ a_{n1}x_1 + a_{n2}x_2 + \cdots + a_{nn}x_n = b_n \end{cases}$

and provided the determinant of the coefficient matrix is not zero, then the solution of the system is given by

$$x_i = \frac{D_i}{D} = \frac{\begin{vmatrix} a_{11} & a_{12} & \ldots & b_1 & \ldots & a_{1n} \\ a_{21} & a_{22} & \ldots & b_2 & \ldots & a_{2n} \\ \multicolumn{6}{c}{\ldots} \\ a_{n1} & a_{n2} & \ldots & b_n & \ldots & a_{nn} \end{vmatrix}}{\begin{vmatrix} a_{11} & a_{12} & \ldots & a_{1i} & \ldots & a_{1n} \\ a_{21} & a_{22} & \ldots & a_{2i} & \ldots & a_{2n} \\ \multicolumn{6}{c}{\ldots} \\ a_{n1} & a_{n2} & \ldots & a_{ni} & \ldots & a_{nn} \end{vmatrix}} \qquad i = 1, 2, \ldots, n$$

*i*th column

EXAMPLE 7

For $n = 3$, that is, three equations in three unknowns, the theorem states that given the system

$$\begin{cases} a_{11}x_1 + a_{12}x_2 + a_{13}x_3 = b_1 \\ a_{21}x_1 + a_{22}x_2 + a_{23}x_3 = b_2 \\ a_{31}x_1 + a_{32}x_2 + a_{33}x_3 = b_3 \end{cases}$$

then the solution to such a system is given by

$$x_1 = \frac{D_1}{D} = \frac{\begin{vmatrix} b_1 & a_{12} & a_{13} \\ b_2 & a_{22} & a_{23} \\ b_3 & a_{32} & a_{33} \end{vmatrix}}{\begin{vmatrix} a_{11} & a_{12} & a_{13} \\ a_{21} & a_{22} & a_{23} \\ a_{31} & a_{32} & a_{33} \end{vmatrix}}, \quad x_2 = \frac{D_2}{D} = \frac{\begin{vmatrix} a_{11} & b_1 & a_{13} \\ a_{21} & b_2 & a_{23} \\ a_{31} & b_3 & a_{33} \end{vmatrix}}{\begin{vmatrix} a_{11} & a_{12} & a_{13} \\ a_{21} & a_{22} & a_{23} \\ a_{31} & a_{32} & a_{33} \end{vmatrix}},$$

$$x_3 = \frac{D_3}{D} = \frac{\begin{vmatrix} a_{11} & a_{21} & b_1 \\ a_{21} & a_{22} & b_2 \\ a_{31} & a_{32} & b_3 \end{vmatrix}}{\begin{vmatrix} a_{11} & a_{12} & a_{13} \\ a_{21} & a_{22} & a_{23} \\ a_{31} & a_{32} & a_{33} \end{vmatrix}}$$

PROBLEM 9

Using Cramer's Rule, solve the system

$$\begin{cases} 4x_1 + x_2 + 2x_3 = 10 \\ 3x_1 + 2x_2 + x_3 = 5 \\ 2x_1 + 3x_2 + 2x_3 = 10 \end{cases}$$

Solution: We first compute D, D_1, D_2, and D_3. Expanding about the first row we get

$$D = \begin{vmatrix} 4 & 1 & 2 \\ 3 & 2 & 1 \\ 2 & 3 & 2 \end{vmatrix} = (4)(1) + (1)(-4) + (2)(5) = 10$$

$$D_1 = \begin{vmatrix} 10 & 1 & 2 \\ 5 & 2 & 1 \\ 10 & 3 & 2 \end{vmatrix} = 0, \text{ because columns one and three are multiples,}$$
and by Property III.

$$D_2 = \begin{vmatrix} 4 & 10 & 2 \\ 3 & 5 & 1 \\ 2 & 10 & 2 \end{vmatrix} = 0, \text{ because columns two and three are multiples,}$$
and by Property III.

$$D_3 = \begin{vmatrix} 4 & 1 & 10 \\ 3 & 2 & 5 \\ 2 & 3 & 10 \end{vmatrix} = (4)(5) + (1)(-20) + (10)(5) = 50$$

Thus, $x_1 = \dfrac{D_1}{D} = \dfrac{0}{10} = 0$, $x_2 = \dfrac{D_2}{D} = \dfrac{0}{10} = 0$, and $x_3 = \dfrac{D_3}{D} = 5$, and the solution is the ordered triple $(0, 0, 5)$.

Although Cramer's Rule is of interest from a theoretical point of view, its practical uses are somewhat limited. It is a good method for systems of two variables and systems of three variables. However, for higher-order systems it is not as practical because of the difficulty in evaluating the determinants of such systems. Moreover, the number of equations must be the same as the number of unknowns. The **elimination method** and **row-equivalent matrices method** do not have these limitations and are the most effective ways of solving a system of linear equations.

Applications of Determinants

One very important elementary application of determinants is for simplifying notation. Consider the equation of a line passing through two points (x_1, y_1) and (x_2, y_2). This can be written in the form of a determinant.

Theorem 1: The equation of a line passing through the points (x_1, y_1) and (x_2, y_2) is given by the equation

$$\begin{vmatrix} x & y & 1 \\ x_1 & y_1 & 1 \\ x_2 & y_2 & 1 \end{vmatrix} = 0$$

Proof. Evaluating the determinant we get

$$\begin{vmatrix} x & y & 1 \\ x_1 & y_1 & 1 \\ x_2 & y_2 & 1 \end{vmatrix} = x(y_1 - y_2) - y(x_1 - x_2) + (x_1y_2 - x_2y_1) = 0$$

or, equivalently, $xy_1 - xy_2 - x_1y + x_2y + x_1y_2 - x_2y_1 = 0$. Adding and subtracting x_1y_1 to the left side gives $xy_1 - xy_2 - x_1y + x_2y + x_1y_2 - x_2y_1 + x_1y_1 - x_1y_1 = 0$. Rearranging, we have

$$x_2y - x_1y - x_2y_1 + x_1y_1 = xy_2 - xy_1 - x_1y_2 + x_1y_1$$
$$y(x_2 - x_1) - y_1(x_2 - x_1) = x(y_2 - y_1) - x_1(y_2 - y_1)$$
$$(y - y_1)(x_2 - x_1) = (y_2 - y_1)(x - x_1)$$

and

$$(y - y_1) = \frac{(y_2 - y_1)}{(x_2 - x_1)} (x - x_1)$$

This last equation is the two-point equation we discussed in a previous section.

EXAMPLE 8

The equation of the line passing through the points $(4, -2)$ and $(1, 3)$ is

$$\begin{vmatrix} x & y & 1 \\ 4 & -2 & 1 \\ 1 & 3 & 1 \end{vmatrix} = 0.$$

Evaluating the determinant we get $x(-5) - y(3) + 1(14) = 0$, that is, $-5x - 3y + 14 = 0$.

A direct result of Theorem 1 is the idea of collinear points—that is, points which are on the same line. Suppose that the three points (x_1, y_1), (x_2, y_2), and (x_3, y_3) are collinear. Then (x_3, y_3) must satisfy the equations in Theorem 1:

$$\begin{vmatrix} x_3 & y_3 & 1 \\ x_1 & y_1 & 1 \\ x_2 & y_2 & 1 \end{vmatrix} = 0$$

If we interchange the first and third rows and then the second and first rows of the resulting determinant, we have

$$\begin{vmatrix} x_1 & y_1 & 1 \\ x_2 & y_2 & 1 \\ x_3 & y_3 & 1 \end{vmatrix} = 0.$$

Then $(y_3 - y_2) = \frac{(y_2 - y_1)}{(x_2 - x_1)} (x_3 - x_2)$, or $\frac{(y_3 - y_2)}{(x_3 - x_2)} = \frac{(y_2 - y_1)}{(x_2 - x_1)}$.

Further manipulation shows that $\frac{(y_3 - y_1)}{(x_3 - x_2)} = \frac{(y_2 - y_1)}{(x_2 - x_1)}$.

Therefore, $\frac{(y_2 - y_1)}{(x_2 - x_1)} = \frac{(y_3 - y_2)}{(x_3 - x_1)} = \frac{(y_3 - y_1)}{(x_3 - x_1)}$, which says that the slopes of the line passing through any pair of the points are the same. Thus, the points (x_1, y_1), (x_2, y_2), and (x_3, y_3) are collinear. These results can be stated as a theorem.

Theorem 2: Three points (x_1, y_1), (x_2, y_2), and (x_3, y_3) are collinear if and only if

$$\begin{vmatrix} x_1 & y_1 & 1 \\ x_2 & y_2 & 1 \\ x_3 & y_3 & 1 \end{vmatrix} = 0$$

EXAMPLE 9

The points $(1, 3)$, $(-2, -3)$, and $(3, 7)$ are collinear. This can be shown by using Theorem 2.

$$\begin{vmatrix} 1 & 3 & 1 \\ -2 & -3 & 1 \\ 3 & 7 & 1 \end{vmatrix} = (1)(-10) + (3)(5) + (1)(-5) = 0$$

PROBLEM 10

We check to see whether the points $(-1, 2)$, $(-3, 5)$, and $(4, 4)$ are collinear.

Solution: Using Theorem 2 we have

$$\begin{vmatrix} -1 & 2 & 1 \\ -3 & 5 & 1 \\ 4 & 4 & 1 \end{vmatrix} = (-1)(1) + (2)(7) + (1)(-32) = -19 \neq 0$$

The points $(-1, 2)$, $(-3, 5)$, and $(4, 4)$ are *not* collinear.

A final application of determinants is finding the area of a triangle in terms of the coordinates of its vertices.

Theorem 3: The area of a triangle with vertices (x_1, y_1), (x_2, y_2), and (x_3, y_3) is the absolute value of

$$\frac{1}{2} \cdot \begin{vmatrix} x_1 & y_1 & 1 \\ x_2 & y_2 & 1 \\ x_3 & y_3 & 1 \end{vmatrix}$$

Proof. Suppose that a triangle ABC has vertices (x_1, y_1), (x_2, y_2), and (x_3, y_3) (see Fig. 89). The area of ABC = area of $AEFC$ + area of $CFGB$ − area of $AEGB$. Recall that the area of a trapezoid equals $\frac{1}{2}$ the product of its altitude and the sum of its parallel sides. Therefore,

$$\text{area of } ABC = \frac{1}{2}(x_3 - x_1)(y_1 + y_3) + \frac{1}{2}(x_2 - x_3)(y_2 + y_3)$$

$$- \frac{1}{2}(x_2 - x_1)(y_1 + y_2)$$

$$= \frac{1}{2}[x_3y_1 + x_3y_3 - x_1y_1 - x_1y_3 + x_2y_2 + x_2y_3 - x_3y_2 - x_3y_3$$

$$- x_2y_1 - x_2y_2 + x_1y_1 + x_1y_2]$$

$$= \frac{1}{2}[(x_1y_2 - x_2y_1) - (x_1y_3 - x_3y_1) + (x_2y_3 - x_3y_2)]$$

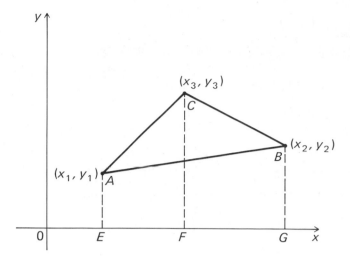

Figure 89

which can be represented by a determinant:

$$\text{area } ABC = \frac{1}{2} \begin{vmatrix} x_1 & y_1 & 1 \\ x_2 & y_2 & 1 \\ x_3 & y_3 & 1 \end{vmatrix}$$

This may be seen by expanding the determinant using the third column. The absolute value must be used because of the placement of the points *ABC*. If the order of *ABC* is traced counterclockwise, then the determinant will be positive. If the points *ABC* are not traced counterclockwise then the determinant will be negative, and since area is not negative we take the absolute value.

PROBLEM 11

Find the area of the triangle with vertices (4, 2), (6, 3), and (5, 5).

Solution: The area is given by the absolute value of

$$\frac{1}{2} \begin{vmatrix} 4 & 2 & 1 \\ 6 & 3 & 1 \\ 5 & 5 & 1 \end{vmatrix} = \frac{1}{2} |(4)(-2) + (-2)(1) + (1)(15)| = \frac{5}{2}$$

3.4 EXERCISES

1. Evaluate the following determinants:

(a) $\begin{vmatrix} 1 & 2 \\ -3 & 1 \end{vmatrix}$ (b) $\begin{vmatrix} \frac{1}{3} & 0 \\ 4 & -6 \end{vmatrix}$

(c) $\begin{vmatrix} 1.7 & -.5 \\ 2.1 & .6 \end{vmatrix}$ (d) $\begin{vmatrix} 7 & -2 \\ -14 & 4 \end{vmatrix}$

2. Solve the following systems using Cramer's Rule:

(a) $\begin{cases} 2x + y = 7 \\ 3x - 2y = 0 \end{cases}$

(b) $\begin{cases} 2x + y = 1 \\ 5x + 3y = 2 \end{cases}$

(c) $\begin{cases} 5x + 3y = -26 \\ 4x - 9y = 2 \end{cases}$

3. Using determinants, determine whether the following systems are consistent and independent, consistent and dependent, or inconsistent and independent. Graph each system.

(a) $\begin{cases} x + y = 5 \\ x - y = 1 \end{cases}$ (c) $\begin{cases} -x + y = 1 \\ x - y = 2 \end{cases}$

(b) $\begin{cases} x - 3y = 2 \\ -2x + 6y = -4 \end{cases}$

4. Given the determinant $\begin{vmatrix} a_{11} & a_{12} & a_{13} \\ a_{21} & a_{22} & a_{23} \\ a_{31} & a_{32} & a_{33} \end{vmatrix}$, find the minor and cofactor of each of the following elements:

(a) a_{31} (b) a_{22} (c) a_{23}

5. Show that

(a) $2\begin{vmatrix} 4 & -1 \\ -3 & 5 \end{vmatrix} = \begin{vmatrix} 4 & -1 \\ -6 & 10 \end{vmatrix}$

(b) $\begin{vmatrix} -1 & 2 \\ 5 & 3 \end{vmatrix} = -\begin{vmatrix} 5 & 3 \\ -1 & 2 \end{vmatrix}$

6. Let $a, b, c, d,$ and k be real numbers. Show that

(a) $\begin{vmatrix} a & b \\ kc & kd \end{vmatrix} = k\begin{vmatrix} a & b \\ c & d \end{vmatrix}$

(b) $\begin{vmatrix} c & d \\ kc & kd \end{vmatrix} = 0$

(c) $\begin{vmatrix} a & b \\ c & d \end{vmatrix} = -\begin{vmatrix} c & d \\ a & b \end{vmatrix}$

(d) $\begin{vmatrix} 0 & 0 \\ c & d \end{vmatrix} = 0$

7. Evaluate the following determinants by the expansion of cofactors:

(a) $\begin{vmatrix} 3 & 0 & 1 \\ 6 & 4 & 2 \\ 0 & 2 & -1 \end{vmatrix}$

(b) $\begin{vmatrix} 16 & 4 & 22 \\ 4 & 2 & -3 \\ 12 & 2 & 25 \end{vmatrix}$

(c) $\begin{vmatrix} -4 & 0 & 0 \\ 0 & 3 & 0 \\ 0 & 0 & -5 \end{vmatrix}$

(d) $\begin{vmatrix} 4 & 3 & -2 & 1 \\ -9 & -2 & 5 & 0 \\ 7 & 7 & 1 & 0 \\ 10 & -7 & 3 & 0 \end{vmatrix}$

(e) $\begin{vmatrix} 1 & 0 & 0 & 0 \\ 0 & 1 & 0 & 0 \\ 0 & 0 & 1 & 0 \\ 0 & 0 & 0 & 1 \end{vmatrix}$

8. Use the properties of determinants to show the following:

(a) $\begin{vmatrix} -6 & -12 & 8 \\ 1 & -1 & 3 \\ 3 & 6 & -4 \end{vmatrix} = 0$

(b) $\begin{vmatrix} ka & kb & kc \\ kd & ke & kf \\ kg & kh & ki \end{vmatrix} = k^3\begin{vmatrix} a & b & c \\ d & e & f \\ g & h & i \end{vmatrix}$

9. For what value of x does

$\begin{vmatrix} x-1 & 7 & 2 \\ 3 & 5 & 0 \\ 0 & -5 & -3 \end{vmatrix} = 18?$

10. Solve the following systems of equations using Cramer's Rule:

(a) $\begin{cases} x_1 - x_2 + x_3 = -2 \\ 2x_1 + x_2 + 3x_3 = 1 \\ x_1 - 2x_2 - 2x_3 = -1 \end{cases}$

(b) $\begin{cases} x_1 + x_2 - x_3 = 3 \\ 5x_1 - x_2 - 2x_3 = -3 \\ 2x_2 + x_3 = 10 \end{cases}$

11. Using determinants, find the equation of the line passing through each of the following pairs of points:
    (a) $(-4, 1)$ and $(3, -5)$
    (b) $(0, 0)$ and $(5, -3)$
    (c) $(5, -3)$ and $(5, 2)$
    (d) $(-5, 2)$ and $(3, 2)$

12. Using determinants, determine whether the following points are collinear:
    (a) $(0, 4)$, $(3, -2)$, and $(-2, 8)$
    (b) $(-2, 1)$, $(3, 2)$, and $(6, 3)$
    (c) $(1, 2)$, $(-3, 10)$, and $(4, -4)$

13. Using determinants, find the area of the triangles whose vertices are:
    (a) $(-3, 4)$, $(6, 2)$, and $(4, -3)$
    (b) $(0, 4)$, $(-8, 0)$, and $(-1, -4)$
    (c) $(\sqrt{2}, 2)$, $(-4, 6)$, and $(4, -2\sqrt{2})$

14. Find the area of the triangle whose vertices are $(0, k)$, $(k, 2k)$, and $(3k, 0)$. Determine if there is any value of k for which the area is zero.

15. We have seen that the determinant can be used to find the area of a triangle in the plane. This use of a determinant also has meaning in three dimensions. Let A, B, C, and D be the vertices of a **tetrahedron** (triangular pyramid) with coordinates (x_1, y_1, z_1), (x_2, y_2, z_2), (x_3, y_3, z_3), and (x_4, y_4, z_4). Geometric arguments can show that the volume of the tetrahedron is the absolute value of

$$\frac{1}{6} \cdot \begin{vmatrix} x_1 & y_1 & z_1 & 1 \\ x_2 & y_2 & z_2 & 1 \\ x_3 & y_3 & z_3 & 1 \\ x_4 & y_4 & z_4 & 1 \end{vmatrix}$$

Find the volume of the tetrahedron with vertices $(1, 0, 0)$, $(0, 1, 0)$, $(0, 0, 1)$, and $(1, 1, 1)$.

3.5 QUADRATIC FUNCTIONS

Consider now polynomials of degree two, $ax^2 + bx + c$. Such polynomials define a class of functions called **quadratic functions.**

Definition: A function f defined by $f(x) = ax^2 + bx + c$, where a, b, and c are real numbers, $a \neq 0$, is called a **quadratic function.**

EXAMPLE 1

The function f defined by $f(x) = x^2 + 2x$ is a quadratic function where $a = 1$, $b = 2$, and $c = 0$.

EXAMPLE 2

The function g defined by $g(x) = x^2$ is a quadratic function where $a = 1$, $b = 0$, and $c = 0$.

EXAMPLE 3

The expression $h(x) = -x^2 + 2x - 2$ defines a quadratic function. Here we have $a = -1$, $b = 2$, and $c = -2$.

The graphs of each of these functions are examples of **parabolas,** as was shown in Section 2.3 of Chapter 2. We shall give a more detailed discussion of parabolas in Chapter 7. The graphs of Examples 1 to 3 are given in Figure 90.

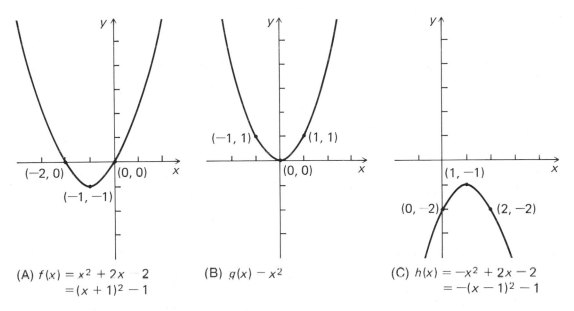

(A) $f(x) = x^2 + 2x - 2$
  $= (x + 1)^2 - 1$

(B) $g(x) - x^2$

(C) $h(x) = -x^2 + 2x - 2$
  $= -(x - 1)^2 - 1$

Figure 90

In this section we are interested in various points that appear on the graph of a quadratic function. First, we shall investigate those points, if any, where the graph intersects the x-axis. For example, the function f in Example 1 intersects the x-axis at points $(-2, 0)$ and $(0, 0)$ as shown in Figure 90(A); the function g intersects at the origin, $(0, 0)$, as shown in Figure 90(B); and the function h does not intersect the x-axis at any point. Such points are called **zeros** of the quadratic function. Secondly, we shall examine the **vertex,** or turning point, of a parabola. The vertices of the parabolas which represent the functions given in Examples 1 to 3 are $(-1, 1)$ for the function f; $(0, 0)$ for the function g; and $(1, -1)$ for the function h. Before we study the concepts of zero(s) and vertex of a quadratic function, we shall discuss the graph of a quadratic function which is a parabola.

Completing the Square

PROBLEM 1

Graph the quadratic function f defined by

$$f(x) = 2x^2 + 4x + 1$$

Solution: The graph could be obtained by plotting several points; however, we shall make use of the graphing techniques discussed in Section 2.3 of Chapter 2. Our goal is to write the expression $f(x) = 2x^2 + 4x + 1$ in a form that will enable us to use the graph of $y = x^2$ to obtain the graph of $f(x) = 2x^2 + 4x + 1$. The technique used is that of completing the square on the right side of the expression $f(x) = 2x^2 + 4x + 1$.

$$
\begin{aligned}
f(x) &= 2x^2 + 4x + 1 \\
&= 2(x^2 + 2x) + 1 && \text{factor 2} \\
&= 2(x^2 + 2x + 1 - 1) + 1 && \text{add and subtract } \left\{\frac{1}{2} \cdot \text{coeff. of } x\right\}^2 \\
& && \left(\text{That is, } \left\{\frac{1}{2} \cdot 2\right\}^2 = \{1\}^2 = 1.\right) \\
&= 2(x^2 + 2x + 1) - 2 + 1 \\
&= 2(x + 1)^2 - 1
\end{aligned}
$$

Thus, $f(x) = 2x^2 + 4x + 1 = 2(x + 1)^2 - 1$ and the graph of f is the graph of $y = x^2$ shifted to the left 1 unit, stretched vertically 2 units, and shifted vertically downward 1 unit (see Fig. 91). Note that the vertex is

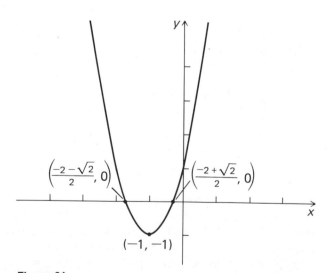

Figure 91

$(-1, -1)$ and the zeros are $\left(\dfrac{-2 - \sqrt{2}}{2}, 0\right)$ and $\left(\dfrac{-2 + \sqrt{2}}{2}, 0\right)$.

We now complete the square in the general quadratic function f defined by $f(x) = ax^2 + bx + c$, where a, b, and c are real numbers, $a \neq 0$.

$$
\begin{aligned}
f(x) &= ax^2 + bx + c \\
&= a\left(x^2 + \frac{b}{a}x\right) + c \\
&= a\left(x^2 + \frac{b}{a}x + \frac{b^2}{4a^2} - \frac{b^2}{4a^2}\right) + c \\
&= a\left(x^2 + \frac{b}{a}x + \frac{b^2}{4a^2}\right) - \frac{b^2}{4a} + c \\
&= a\left(x + \frac{b}{2a}\right)^2 + \frac{4ac - b^2}{4a}
\end{aligned}
$$

Therefore, if we set $h = -\dfrac{b}{2a}$ and $k = \dfrac{4ac - b^2}{4a}$, then we have

$$
\begin{aligned}
f(x) &= ax^2 + bx + c \\
&= a(x - h)^2 + k
\end{aligned}
$$

Graphically, we get a parabola whose graph is the graph of $y = x^2$ stretched or contracted vertically by a factor of $|a|$, shifted horizontally h units, and shifted vertically k units, and if $a < 0$, then reflected about the x-axis. Also, the **vertex** is at (h, k).

PROBLEM 2

Let f, g, and p be quadratic functions defined by $f(x) = x^2 - 4x + 3$, $g(x) = \dfrac{1}{2}x^2 - x + \dfrac{1}{2}$, and $p(x) = x^2 - 6x + 10$. Graph f, g, and p. Indicate the vertices.

Solution:
For f: If $f(x) = x^2 - 4x + 3$, we have $a = 1$, $b = -4$, and $c = 3$. Thus, $h = \dfrac{-b}{2a} = \dfrac{4}{2} = 2$ and $k = \dfrac{4ac - b^2}{4a} = \dfrac{12 - 16}{4} = -1$. The function f takes the form $f(x) = (x - 2)^2 - 2$. The vertex $= (2, -2)$.

For g: Similarly, for g we have $a = \dfrac{1}{2}$, $b = -1$, and $c = \dfrac{1}{2}$. Hence, $h = 1$ and $k = 0$ and the function g takes the form $g(x) = \dfrac{1}{2}(x - 1)^2$. The vertex $= (1, 0)$.

For p: Finally, for the function p we have $a = 1$, $b = -6$, and $c = 10$. This yields $h = 3$ and $k = 1$. The form p takes is $p(x) = (x - 3)^2 + 1$. The vertex $= (3, 1)$.

The graphs of f, g, and p are given in Figure 92.

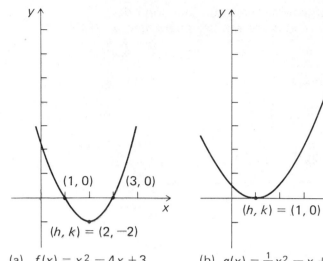

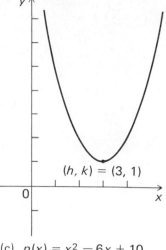

(a) $f(x) = x^2 - 4x + 3$

$= (x - 2)^2 - 2$

(b) $g(x) = \frac{1}{2}x^2 - x + \frac{1}{2}$

$= \frac{1}{2}(x - 1)^2$

(c) $p(x) = x^2 - 6x + 10$

$= (x - 3)^2 + 1$

Figure 92

Zeros of a Quadratic Function

Consider the quadratic function f defined by $f(x) = ax^2 + bx + c$ or, equivalently, $y = ax^2 + bx + c$. If we set $y = 0$, then we have a **quadratic equation,** $ax^2 + bx + c = 0$. This equation is said to have solutions $x = r_1$ and $x = r_2$ if and only if $f(r_1) = 0$ and $f(r_2) = 0$. Geometrically these points have coordinates $(r_1, 0)$ and $(r_2, 0)$. The values r_1 and r_2 are called **zeros** of the quadratic function f, since they are the values at which the function is zero. That is, they represent the points where the parabola crosses the x-axis. Algebraically we can find the zeros of the quadratic function f, defined by $f(x) = ax^2 + bx + c$, by solving the quadratic equation $ax^2 + bx + c = 0$. Once again we use the technique of completing the square.

$ax^2 + bx + c = 0$ $ax^2 + bx = -c$ $x^2 + \dfrac{b}{a}x = -\dfrac{c}{a}$	Take constant c to right side. Divide each side by a.

$x^2 + \dfrac{b}{a}x + \dfrac{b^2}{4a^2} = \dfrac{b^2}{4a^2} - \dfrac{c}{a}$	Add $\left\{\dfrac{1}{2}\cdot \text{coeff. of } x\right\}^2$ to each side.

$\left(x + \dfrac{b}{2a}\right)^2 = \dfrac{b^2 - 4ac}{4a^2}$	Write the left side as the square of a binomial.

$$x + \frac{b}{2a} = \frac{\pm\sqrt{b^2 - 4ac}}{2a} \qquad \text{Take square roots of both sides.}$$

$$x = \frac{-b \pm \sqrt{b^2 - 4ac}}{2a} \qquad \text{Solve for } x.$$

Quadratic Formula:

$$x = \frac{-b \pm \sqrt{b^2 - 4ac}}{2a}$$

This last equation is called the **quadratic formula.** Of special interest is the expression $b^2 - 4ac$, called the **discriminant.** The discriminant plays an important role in determining the nature of the zeros (roots) of a quadratic function. If $b^2 - 4ac \geq 0$, then $\sqrt{b^2 - 4ac}$ is real and x is real. If $b^2 - 4ac < 0$, then $\sqrt{b^2 - 4ac}$ is not real and x is not real, and the graph does *not* cross the x-axis.

PROBLEM 3

Using the discriminant determine the nature of the roots of each of the following quadratic functions:

(a) $f(x) = x^2 - 4x + 3$

(b) $g(x) = \frac{1}{2} x^2 - x + \frac{1}{2}$

(c) $p(x) = x^2 - 6x + 10$

Solution:
(a) $f(x) = x^2 - 4x + 3$. The discriminant $b^2 - 4ac = (-4)^2 - 4(1)(3) = 16 - 12 = 4 > 0$. Therefore, there are two real zeros for the function f (see Fig. 92A).

(b) $g(x) = \frac{1}{2} x^2 - x + \frac{1}{2}$. The discriminant $b^2 - 4ac = (-1)^2 - 4\left(\frac{1}{2}\right)\left(\frac{1}{2}\right) = 1 - 1 = 0$. The function g has one real zero (see Fig. 92B).

(c) $p(x) = x^2 - 6x + 10$. The discriminant $b^2 - 4ac = (-6)^2 - 4(1)(10) = 36 - 40 = -4 < 0$. Thus, the function p has no real zeros (see Fig. 92C).

Table 1 summarizes the three possible cases concerning the zeros of a quadratic function.

TABLE 1.   The Nature of the Zeros of a Quadratic Function

DISCRIMINANT: $b^2 - 4ac$	NATURE OF ZEROS	GEOMETRIC INTERPRETATION
1. $b^2 - 4ac > 0$	Two real zeros $$x_1 = \frac{-b + \sqrt{b^2 - 4ac}}{2a}$$ and $$x_2 = \frac{-b - \sqrt{b^2 - 4ac}}{2a}$$	The graph of the function intersects the x-axis in two points:
2. $b^2 - 4ac = 0$	One real zero $$x = -\frac{b}{2a}$$	The graph of the function intersects the x-axis in one point:
3. $b^2 - 4ac < 0$	No real zero	The graph of the function does *not* intersect the x-axis:

The Vertex: Minimum and Maximum Points

We have seen that the graph of a quadratic function is a parabola and that the parabola either opens upward or downward. It can be shown that the graph of the quadratic function f, defined by $f(x) = ax^2 + bx + c$, is a parabola which

(1) opens upward if $a > 0$.
(2) opens downward if $a < 0$.

Of special interest is the point on the curve called the **vertex.** This point represents the *lowest* point of the parabola if the parabola opens upward and represents the *highest* point on the parabola if the parabola opens downward. If the vertex is the lowest point of the parabola, then the vertex is called the **minimum point.** If the vertex is the highest point of the parabola, then the vertex is called the **maximum point.** Recall that the vertex of the quadratic function f, defined by

$f(x) = ax^2 + bx + c$, is the point (h, k) where $h = \dfrac{-b}{2a}$ and $k = \dfrac{4ac - b^2}{4a}$.

Thus we have the following theorem.

Theorem: The quadratic function f, defined by $f(x) = ax^2 + bx + c$, has

(i) a maximum at $x = -\dfrac{b}{2a}$ if $a < 0$.

(ii) a minimum at $x = -\dfrac{b}{2a}$ if $a > 0$.

In either case the value of the function f is

$$\frac{4ac - b^2}{4a} = -\frac{(b^2 - 4ac)}{4a}$$

EXAMPLE 4

Given $f(x) = -4x^2 - 3x + 1$, we have $h = \dfrac{-b}{2a} = -\dfrac{3}{8}$ and $k = \dfrac{4ac - b^2}{4a} = \dfrac{25}{16}$. Therefore, the vertex is the point $\left(-\dfrac{3}{8}, \dfrac{25}{16}\right)$. Since $a = -4 < 0$, the vertex is a maximum.

EXAMPLE 5

Given $g(x) = x^2 + 1$, we have $h = -\dfrac{b}{2a} = \dfrac{0}{2} = 0$ and $k = \dfrac{4ac - b^2}{4a} = \dfrac{4}{4} = 1$. Thus, the vertex is the point $(0, 1)$ and, since $a > 0$, the vertex is a minimum.

We summarize the information concerning quadratic functions using the following graphs (Fig. 93):

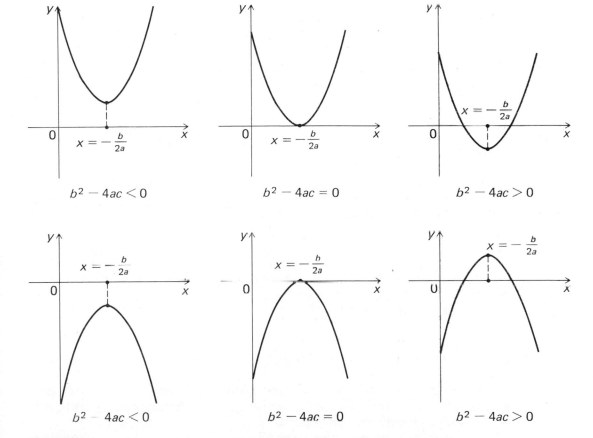

Figure 93

Case 1: $a > 0$; the vertex $\left(-\dfrac{b}{2a}, \dfrac{4ac - b^2}{4a}\right)$ is a minimum.

Case 2: $a < 0$; the vertex $\left(-\dfrac{b}{2a}, \dfrac{4ac - b^2}{4a}\right)$ is a maximum.

Applications of Quadratic Functions

We now consider a few of the many applications of quadratic functions.

PROBLEM 4

The length of a rectangle is twice its width. If the width is diminished by 1 foot and the length increased by 2 feet, the area will be 100 square feet. Find the dimensions of the original rectangle.

Solution: Let x be the original width. Then $2x$ is the length of the original rectangle. The dimensions of the new rectangle are width $= x - 1$ feet and length $= 2x + 2$ feet. Thus, the area of the new rectangle can be expressed by the quadratic equation

$$(2x + 2)(x - 1) = 100$$
$$2(x + 1)(x - 1) = 100$$
$$(x + 1)(x - 1) = 50$$
$$x^2 - 1 = 50$$
$$x = \pm\sqrt{51}$$

Since distance cannot be negative, we have the dimensions of the original rectangle being width $= \sqrt{51}$ and length $= 2\sqrt{51}$.

PROBLEM 5

An object is projected vertically upward with initial velocity of v_0 feet per second. The distance s above the ground at any time t in seconds is described by the quadratic function s defined by

$$s(t) = v_0 t - 16t^2$$

If a ball is thrown vertically upward from the ground with an initial velocity of 64 feet per second, then
   (a) how many seconds will it take the ball to reach its maximum height?
   (b) what is the maximum height?
   (c) how many seconds will it take the ball to reach the ground?

Solution: The equation which describes the motion of the ball is $s(t) = 64t - 16t^2$. Since the coefficient of t^2 is -16, the graph is a parabola which opens downward (see Fig. 94).
   (a) The fact that the graph of the equation $s = 64t - 16t^2$ is a parabola which opens downward means that the vertex is a maximum. The time it takes the ball to reach its maximum height is given by the first coordinate of the vertex—that is, $t = -\dfrac{b}{2a} = \dfrac{-64}{-32} = 2$ seconds. It takes the ball 2 seconds to reach its maximum height.

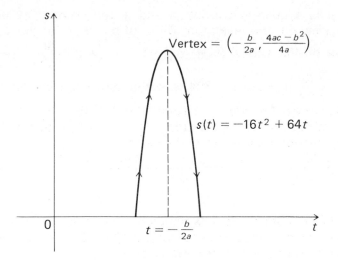

Figure 94

(b) It follows that the maximum height is given by the second coordinate of the vertex—that is,

$$s = \frac{4ac - b^2}{4a} = \frac{4(-16)(0) - (64)^2}{4(-16)} = \frac{-(64)^2}{-64} = 64 \text{ feet.}$$

The maximum height is 64 feet.

(c) When the ball returns to the ground the distance will be zero. Therefore we are looking for the value(s) for which $s(t) = 0$—that is,

$$64t - 16t^2 = 0$$
$$16t(4 - t) = 0$$
$$t = 0$$
$$\text{and } t = 4$$

There are two times at which the distance is zero, when the ball is thrown and when it reaches the ground after its flight. The former is represented by the time $t = 0$. Hence, $t = 4$ is the number of seconds it will take the ball to reach the ground after the ball is thrown.

PROBLEM 6

The number 36 is written as the sum of two numbers so that the product of the numbers is maximized. Find the numbers.

Solution: Let x be one of the numbers. Then $36 - x$ is the other number. The product p is defined by

$$p(x) = x(36 - x)$$
$$p(x) = 36x - x^2$$

Since p is a quadratic function and is represented by a parabola opening downward, the maximum value of p occurs at the vertex. The coordinates of the vertex (h, k) are $h = -\dfrac{b}{2a} = \dfrac{-36}{-2} = 18$ and $k = \dfrac{4ac - b^2}{4a} = \dfrac{-(36)^2}{-4} = (18)^2 = 324$, and the vertex is $(18, 324)$. Thus, the numbers we were to find are $x = 18$ and $36 - x = 36 - 18 = 18$. The value of the maximized product is $(18)(18) = (18)^2 = 324$. Note that this is the value of the second coordinate of the vertex.

PROBLEM 7

The total cost of producing x radios per day is $\$(x^2 + 4x + 5)$ and the price per set at which they may be sold is $\$(100 - 2x)$. What should be the daily output to obtain a maximum total profit?

Solution: Profit $=$ *(Revenue)* $-$ *(Cost)*. If x radios are sold per day at a price of $\$(100 - 2x)$ each, then the revenue generated is $x(100 - 2x)$. Thus, the profit on the sale of x sets per day is

$$
\begin{aligned}
P &= x(100 - 2x) - (x^2 + 4x + 5) \\
&= 100x - 2x^2 - x^2 - 4x - 5 \\
&= -3x^2 + 96x - 5.
\end{aligned}
$$

The profit is represented by the quadratic function whose graph is a parabola opening downward. The maximum value of the profit P is given by the first coordinate of the vertex. The first coordinate of the vertex $h = -\dfrac{b}{2a} = \dfrac{-96}{-6} = 16$. Thus the production required to yield maximum profit is 16 sets per day and the radios sell for $68.

PROBLEM 8

A wire, 40 inches long, is cut into two pieces. One piece is to be used to form a circle. The other piece is used to form a square. How should the wire be cut if the sum of the areas of the square and the circle is to be a minimum?

Solution: Let x be the length of the wire used to form the circle and $40 - x$ the length used to form the square. Since x will be the circumference of the circle we have $x = 2\pi r$, where r is the radius, and $r = \dfrac{x}{2\pi}$. Substituting this into the area formula for a circle, $A = \pi r^2$, we have

$$
\text{area of circle} = \dfrac{x^2}{4\pi}
$$

The area of the square $= \left(\dfrac{40 - x}{4}\right)^2 = \dfrac{1600 - 80x + x^2}{16}$. Thus, the sum of the areas $= \dfrac{x^2}{4\pi} + \dfrac{1600 - 80x + x^2}{16}$, or equivalently, the sum of the areas $= \left(\dfrac{4 + \pi}{16\pi}\right) x^2 - 5x + 100$. We have a quadratic function whose graph is a parabola opening upward because the coefficient of x^2 is greater than zero. The value of x is given by the first coordinate of the vertex. The vertex is given by the coordinates $h = -\dfrac{b}{2a} = \dfrac{5}{\dfrac{4 + \pi}{8\pi}} =$

$\dfrac{40\pi}{4 + \pi}$ and $k = \dfrac{4ac - b^2}{4a} = \dfrac{\dfrac{100}{\pi}}{\dfrac{4 + \pi}{4\pi}} = \dfrac{400}{4 + \pi}$. Thus, when $x = \dfrac{40\pi}{4 + \pi}$ and

$40 - x = 40 - \dfrac{40\pi}{4 + \pi}$, we have the sum of the areas, $\dfrac{40\pi}{4 + \pi}$, which is a minimum.

---

1. Put the following quadratic expressions into the form $p(x) = a(x - h)^2 + k$ using the technique of completing the square. Graph.
   (a) $p(x) = x^2 - 6x - 3$
   (b) $p(x) = -2x^2 - 4x + 1$
   (c) $p(x) = 3x^2 + 5x - 4$
   (d) $p(x) = 8x^2 + 2x \quad 3$

2. Find the roots of the following quadratic equations using the Quadratic Formula:
   (a) $x^2 - 3x - 4 = 0$
   (b) $-x^2 + 10x + 8 = 0$
   (c) $-3x^2 - 9x = -4$
   (d) $\dfrac{1}{3}x^2 - \dfrac{1}{2}x - \dfrac{3}{2} = 0$

3. In each of the following, find the discriminant, $b^2 - 4ac$, and determine the nature of the roots. That is, determine whether the function has two real zeros, one real zero, or no real zeros.
   (a) $f(x) = x^2 + 5x - 6$
   (b) $f(x) = 5x - 2x^2 - 2$
   (c) $f(x) = 1 + 2x + 2x^2$
   (d) $f(x) = -4x^2 + 4\sqrt{3}x - 3$

4. Determine the vertex for each of the following:
   (a) $g(x) = -5x^2 - 14x - 8$
   (b) $g(x) = 4x^2 - 9$
   (c) $g(x) = 2x^2 - 6x + 3$
   (d) $g(x) = \sqrt{7}x^2 - 4x + \sqrt{5}$

5. Determine whether the graph of each of the following quadratic expression opens upward or downward. Find the maximum or minimum point of each graph.
   (a) $q(x) = 1 + 3x - 4x^2$
   (b) $q(x) = 3x^2 - x - 2$
   (c) $q(x) = x^2$
   (d) $q(x) = 6x + 1 - 3x^2$

6. Explain how the graph of the following quadratic equations can be obtained from the graph of $f(x) = x^2$:
   (a) $h(x) = (x - 1)^2 + 1$
   (b) $h(x) = -x^2 + 3$
   (c) $h(x) = 2(x + 2)^2 - 3$
   (d) $h(x) = -(x - 1)^2 - 1$
   (e) $h(x) = -x^2$
   (f) $h(x) = 3(x - \sqrt{2})^2 + 1$

7. A stone is thrown upward with an initial velocity of 48 feet per second. (a) How many seconds will it take the stone to reach its maximum height? (b) What is the maximum height of the stone? (c) When does the stone hit the ground?

8. Find two positive numbers, the sum of which is 20, and whose product is maximized.

9. A boat manufacturer can sell x boats per week at $\$(2460 - 5x)$ per boat. It costs $\$(2.4 + 1000x)$ to produce x boats per week. How many boats per week should be sold to maximize profits?

10. A gardener has 100 feet of wire fencing to use around a garden. One side of the garden will be formed by the side of the house, so no wire will be needed for that side of the garden. What are the dimensions of the garden if it is to be rectangular in shape and has maximum area?

3.6 GRAPHING POLYNOMIALS

So far we have studied polynomials of degrees one and two. In this section we shall study polynomials of degree greater than two. The rational zeros and graphs of these polynomial functions shall be discussed. There is no general method of finding the zeros; however, we shall present several theorems that are useful in finding some zeros. These methods will aid us in graphing polynomial functions of degree greater than two.

Rational Zeros of Polynomials

INTEGER COEFFICIENTS

If all the coefficients of a polynomial function p, defined by $p(x)$, are integers, then the following theorem will help us find all rational zeros of p, if they exist.

Theorem: Rational Roots
Let p be a polynomial function defined by

$$p(x) = a_n x^n + a_{n-1} x^{n-1} + \cdots + a_1 x + a_0,$$

where the coefficients $a_0, a_1, \ldots, a_n$ are all integers with $a_n \neq 0$. Let $\dfrac{b}{c}$ be a rational number in its lowest terms. If $\dfrac{b}{c}$ is a root of p, then b is a factor of a_0 and c is a factor of a_n.

Proof. Assuming $\dfrac{b}{c}$ is a root we have

$$a_n \left(\frac{b}{c}\right)^n + a_{n-1} \left(\frac{b}{c}\right)^{n-1} + \cdots + a_1 \left(\frac{b}{c}\right) + a_0 = 0 \tag{I}$$

Multiplying both sides by c^n we get

$$a_nb^n + a_{n-1}b^{n-1}c + \cdots + a_1bc^{n-1} + a_0c^n = 0 \qquad \text{(II)}$$

which can be written

$$a_nb^n = c(-a_{n-1}b^{n-1} - \cdots - a_0c^{n-1})$$

Therefore c is a factor of a_nb^n, since the expression inside the parentheses is an integer. Also, c is not a factor of b because c and b are in their lowest form, which means they have only 1 and -1 as a common factor. Thus, c is not a factor of b^n and hence must be a factor of a_n. If we now write equation (II) in the form

$$a_0c^n = b(-a_nb^{n-1} - \cdots - a_1c^{n-1})$$

we can similarly show that b is a factor of a_0c^n, and thus a factor of a_0.

If the leading coefficient $a_n = 1$, then because c is a factor of a_n, c must be equal to 1 or -1. Therefore the rational root $\dfrac{b}{c}$ is an integer.

Corollary: If p is a polynomial function, with a leading coefficient of 1, and p is defined by

$$p(x) = x^n + a_{n-1}x^{n-1} + \cdots + a_1x + a_0,$$

then any rational root is an integer and a factor of a_0.

We must emphasize that the Rational Root Theorem does not say that a polynomial function defined by a polynomial with integer coefficients has rational roots. It merely says that if the polynomial function does have rational roots, then they are included in the list of possibilities given by the theorem.

PROBLEM 1

Let p be the polynomial function defined by $p(x) = 3x^3 - 7x^2 - 3x + 2$. Find the possible rational roots.

Solution: The Rational Root Theorem states that if $\dfrac{b}{c}$ is a rational root of $p(x)$, then b must be a factor of 2 and c must be a factor of 3. Therefore, the possibilities for b and c are

$$b: 1, -1, 2, -2$$

$$c: 1, -1, 3, -3$$

Thus, the possible rational roots are

$$\frac{b}{c}: 1, -1, 2, -2, \frac{1}{3}, -\frac{1}{3}, \frac{2}{3}, -\frac{2}{3}$$

To find which, if any, are roots we could evaluate $p(x)$ at each possibility and see which give $p(x) = 0$. However, synthetic division is generally a better method.

Try 1:

$$
\begin{array}{r|rrrr}
1 & 3 & -7 & -3 & 2 \\
  &   & 3 & -4 & -7 \\
\hline
  & 3 & -4 & -7 & \boxed{-5} \rightarrow p(1)
\end{array}
$$

Try $-\frac{2}{3}$:

$$
\begin{array}{r|rrrr}
-\dfrac{2}{3} & 3 & -7 & -3 & 2 \\
              &   & -2 & 6 & -2 \\
\hline
              & 3 & -9 & 3 & \boxed{0} \rightarrow p\left(-\dfrac{2}{3}\right)
\end{array}
$$

Thus $p(1) = -5$ and 1 is not a root while $p\left(-\frac{2}{3}\right) = 0$ and $-\frac{2}{3}$ is a root. The Factor Theorem says we can write $p(x)$ as

$$p(x) = \left(x + \frac{2}{3}\right)(3x^2 - 9x + 3)$$

We could use $3x^2 - 9x + 3$ to check the other possibilities. However, since this is a quadratic polynomial, we can use the Quadratic Formula to find the other roots: $3x^2 - 9x + 3 = 0$ or, equivalently, $x^2 - 3x + 1 = 0$ if and only if $x = \dfrac{3 \pm \sqrt{9 - 4}}{2} = \dfrac{3 \pm \sqrt{5}}{2}$.

PROBLEM 2

Let $p(x) = x^4 + 4x^3 + 3x^2 - 4x - 4$. Find the rational roots of $p(x)$, if they exist.

Solution: The leading coefficient $a_n = 1$. Thus, by the corollary, the only possible rational roots are integers. This is the result of b being a factor of -4 and c being a factor of 1.

$$b: 1, -1, 2, -2, 4, -4$$
$$c: 1, -1$$

Therefore, the possible rational roots are

$$\frac{b}{c}: 1, -1, 2, -2, 4, -4$$

which are factors of the constant term, -4, in the polynomial $p(x)$. Using synthetic division we have

$$\begin{array}{r|rrrr} 1 & 1 & 4 & 3 & -4 & -4 \\ & & 1 & 5 & 8 & 4 \\ \hline & 1 & 5 & 8 & 4 & 0 \rightarrow p(1) \end{array}$$

Thus 1 is a root and we can express $p(x)$ as

$$p(x) = (x - 1)(x^3 + 5x^2 + 8x + 4)$$

We now use $x^3 + 5x^2 + 8x + 4$ to check for other rational roots. We try -1:

$$\begin{array}{r|rrrr} -1 & 1 & 5 & 8 & 4 \\ & & -1 & -4 & -4 \\ \hline & 1 & 4 & 4 & 0 \rightarrow p(-1) \end{array}$$

Thus $p(-1) = 0$, -1 is another root, and $p(x)$ is of the form

$$p(x) = (x - 1)(x + 1)(x^2 + 4x + 4)$$

Note that the last factor is quadratic and the last two roots can be obtained by using the Quadratic Formula or by factoring:

$$x^2 + 4x + 4 = 0 \text{ if and only if } (x + 2)(x + 2) = 0$$

Therefore, the last two roots are -2 and -2. We see that all the roots of $p(x)$ are rational.

PROBLEM 3

Find the rational roots, if any, of the quartic function defined by

$$p(x) = 2x^4 - x^3 + 2x^2 - 2x - 4$$

Solution: The possible rational roots are

$$\frac{b}{c}: 1, -1, 2, -2, 4, -4, \frac{1}{2}, -\frac{1}{2}$$

If we try these values we would find that there are no rational roots.

The real roots, which are irrational, can be approximated by graphical methods.

RATIONAL COEFFICIENTS

If the coefficients of the polynomial are rational, but not necessarily integers, then the Rational Root Theorem can still be used after a simple modification. Before using the theorem we must multiply by the Least Common Multiple of the denominators appearing in the polynomial.

PROBLEM 4

Let p be the polynomial function defined by $p(x) = \frac{1}{6}x^4 - \frac{1}{3}x^2 - \frac{1}{2}x - \frac{1}{3}$. Find the rational roots.

Solution: Multiplying by 6 gives $6p(x) = x^4 - 2x^2 - 3x - 2$. Note that any number which is a root of $p(x)$ is also a root of $6p(x)$. Thus, the two equations are equivalent. However, $6p(x)$ has all integers as coefficients and we can use the Rational Root Theorem. The possible rational roots are factors of -2. Therefore, the possibilities are

$$1, -1, 2, -2$$

We try -1:

$$
\begin{array}{r|rrrr}
-1 & 1 & 0 & -2 & -3 & -2 \\
   &   & -1 & 1 & 1 & 2 \\
\hline
   & 1 & -1 & -1 & -2 & 0 \rightarrow p(-1)
\end{array}
$$

Thus, -1 is a root and $p(x) = (x + 1)(x^3 - x^2 - x - 2)$.
We try 2:

$$
\begin{array}{r|rrrr}
2 & 1 & -1 & -1 & -2 \\
  &   & 2 & 2 & 2 \\
\hline
  & 1 & 1 & 1 & 0 \rightarrow p(2)
\end{array}
$$

So, 2 is also a root and $p(x) = (x + 1)(x - 2)(x^2 + x + 1)$. Further calculations show that these are the only rational roots.

Locating Real Roots

One of the most widely used methods to find the real zeros of a polynomial function p with real coefficients is by graphing the function and determining the values of x at the points at which the graph intersects the x-axis. Algebraically, this is the point where the function takes the value zero—that is, $p(x) = 0$. The following theorem will be of great use not only in finding roots, but also in graphing polynomials. The theorem is based on the continuity or smoothness of the graph.

Theorem: Let p be a polynomial function defined by $p(x)$. If there are values $x = a$ and $x = b$ such that

(i) $p(a)$ and $p(b)$ have **opposite signs,** then there is either **one** real root or an **odd** number of real roots between $x = a$ and $x = b$.

(ii) $p(a)$ and $p(b)$ have the **same sign,** then there is either **no** real root or an **even** number of real roots between $x = a$ and $x = b$.

The reason why this theorem is true can be seen by the graphs in Figure 95.

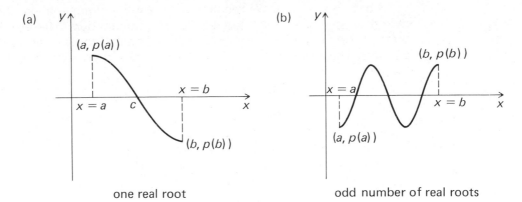

(a) one real root

(b) odd number of real roots

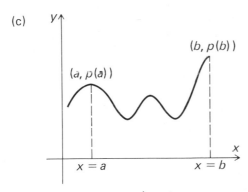

(c) no real roots

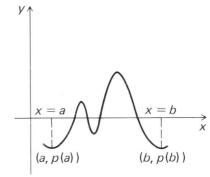

(d) even number of real roots

Figure 95

For example, Figure 95(a) shows that since the graph of the polynomial function connects points $(a, p(a))$, above the x-axis, and $(b, p(b))$, below the x-axis, the graph must cross the x-axis at a point between $x = a$ and $x = b$; say at $x - c$. Thus, at $x - c$ we have $p(c) = 0$. Similar arguments can be given for the other cases.

EXAMPLE 1

Consider the equation $p(x) = x^3 - 3x^2 - 13x + 15$. We evaluate this equation at $x = -4, 0,$ and 6:

$$p(-4) = (-4)^3 - 3(-4)^2 - 13(-4) + 15 = -45$$
$$p(0) = (0)^3 - 3(0)^2 - 13(0) + 15 = 15$$
$$p(6) = (6)^3 - 3(6)^2 - 13(6) + 15 = 45$$

Since $p(-4)$ and $p(0)$ have opposite signs, there is either a real root or an odd number of roots between $x = -4$ and $x = 0$. Since $p(-4)$ and $p(6)$ have the same sign, there is either no real root or an even number of roots between $x = 0$ and $x = 6$. Using previous methods it can be shown that the function has zeros at $x = -3, 1,$ and 5.

EXAMPLE 2

Let $p(x) = x^2 + 1$. We recall that the square of any real number is positive; thus, $p(x) = x^2 + 1$ is positive for all real numbers. Therefore, $p(x)$ has the same sign for all x, and $p(x)$ has either no real roots or an even number of roots. The graph of $p(x) = x^2 + 1$ shows that there are no real roots (see Fig. 96).

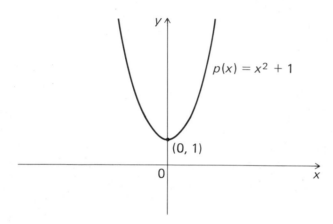

Figure 96

Graphing Polynomial Functions

To graph a polynomial function we shall make use of the techniques and information presented in previous sections. However, a great deal of graphing polynomials depends on a table of values. Our first step will be to find as many roots of the polynomial as possible.

PROBLEM 5

Graph $p(x) = x^3 - x^2$.

Solution: Factoring $p(x)$ we get $p(x) = x^2(x - 1)$. Thus $p(x)$ has roots at $x = 0$ and $x = 1$. We now test for the sign of $p(x)$ on the intervals determined by the roots 0 and 1.

Interval	Sign of $p(x)$	Graph of $p(x)$
$x < 0$	$-$	graph below x-axis
$x = 0$	$p(0) = 0$	graph tangent to x-axis
$0 < x < 1$	$-$	graph below x-axis
$x = 1$	$p(1) = 0$	graph crosses x-axis
$x > 1$	$+$	graph above x-axis

We now evaluate $p(x)$ at values in each of the three intervals.

x	x^2	$x - 1$	$p(x)$
-1	1	-2	-2
0	0	-1	0
$\dfrac{2}{3}$	$\dfrac{4}{9}$	$-\dfrac{1}{3}$	$-\dfrac{4}{27}$
1	1	0	0
1.5	2.25	.5	1.125

The graph of $p(x) = x^3 - x^2$ is given in Figure 97.

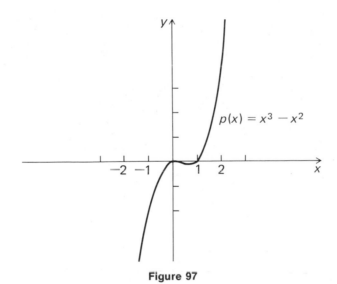

Figure 97

PROBLEM 6

Graph the function defined by $f(x) = x^4 + 2x^3 - 3x^2 - 4x + 4$.

Solution: The Rational Root Theorem states that the only possible rational roots are factors of 4. Using synthetic division we obtain the roots -2 and 1. Determining the signs of $f(x)$ we obtain:

Interval	Sign of $f(x)$	Graph of $f(x)$
$x < -2$	$+$	graph above x-axis
$x = -2$	$f(x) = 0$	graph tangent to x-axis
$-2 < x < 1$	$+$	graph above x-axis
$x = 1$	$f(x) = 0$	graph tangent to x-axis
$x > 1$	$+$	graph above x-axis

Before evaluating $f(x)$ we use synthetic division to express $f(x)$ in a factored form:

$$
\begin{array}{r}
-2 \,\rvert\ 1 \quad 2 \ -3 \ -4 \quad 4 \\
\underline{\quad\ -2 \quad 0 \quad 6 \ -4} \\
1 \,\rvert\ 1 \quad 0 \ -3 \quad 2 \quad \rvert\,0 \ \to f(x) = (x + 2)(x^3 - 3x + 2) \\
\underline{\quad\quad 1 \quad 1 \ -2} \\
1 \quad 1 \ -2 \quad \rvert\,0
\end{array}
$$

Thus $f(x) = (x + 2)(x - 1)(x^2 + x - 2)$. We now evaluate $f(x)$ at several values.

x	$x + 2$	$x - 1$	$x^2 + x - 2$	$f(x)$
-3	-1	-4	4	16
-2				0
$-\dfrac{1}{2}$	$\dfrac{3}{2}$	$-\dfrac{3}{2}$	$-\dfrac{9}{4}$	$\dfrac{81}{16}$
0	2	-1	-2	4
1				0
2	4	1	4	16

The graph of $f(x) = x^4 + 2x^3 - 3x^2 - 4x + 4$ is given in Figure 98.

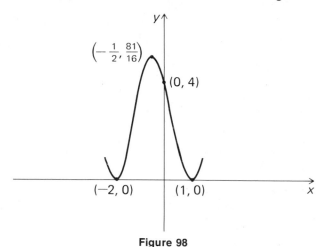

Figure 98

PROBLEM 7

Graph $p(x) = x^3 - 3x^2 + 3$.

Solution: The Rational Root Theorem shows us that there are no rational roots. Thus, we must rely on a table of values. We use synthetic division to aid us in evaluating $p(x)$ at several values of x. Recall that the Remainder Theorem says that if we divide a polynomial p synthetically by $x - a$, then the remainder is equal to $p(a)$.

$$
\begin{array}{r|rrrr}
-2 & 1 & -3 & 0 & 3 \\
& & -2 & 10 & -20 \\
\hline
& 1 & -5 & 10 & \boxed{-17} \rightarrow p(-2)
\end{array}
$$

$$
\begin{array}{r|rrrr}
-1 & 1 & -3 & 0 & 3 \\
& & -1 & 4 & -4 \\
\hline
& 1 & -4 & 4 & \boxed{-1} \rightarrow p(-1)
\end{array}
$$

$$
\begin{array}{r|rrrr}
-\dfrac{1}{2} & 1 & -3 & 0 & 3 \\[2ex]
& & -\dfrac{1}{2} & \dfrac{7}{4} & -\dfrac{7}{8} \\[2ex]
\hline
& 1 & -\dfrac{7}{2} & \dfrac{7}{4} & \boxed{\dfrac{17}{8}} \rightarrow p\left(-\dfrac{1}{2}\right)
\end{array}
$$

$$
\begin{array}{r|rrrr}
0 & 1 & -3 & 0 & 3 \\
& & 0 & 0 & 0 \\
\hline
& 1 & -3 & 0 & \boxed{3} \rightarrow p(0)
\end{array}
$$

Continuing in this way, we can obtain the graph of $p(x) = x^3 - 3x^2 + 3$ (see Fig. 99).

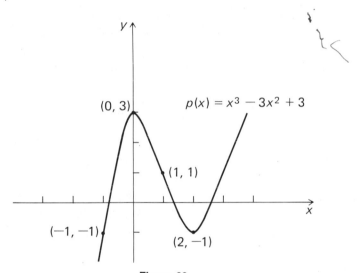

Figure 99

3.6 EXERCISES

1. List all possible rational roots for the following polynomial expressions:
   (a) $f(x) = 3x + 1$
   (b) $f(x) = x^2 - 7x + 9$
   (c) $f(x) = 3x^2 + 8x + 12$
   (d) $f(x) = x^3 - x^2 + x + 24$
   (e) $f(x) = 4x^3 + 1$
   (f) $f(x) = 5x^4 - x^2 - 3x + 5$
   (g) $f(x) = x^5 + x^4 + 3x^3 - x^2 + x - 10$
   (h) $f(x) = 3x^5 + x^3 + 2x^2 + 5x$
   (i) $f(x) = \dfrac{1}{2}x^3 - 6x^2 + 4x - \dfrac{1}{3}$
   (j) $f(x) = \dfrac{3}{4}x^4 - \dfrac{7}{3}$

2. Find all rational roots of the following polynomial expressions:
   (a) $p(x) = x + 4$
   (b) $p(x) = 3x - 5$
   (c) $p(x) = x^2 - x - 2$
   (d) $p(x) = x^2 + 4$
   (e) $p(x) = x^3 - 2x^2 + 9x - 18$
   (f) $p(x) = 4x^4 + 27x^3 + 46x^2 + 61x + 30$
   (g) $p(x) = x^3 + x^2 + x + 13$
   (h) $p(x) = 2x^3 - 5x^2 - 8x + 6$
   (i) $p(x) = 12x^3 - 16x^2 - 5x + 3$
   (j) $p(x) = x^4 - 2x^2 - 3x - 2$

3. Find all the roots of the following polynomial expressions by finding the rational roots first:
   (a) $h(x) = x^3 - 6x^2 + 11x - 6$
   (b) $h(x) = x^3 - 5x^2 - 2x + 24$
   (c) $h(x) = 2x^3 - 9x^2 + 8x - 2$
   (d) $h(x) = x^3 + 7x^2 + 12x$
   (e) $h(x) = x^4 - 4x^3 - 3x^2 + 16x - 4$

5. Graph:
   (a) $f(x) = x^3 - 2x^2 - 5x + 6$
   (b) $f(x) = x^3 - 3x - 2$
   (c) $f(x) = (x - 2)^3$
   (d) $f(x) = (x + 1)(x - 2)(x - 3)$
   (e) $f(x) = -x^3 - 2x^2 + 5x + 6$

7. As in Exercise 6, show that each of the
   following is not rational:
   (a) $\sqrt{7}$ (b) $\sqrt[3]{4}$ (c) $\sqrt[4]{5}$ (d) $\sqrt[5]{6}$

4. Show that each of the following polynomials has at least one real root between
   the given values:
   (a) $p(x) = x^3 - 3x + 1$; $x = 1$ and $x = 3$
   (b) $p(x) = x^3 - 2x^2 - 5x + 6$; $x = -3$
   and $x = 4$
   (c) $p(x) = 2x^3 + x^2 - 8x - 4$; $x = -1$ and
   $x = 0$

6. Consider the function p defined by
   $p(x) = x^2 - 2$. Show that $p(x)$ has no
   rational roots. What does this say about
   $\sqrt{2}$?

8. Consider the function f defined by $f(x) =$
   $x^n + a$, where n is a positive even integer
   and a is a positive real number. Show
   that f has no rational roots.

3.7   COMPLEX ROOTS

In the previous section we saw that certain polynomials had no real roots. In
particular, if p is a quadratic defined by $p(x) = ax^2 + bx + c, a \neq 0$, and if the discriminant $b^2 - 4ac < 0$, then the polynomial p has no real roots. In this section
we introduce a new set of numbers such that all polynomials will have roots.

EXAMPLE 1

Consider the polynomial p defined by $p(x) = x^2 + 1$. The discriminant
$b^2 - 4ac = 0 - 4 = -4 < 0$ indicates that this polynomial has no real
roots. This can be seen by using the Quadratic Formula:

$$x = \frac{-b \pm \sqrt{b^2 - 4ac}}{2a}$$

$$= \frac{0 \pm \sqrt{-4}}{2}$$

$$= \frac{\pm \sqrt{-4}}{2}$$

Algebraically we cannot solve this equation using the real numbers
since there is no real number which is the $\sqrt{-4}$.

We now introduce the set of **complex numbers** to handle the problem of
finding roots of such equations. Such a set of numbers will enable us to include
numbers whose square could be negative, and thus we can take the square root
of a negative number, such as $\sqrt{-4}$. Complex numbers, as we shall see, will
contain the real numbers and many of the properties of real numbers.

> **Definition:** A number of the form $a + bi$, with a and b real constants and $i = \sqrt{-1}$, is called a **complex number.** The number a is called the **real part** and b is called the **imaginary part.**

EXAMPLE 2

The number $z = 4 + 3i$ is a complex number with real part $a = 4$ and imaginary part $b = 3$.

EXAMPLE 3

The number $z = -2i$ is a complex number with real part $a = 0$ and imaginary part $b = -2$.

EXAMPLE 4

The number $z = 5$ is a complex number with real part $a = 5$ and imaginary part $b = 0$.

Example 4 shows that the real numbers are contained in the set of complex numbers. By assigning the imaginary part $b = 0$, we get the set of real numbers.

EXAMPLE 5

The number $\sqrt{-4}$ is a complex number. To see this we use the definition $i = \sqrt{-1}$. $\sqrt{-4} = \sqrt{(4)(-1)} = \sqrt{4} \cdot \sqrt{-1} = 2i$. Thus we can find the roots of the polynomial $p(x) = x^2 + 1$ given in Example 1. Using the Quadratic Formula we get

$$x = \frac{\pm\sqrt{-4}}{2} = \pm\frac{2i}{2} = \pm i$$

To check:
(a) If $x = i$, then $x^2 + 1 = (i)^2 + 1 = (-1) + 1 = 0$
(b) If $x = -i$, then $x^2 + 1 = (-i)^2 + 1 = i^2 + 1$
$$= -1 + 1 = 0.$$

PROBLEM 1

Find the roots of the polynomial p defined by the expression $p(x) = x^2 + 6x + 13$.

Solution: The Quadratic Formula yields

$$x = \frac{-6 \pm \sqrt{36 - 52}}{2} = \frac{-6 \pm \sqrt{-16}}{2}$$

$$= \frac{-6 \pm 4i}{2} = -3 \pm 2i$$

Thus, the roots are $-3 + 2i$ and $-3 - 2i$.

In Problem 1, the roots $-3 + 2i$ and $-3 - 2i$, and in Example 5, the roots i and $-i$, differ only in the sign preceding the imaginary parts. Such pairs of complex numbers are said to be **conjugates** of one another.

Definition: If $z = a + bi$ is a complex number, then the **conjugate** of z, denoted by $\bar{z}$, is $\bar{z} = a - bi$.

EXAMPLE 6

If $z = -1 + 5i$, then the conjugate $\bar{z} = -1 - 5i$.

EXAMPLE 7

If $z = 4 - 3i$, then the conjugate $\bar{z} = 4 + 3i$.

EXAMPLE 8

If $z = 7 = 7 + 0i$, then the conjugate $\bar{z} = 7 - 0i = 7$.

Real numbers are their own conjugates.

PROBLEM 2

Find the roots of the polynomial function p defined by $p(x) = 3x^3 + 2x^2 + 2x - 1$.

Solution: We first try to find all possible rational roots. The Rational Root Theorem gives the following possible rational roots: 1, -1, $\frac{1}{3}$ and $-\frac{1}{3}$. Synthetic division shows that $\frac{1}{3}$ is a root.

$$
\begin{array}{r|rrrr}
\frac{1}{3} & 3 & 2 & 2 & -1 \\
 & & 1 & 1 & 1 \\
\hline
 & 3 & 3 & 3 & 0
\end{array}
$$

Thus, $p(x) = \left(x - \frac{1}{3}\right)(3x^2 + 3x + 3) = (3x - 1)(x^2 + x + 1)$. To solve the polynomial $3x^2 + 3x + 3$ we use the Quadratic Formula:

$$x = \frac{-3 \pm \sqrt{9 - 36}}{6} = \frac{-3 \pm \sqrt{-27}}{6}$$

$$\frac{-3 \pm 3i\sqrt{3}}{6} = -\frac{1}{2} \pm \frac{\sqrt{3}}{2}i$$

Therefore the roots are

$$\frac{1}{3}, \quad -\frac{1}{2} + \frac{\sqrt{3}}{2}\, i, \text{ and } -\frac{1}{2} - \frac{\sqrt{3}}{2}\, i.$$

A closer examination of the polynomials solved in this section shows that complex numbers appear as conjugate roots. This results in the following theorem.

Theorem 1: If a complex number $z = a + bi$, $b \neq 0$, is a root of a polynomial function with real coefficients, then its conjugate $\bar{z} = a - bi$ is also a root.

PROBLEM 3

Given that $1 - 3i$ is a root of $p(x) = x^3 - 4x^2 + 14x - 20$, what can be said about the nature of the other roots of p?

Solution: Theorem 1 says that if $1 - 3i$ is a root of $p(x)$, then $1 + 3i$ is also a root. Also, note that since complex numbers occur only in conjugate pairs, the third root of $p(x)$ must be real.

We are now interested in finding the roots of polynomial functions in general. First, we must consider polynomials with complex coefficients and complex variables. Such polynomials are called **complex polynomials.**

Definition: A **complex polynomial function** is a function f defined by

$$f(z) = a_n z^n + a_{n-1} z^{n-1} + \cdots + a_1 z + a_0,$$

where $a_n, a_{n-1}, \ldots, a_1, a_0$ are complex numbers.

EXAMPLE 9

The polynomial defined by $f(z) = 4z^3 + (1 - 2i)z^2 + z + i$ is a complex polynomial.

EXAMPLE 10

The polynomial defined by $p(x) = 3x^2 + 2x - 5$ is a complex polynomial. This follows from the fact that the real numbers are contained in the complex numbers. Thus, real polynomial functions are also complex polynomial functions.

The following theorems answer the question of roots of any polynomial. The theorems tell us exactly how many roots exist for a polynomial function of a given degree. The first theorem, the Fundamental Theorem of Algebra, is stated without proof since the proof requires techniques beyond this text.

The Fundamental Theorem of Algebra: Every polynomial of degree $n \geq 1$, with real or complex coefficients, has at least one real or complex root.

While the Fundamental Theorem assures us that a root exists, it does not tell us how to find the roots. The following theorems give some insight on how to find roots.

Factorization Theorem:

Every polynomial of degree $n \geq 1$, with real or complex coefficients, can be expressed as the product of n linear factors of the type $x - r$, where r is a real or complex number. That is, if p is a polynomial function defined by $p(x) = a_n x^n + a_{n-1} x^{n-1} + \cdots + a_1 x + a_0$, then

$$p(x) = a_n(x - r_1)(x - r_2) \cdots (x - r_n)$$

Proof. Let $p(x) = a_n x^n + a_{n-1} x^{n-1} + \cdots + a_1 x + a_0$

$$a_n \neq 0$$

The Fundamental Theorem of Algebra says that $p(x)$ has at least one root, say r_1. Thus $p(r_1) = 0$ and the Factorization Theorem gives

$$p(x) = (x - r_1)q_1(x)$$

where $q_1(x)$ is a polynomial of degree $n - 1$. If $n - 1 = 0$, then $q_1(x) = a_n$ and the proof is complete.

If $n - 1 \geq 1$, we proceed as above. Since $q_1(x)$ is of degree $n - 1 \geq 1$, the Fundamental Theorem of Algebra assures us that $q_1(x)$ has at least one root, say r_2. Thus,

$$p(x) = (x - r_1)(x - r_2)q_2(x)$$

where $q_2(x)$ is of degree $n - 2$. Continuing the process, we get

$$p(x) = (x - r_1)(x - r_2) \cdots (x - r_n)q_n(x)$$

where $q_n(x)$ has degree 0—that is, $q_n(x)$ is a constant. If we multiply out the expression for $p(x)$, then it can be seen that $q_n(x) = a_n$.

EXAMPLE 11

Consider the polynomial function p defined by

$$p(x) = 2x^3 + 3x^2 - 11x - 6$$

The linear factorization of $p(x)$ is

$$p(x) = 2\left(x + \frac{1}{2}\right)(x + 3)(x - 2)$$

and the roots are $-\dfrac{1}{2}, -3, 2$.

EXAMPLE 12

The polynomial function p defined by

$$p(x) = x^4 - 4x^3 + 13x^2 - 36x + 36$$

has the linear factorization

$$\begin{aligned} p(x) &= (x + 3i)(x - 3i)(x - 2)(x - 2) \\ &= (x + 3i)(x - 3i)(x - 2)^2 \end{aligned}$$

The roots are $3i, -3i, 2, 2$.

Note that the polynomial in Example 12 has four factors but only three roots. The root 2 appears twice and is said to be a **root of multiplicity** two.

PROBLEM 4

Write a polynomial function which has 4 as a root of multiplicity three, -3 as a root of multiplicity two, and 1 as a root of multiplicity one.

Solution: $p(x) = (x - 4)^3(x + 3)^2(x - 1)$

The following theorem summarizes the results concerning roots of polynomials.

> **Theorem:** Every polynomial of degree $n \geq 1$ has exactly n roots. Not all of the roots are necessarily distinct.

In an earlier section we saw that a quadratic equation with **real coefficients** which has the complex number $z = a + bi$, $b \neq 0$, as a root also has $z = a - bi$ as a root. In other words, complex roots appear as conjugate pairs. The following theorem generalizes this relationship between complex roots of polynomials with real coefficients.

> **The Complex Conjugates Theorem:** If the complex number $z = a + bi$ is a root of a polynomial $p(x)$ of degree $n \geq 1$ with **real coefficients,** then its conjugate $z = a - bi$ is a root.

The condition that the coefficients be real can be seen in the following example.

EXAMPLE 13

If p is a polynomial function with complex coefficients defined by $p(x) = x + 3i$, then the only root is $x = -3i$.

The following problem illustrates the use of the Conjugate Theorem.

PROBLEM 5

Let $p(x) = x^4 - 4x^3 + 13x^2 - 36x + 36$ and let $r = -3i$ be one root of $p(x)$. Find the other roots.

Solution: By the Conjugate Theorem we know that if $-3i$ is a root, then $3i$ is also a root. Using synthetic division we have $p(x) = (x + 3i)(x - 3i)$ $(x^2 - 4x + 4)$ or $p(x) = (x + 3i)(x - 3i)(x - 2)^2$. Thus the roots are $-3i$, $3i$, and 2 as a root of multiplicity two.

Operations with Complex Numbers

Equality. Two complex numbers are **equal** if and only if their real parts are equal and their imaginary parts are equal. That is, $a + bi = c + di$ if and only if $a = c$ and $b = d$.

EXAMPLE 14

$a + 7i = -4 + di$ if and only if $a = -4$ and $d = 7$.

Addition. To add two complex numbers, add the real parts and the imaginary parts separately. That is,

$$(a + bi) + (c + di) = (a + c) + (b + d)i$$

EXAMPLE 15

Let $z_1 = -2 + 3i$ and $z_2 = 4 - 7i$.
Then $z_1 + z_2 = (-2 + 3i) + (4 - 7i) = (-2 + 4) + (3 - 7)i = 2 - 4i$.

Subtraction. To subtract two complex numbers, subtract the real parts and the imaginary parts separately. That is,

$$(a + bi) - (c + di) = (a - c) + (b - d)i$$

EXAMPLE 16

Let $z_1 = \sqrt{3} - 4i$ and $z_2 = 5 - i$.
Then $z_1 - z_2 = (\sqrt{3} - 4i) - (5 - i) = (\sqrt{3} - 5) + (-4 + 1)i = (\sqrt{3} - 5) - 3i$.

Multiplication. To multiply two complex numbers, multiply as you would two binomial expressions and use the fact that $i^2 = -1$. That is,

$$
\begin{aligned}
(a + bi)(c + di) &= ac + adi + bci + bdi^2 \\
&= ac + (ad + bc)i + bd(-1) \\
&= (ac - bd) + (ad + bc)i
\end{aligned}
$$

Thus,

$$(a + bi)(c + di) = (ac - bd) + (ad + bc)i$$

EXAMPLE 17

Let $z_1 = 1 + 2i$ and $z_2 = 5 - 3i$. Then $z_1 z_2 = (1 + 2i)(5 - 3i) = 5 - 3i + 10i - 6i^2 = 5 + 7i - 6(-1) = 11 + 7i$.

PROBLEM 6

Evaluate: (a) i^3; (b) i^4; (c) i^5.

Solution: (a) $i^3 = i^2 \cdot i = (-1)(i) = -i$
           (b) $i^4 = i^2 \cdot i^2 = (-1)(-1) = 1$
           (c) $i^5 = i^2 \cdot i^2 \cdot i = (-1)(-1)i = i$

Division: To divide two complex numbers, multiply the numerator and denominator of the fraction by the conjugate of the denominator. That is,

$$\frac{a + bi}{c + di} = \frac{(a + bi)(c - di)}{(c + di)(c - di)} = \frac{(a + bi)(c - di)}{c^2 + d^2}$$

Thus,

$$\frac{a + bi}{c + di} = \frac{(a + bi)(c - di)}{c^2 + d^2}$$

EXAMPLE 18

Let $z_1 = 2 - 3i$ and $z_2 = 1 + i$. Then

$$\frac{z_1}{z_2} = \frac{(2 - 3i)}{(1 + i)} = \frac{(2 - 3i)(1 - i)}{(1 + i)(1 - i)} = \frac{(2 - 3) + (-2 - 3)i}{1^2 - (-1)} = \frac{-1 - 5i}{2}$$

> **Definition:** Let $z = a + bi$. Then $|z|$ is called the **magnitude** of z and $|z| = |a + bi| = \sqrt{a^2 + b^2}$.

EXAMPLE 19

If $z = 4 - 5i$, then $|z| = |4 - 5i| = \sqrt{4^2 + (-5)^2} = \sqrt{41}$.

EXAMPLE 20

Let $z = a + bi$. Then $\bar{z} = a - bi$ and $z \cdot \bar{z} = (a + bi)(a - bi) = a^2 - b^2 i^2 = a^2 + b^2 = |z|^2$. Thus,

$$z \cdot \bar{z} = |z|^2$$

3.7 EXERCISES

1. Express each of the following in terms of i:
   (a) $\sqrt{-9}$ (b) $\sqrt{-25}$
   (c) $\sqrt{-36}$ (d) $\sqrt{-81}$
   (e) $\sqrt{-2}$ (f) $\sqrt{-5}$
   (g) $\sqrt{-12}$ (h) $\sqrt{-28}$
   (i) $\sqrt{-\dfrac{1}{9}}$ (j) $\sqrt{-\dfrac{16}{49}}$
   (k) $\sqrt{-\dfrac{1}{20}}$ (l) $2 \cdot \sqrt{-\dfrac{1}{8}}$

2. Determine the real and imaginary parts of each of the following complex numbers:
   (a) $3 + 4i$ (b) $1 + i$
   (c) $-5i$ (d) $-4 - i$
   (e) 13 (f) $\sqrt{2} + 5i$
   (g) $14 - \sqrt{2}i$ (h) i

3. Find the conjugate for each of the following complex numbers:
   (a) $2i$ (b) $1 - 4i$
   (c) $\sqrt{2} + 6i$ (d) $-3 - 8i$
   (e) 14 (f) $-\sqrt{3}i$
   (g) $\dfrac{1}{3} - \dfrac{1}{5}i$ (h) $\sqrt{-16}$

4. Solve the following quadratic equations using the Quadratic Formula:
   (a) $3x^2 + 27 = 0$
   (b) $-5x^2 + 4x + 12 = 0$
   (c) $x^2 + 2x + 2 = 0$
   (d) $4x^2 - 16x + 25 = 0$
   (e) $x^2 + 2mx + m^2 + n^2 = 0$, where m and n are constants

5. Find the roots of each polynomial and state the multiplicity of each root:
   (a) $p(x) = (x - 1)(x - 7)$
   (b) $p(x) = (x - 2)^2(x + 1)$
   (c) $p(x) = 7(x - 3)(x - 5)^3$
   (d) $p(x) = -5x(x + 9)^2(x - 8)(x + 1)^3$
   (e) $p(x) = 4(x - 1)(x^2 + 2x + 2)$
   (f) $p(x) = (x - \sqrt{2})^2(x^2 + 9)$

6. Given that the polynomial has the given root, find the other roots:
   (a) $p(x) = x^3 - 2x^2 + 26x; r = 0$
   (b) $p(x) = x^4 - 5x^3 + 6x^2 + 4x - 8; r = 2$
   (c) $p(x) = x^4 - 2x^3 + 2x^2 - 10x + 25; r = 2 + i$
   (d) $p(x) = x^3 - 27; r = 3$

7. Perform the operations indicated and simplify:
   (a) $(4 - 2i) + (-5 + 6i)$
   (b) $(1 - i) - (1 + i)$
   (c) $\left(\dfrac{1}{3} + \dfrac{1}{5}i\right) - \left(\dfrac{2}{3} - \dfrac{4}{5}i\right)$
   (d) $(\sqrt{8} + \sqrt{-1}) + (3\sqrt{2} - 5i)$
   (e) $(a + bi) + (a - bi)$
   (f) $(a + bi) - (a - bi)$

8. Multiply and simplify:
   (a) $(5 - 3i)(i + 2)$
   (b) $(4 + 7i)^2$
   (c) $(1 + i)(4 - i)(3 + 2i)$
   (d) $(\sqrt{2} + 5i)(\sqrt{2} - 5i)$
   (e) $(1 + i)^3$
   (f) $(1 + 2i)^4$

9. Divide and simplify:
   (a) $\dfrac{2}{i}$
   (b) $\dfrac{1 + 2i}{3 - i}$
   (c) $\dfrac{2 + \sqrt{3}i}{5 - \sqrt{3}i}$
   (d) $\dfrac{4\sqrt{3} - 2\sqrt{5}i}{\sqrt{3} + \sqrt{5}i}$

10. Find the magnitude of each of the following complex numbers:
    (a) $z = 4 - i$
    (b) $z = -1 + i$
    (c) $z = \sqrt{3} + 5i$
    (d) $z = 7$
    (e) $z = \sqrt{5} - 3\sqrt{2}i$
    (f) $z = \dfrac{1}{2} + \dfrac{\sqrt{2}}{3}i$

3.8 RATIONAL FUNCTIONS AND PARTIAL FRACTIONS

In previous sections we discussed how addition, subtraction, and multiplication of polynomial functions generated other polynomial functions. In this section we will see that the division of polynomial functions generates a new class of functions called **rational functions.** We will discuss the graphs of such functions, and will end the section by showing how rational functions can be expressed by the sum of two or more simpler rational functions called **partial fractions.**

Rational Functions

Definition: Let g and h be polynomial functions defined by $g(x)$ and $h(x)$, respectively. Then, the function f defined by

$$f(x) = \frac{g(x)}{h(x)}, \; h(x) \neq 0$$

is called a **rational function.**

The rational functions differ from polynomial functions in that polynomial functions are defined for all real numbers, whereas rational functions may be undefined for particular real numbers.

The **domain of a rational function** $f(x) = \dfrac{g(x)}{h(x)}$ is the set of real numbers x such that $h(x) \neq 0$.

EXAMPLE 1

The function f defined by $f(x) = \dfrac{1}{x^2}$ is a rational function. The domain of f is all real numbers $x \neq 0$.

EXAMPLE 2

Let $f(x) = \dfrac{x^2 - x}{x}$. The domain of f is the set of all numbers for which $x \neq 0$. If $x \neq 0$, then we note that $f(x) = \dfrac{x^2 - x}{x} = \dfrac{x(x - 1)}{x} = x - 1$. That is, the graph of the rational function f defined by $f(x) = \dfrac{x^2 - x}{x}$ is the same as the graph of the polynomial function defined by $y = x - 1$ except the point $(0, -1)$.

EXAMPLE 3

The rational function f defined by

$$f(x) = \frac{-2x^2}{x^2 + 1}$$

has all the real numbers as its domain. This is true, since no real numbers make $x^2 + 1 = 0$.

As shown in Example 2, we will put all rational functions in their *lowest form;* that is, in the definition $f(x) = \dfrac{g(x)}{h(x)}$, $g(x)$ and $h(x)$ have *no common factors.* To understand rational functions better, we shall investigate the graphs of rational functions. Of particular interest will be the behavior of the rational function f "close to" the values of x which make $h(x) = 0$ in the expression $f(x) = \dfrac{g(x)}{h(x)}$. Also, we will investigate the "asymptotic behavior" of rational functions f for very large values of $|x|$. We begin by graphing the rational functions in Examples 1 to 3.

PROBLEM 1

Graph $f(x) = \dfrac{1}{x^2}$.

Solution: The function is defined for all values of x, except for $x = 0$. Also, we note that since $x^2 > 0$ for all $x \neq 0$, the function f is positive for all $x \neq 0$; that is, $f(x) > 0$ for $x \neq 0$. As x gets "closer to" $x = 0$, the value $f(x)$ gets larger and the graph of f comes closer to the y-axis. The y-axis is called a **vertical asymptote** of $f(x) = \dfrac{1}{x^2}$. As the values of x get large in a positive or negative direction—that is, as $|x|$ gets larger—the graph of f comes closer to the x-axis. The x-axis is called a **horizontal**

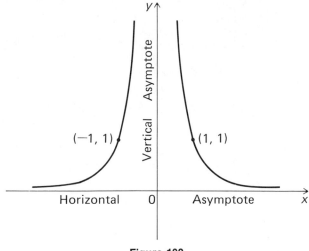

Figure 100

asymptote of $f(x) = \dfrac{1}{x^2}$ (see Fig. 100). Generally, if f is the rational function defined by

$$f(x) = \frac{1}{(x - a)^2}$$

then the graph of f has the line $x = a$ as a **vertical asymptote** and the x-axis as a **horizontal asymptote** (see Fig. 101).

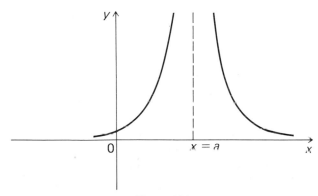

Figure 101

PROBLEM 2

Graph $f(x) = \dfrac{x^2 - x}{x}$.

Solution: As noted in Example 2, $f(x) = \dfrac{x^2 - x}{x} = \dfrac{x(x - 1)}{x} = x - 1$ if $x \neq 0$. Thus, the graph of f is the straight line $y = x - 1$, except the point where $x = 0$—that is, the point $(0, -1)$ (see Fig. 102).

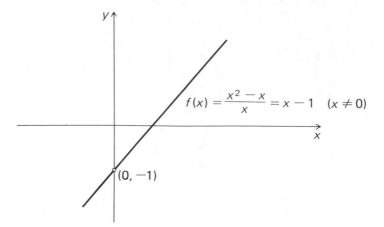

$$f(x) = \frac{x^2 - x}{x} = x - 1 \quad (x \neq 0)$$

$(0, -1)$

Figure 102

PROBLEM 3

Graph $f(x) = \dfrac{-2x^2}{x^2 + 1}$.

Solution: Dividing numerator and denominator by x^2 gives $f(x) =$ $\dfrac{-2x^2}{x^2 + 1} = \dfrac{-2}{1 + \dfrac{1}{x^2}}$. For large values of $|x|$ the term $\dfrac{1}{x^2}$ approaches zero. Thus, the graph of f approaches the line $y = -2$. The line $y = -2$ is a **horizontal asymptote** for the function f. As x gets smaller, the value of $f(x)$ gets smaller. For example, when $x = 0$, $f(x) = 0$. The graph of $f(x) = \dfrac{-2x^2}{x^2 + 1}$ is given in Figure 103. *Note that the denominator, $x^2 + 1$, has no real roots and the function f has no vertical asymptotes.*

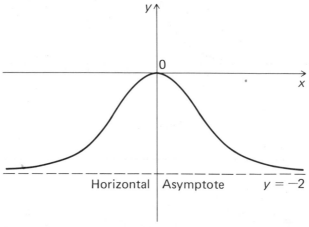

Horizontal │ Asymptote $y = -2$

Figure 103

As the preceding problems illustrate, one of the first steps in graphing a rational function is to examine whether the rational function has asymptotes. We now give the rules for determining the vertical and horizontal asymptotes.

> **Rule for Vertical Asymptotes:** If f is a rational function defined by $f(x) = \dfrac{g(x)}{h(x)}$, where $\dfrac{g(x)}{h(x)}$ have no common factors, then the line $x = a$ is a vertical asymptote of f if $h(a) = 0$ and $g(a) \neq 0$. That is, the vertical asymptotes occur at all the real roots of $h(x)$.

PROBLEM 4

Find all vertical asymptotes of $f(x) = \dfrac{3x - 6}{x^2 - x - 2}$.

Solution: First we must make sure that the numerator and denominator have no factors in common. Thus, $f(x) = \dfrac{3(x - 2)}{(x + 1)(x - 2)} = \dfrac{3}{x + 1}$. The vertical asymptote is given by $x + 1 = 0$—that is, by the line $x = -1$, and the graph of f has a hole at $(2, 1)$.

PROBLEM 5

Find all vertical asymptotes of $f(x) = \dfrac{4x + 5}{2x^2 - 5x - 3}$.

Solution: The vertical asymptotes occur at those values such that $2x^2 - 5x - 3 = (2x + 1)(x - 3) = 0$. Therefore, the lines $x = -\dfrac{1}{2}$ and $x = 3$ are vertical asymptotes.

PROBLEM 6

Given $f(x) = \dfrac{4x}{x^2 + 7}$, find, if any, all vertical asymptotes.

Solution: The denominator $x^2 + 7$ has no real roots. Thus, $f(x) = \dfrac{4x}{x^2 + 7}$ has no vertical asymptotes.

> **Rule for Horizontal Asymptotes:** Let f be a rational function defined by $f(x) = \dfrac{g(x)}{h(x)}$, where $g(x) = a_n x^n + a_{n-1} x^{n-1} + \cdots + a_1 x + a_0$, $a_n \neq 0$, and $h(x) = b_m x^m + b_{m-1} x^{m-1} + \cdots + b_1 x + b_0$, $b_m \neq 0$.
> (1) If $m > n$, the x-axis is a horizontal asymptote of f.
> (2) If $m = n$, then the line $y = \dfrac{a_n}{b_m}$ is a horizontal asymptote of f.
> (3) If $m < n$, then the rational function f has no horizontal asymptote.

PROBLEM 7

Find the horizontal asymptote for each of the rational functions $f(x) = \dfrac{x^2 + 2}{2x^3 + 3x^2 - 4x + 1}$, $p(x) = \dfrac{-7x^4 + 9x^2 - 1}{2x^4 + 1}$, and $q(x) = \dfrac{3x^3 - x^2 + 1}{x^2 - 7}$.

Solution:

(1) For $f(x) = \dfrac{x^2 + 2}{2x^3 + 3x^2 - 4x + 1}$ we have $m = 3$ and $n = 2$. Thus, $m > n$ and the x-axis is the horizontal asymptote.

(2) For $p(x) = \dfrac{-7x^4 + 9x^2 - 1}{2x^4 + 1}$, we have $m = n = 4$, and $a_n = -7$ and $b_m = 2$. Thus, the line $y = \dfrac{a_n}{b_m} = \dfrac{-7}{2}$ is the horizontal asymptote.

(3) The rational function q defined by $q(x) = \dfrac{3x^3 - x^2 + 1}{x^2 - 7}$ gives $m = 2$ and $n = 3$. So, $m < n$ and there is no horizontal asymptote.

We now use the information developed in this section to graph rational functions.

PROBLEM 8

Graph $f(x) = \dfrac{5}{(x + 1)^2}$.

Solution: We first search for vertical and horizontal asymptotes. The vertical asymptote occurs at the point $x + 1 = 0$, and thus the line $x = -1$ is the vertical asymptote. For $f(x) = \dfrac{5}{(x + 1)^2}$ we have $n = 0$ and $m = 2$ and $m > n$. Therefore the x-axis is the horizontal asymptote. Finally, we note that $f(x)$ is always positive. The graph of $f(x)$ is given in Figure 104.

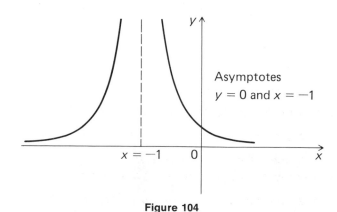

Asymptotes
$y = 0$ and $x = -1$

$x = -1$ 0

Figure 104

PROBLEM 9

Graph $f(x) = \dfrac{2x^2 + 1}{3x^2 + x - 2}$.

Solution: For f we have $m = n = 2$ and the horizontal asymptote is given by $y = \dfrac{a_n}{b_m} = \dfrac{2}{3}$. Factoring the denominator gives $f(x) = \dfrac{2x^2 + 1}{(3x - 2)(x + 1)}$. Thus, the vertical asymptotes are $x = -1$ and $x = \dfrac{2}{3}$. We now construct the following table to aid in the graphing of $f(x)$.

	$3x - 2$	$x + 1$	$2x^2 + 1$	$f(x)$
$x < -1$	$-$	$-$	$+$	$+$
$-1 < x < \dfrac{2}{3}$	$-$	$+$	$+$	$-$
$x > \dfrac{2}{3}$	$+$	$+$	$+$	$+$

The graph of $f(x) = \dfrac{2x^2 + 1}{3x^2 + x - 2}$ is given in Figure 105.

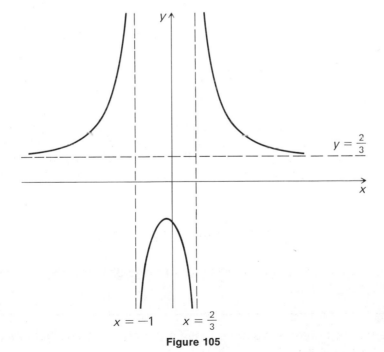

Figure 105

3.8 (A) EXERCISES

1. Determine the vertical and horizontal asymptotes for each of the following:

   (a) $f(x) = \dfrac{2}{x - 4}$

   (b) $f(x) = \dfrac{4}{(x + 7)^2}$

   (c) $f(x) = \dfrac{1}{(x - 1)^3}$

   (d) $f(x) = \dfrac{2x^2}{x^2 + x - 6}$

2. Graph the rational functions in Exercise 1.

3. Graph $f(x) = \dfrac{x}{x^2 - 1}$.

4. Graph $f(x) = \dfrac{x^2 - x}{x^2 - 2x - 3}$.

Partial Fractions

As mentioned in the beginning of this section, we may want to express a rational function as the sum of simpler rational functions. The resulting simple rational functions are called **partial fractions.** We shall consider rational expressions of the form $f(x) = \dfrac{g(x)}{h(x)}$, where the degree of $h(x)$ is greater than the degree of $g(x)$. If the degree of $g(x)$ were greater than the degree of $h(x)$, then we have only to divide $g(x)$ by $h(x)$ to obtain $\dfrac{g(x)}{h(x)} = q(x) + \dfrac{r(x)}{h(x)}$. For example,

$$\frac{x^5 + 3x^3 - 1}{x^3 + 2} = x^2 + 3 + \frac{-2x^2 - 7}{x^3 + 2}$$

In the case where the degree of $g(x)$ is greater than the degree of $h(x)$, the fraction is called an **improper fraction.** However, we are interested in rational expressions in which the degree of $h(x)$ is greater than the degree of $g(x)$. Such a fraction is called a **proper fraction.** For example, the type of decomposition of proper fractions is given by

$$\frac{-x + 8}{x^2 - 7x + 12} = \frac{4}{x - 4} - \frac{5}{x - 3}$$

The following two theorems will be used to aid us in finding such decompositions of proper fractions.

Theorem: Two polynomials are equal if and only if the coefficients of like-degree terms are equal.

Theorem: Partial Fraction Decomposition

Let f be a proper rational function in its reduced form, defined by

$$f(x) = \frac{g(x)}{h(x)}.$$

(1) Corresponding to each *nonrepeating linear factor* $ax + b$ of the denominator $h(x)$ there will be a partial fraction

$$\frac{A}{ax + b}$$

where A is constant.

(2) Corresponding to each *repeated linear factor* $(ax + b)^k$ of the denominator $g(x)$ there will be k partial fractions

$$\frac{A_1}{(ax + b)} + \frac{A_2}{(ax + b)^2} + \cdots + \frac{A_k}{(ax + b)^k}$$

where $A_1, A_2, \ldots, A_k$ are constants.

(3) Corresponding to a *nonrepeating quadratic factor* $(ax^2 + bx + c)$ there will be a partial fraction

$$\frac{Ax + B}{ax^2 + bx + c}$$

where A and B are constants.

(4) Corresponding to each *repeated quadratic factor* $(ax^2 + bx + c)^k$ there will be k partial fractions

$$\frac{A_1 x + B_1}{(ax^2 + bx + c)} + \frac{A_2 x + B_2}{(ax^2 + bx + c)^2} + \cdots + \frac{A_k x + B_k}{(ax^2 + bx + c)^k}$$

where $A_1, B_1, A_2, B_2, \ldots, A_k, B_k$ are constants.

We now illustrate each of the four types.

NONREPEATING LINEAR FACTORS

_____

PROBLEM 10

Decompose $\dfrac{x + 1}{x^2 - x - 6}$ into partial fractions.

Solution by substitution: First we factor the denominator, if possible, and use the Decomposition Theorem. Thus,

$$\frac{x + 1}{(x + 2)(x - 3)} = \frac{A}{x + 2} + \frac{B}{x - 3} = \frac{A(x - 3) + B(x + 2)}{(x + 2)(x - 3)}$$

We must find coefficients A and B such that $x + 1 = A(x - 3) + B(x + 2)$ identically; that is, it must hold for all values of x.

For $x = 3$: $3 + 1 = A(0) + B(3 + 2)$, $B = \dfrac{4}{5}$

For $x = -2$: $-2 + 1 = A(-2 - 3) + B(0)$, $A = \dfrac{1}{5}$

Hence $\dfrac{x + 1}{x^2 - x - 6} = \dfrac{1}{5} \cdot \dfrac{1}{(x + 2)} + \dfrac{4}{5} \cdot \dfrac{1}{(x - 3)}$

Solution by equating coefficients: To determine the constants A and B by equating coefficients of like powers of x, we expand the right side of the equation $x + 1 = A(x - 3) + B(x + 2)$ to get

$$x + 1 = (A + B)x - 3A + 2B$$

Using Theorem 1 and equating the coefficients of x and the constant term, we obtain

$$\begin{cases} A + B = 1 \\ -3A + 2B = 1 \end{cases}$$

Solving this system simultaneously gives $A = \dfrac{1}{5}$, $B = \dfrac{4}{5}$. Therefore,

$$\frac{x + 1}{x^2 - x - 6} = \frac{1}{5} \cdot \frac{1}{(x + 2)} + \frac{4}{5} \cdot \frac{1}{(x - 3)}$$

REPEATED LINEAR FACTORS

PROBLEM 11

Express $\dfrac{x^2 - x + 1}{x^3 + 2x^2 + x}$ as the sum of partial fractions.

Solution by substitution:

$$\frac{x^2 - x + 1}{x^3 + 2x^2 + x} = \frac{x^2 - x + 1}{x(x + 1)^2} = \frac{A}{x} + \frac{B}{(x + 1)} + \frac{C}{(x + 1)^2}$$

Thus, $x^2 - x + 1 = A(x + 1)^2 + Bx(x + 1) + Cx$
For $x = 0$: $1 = A$
For $x = -1$: $3 = -C$ or $C = -3$
For $x = 1$: $1 = 4A + 2B + C$, but $A = 1$ and $C = -3$, so $B = 0$

Therefore, $\dfrac{x^2 - x + 1}{x^3 + 2x^2 + x} = \dfrac{1}{x} - \dfrac{3}{(x + 1)^2}$

Solution by equating coefficients:

$$x^2 - x + 1 = A(x + 1)^2 + Bx(x + 1) + Cx$$
$$= (A + B)x^2 + (2A + B + C)x + A$$

Equating coefficients of like powers of x, we get

$$\begin{cases} A + B & = & 1 \\ 2A + B + C & = & -1 \\ A & = & 1 \end{cases}$$

Solving simultaneously gives $A = 1$, $B = 0$, and $C = -3$.

NONREPEATING QUADRATIC FACTORS

PROBLEM 12

Decompose $\dfrac{x^2 - 2x - 3}{(x - 1)(x^2 + 2x + 2)}$ into partial fractions.

Solution by substitution:

$$\frac{x^2 - 2x - 3}{(x - 1)(x^2 + 2x + 2)} = \frac{A}{x - 1} + \frac{Bx + C}{x^2 + 2x + 2}$$

Thus, $x^2 - 2x - 3 = A(x^2 + 2x + 2) + (Bx + C)(x - 1)$

For $x = 1$: $-4 = 5A$ or $A = -\dfrac{4}{5}$

For $x = 0$: $-3 = 2A - C$, but $A = -\dfrac{4}{5}$ gives $C = \dfrac{7}{5}$

For $x = -1$: $0 = A + 2B - 2C$ and $B - \dfrac{9}{5}$

Hence $\dfrac{x^2 - 2x - 3}{(x - 1)(x^2 + 2x + 2)} = \dfrac{-4}{5(x - 1)} + \dfrac{9x + 7}{5(x^2 + 2x + 2)}$

REPEATING QUADRATIC FACTORS

PROBLEM 13

Express $\dfrac{2x^2 + 3}{(x^2 + 1)^2}$ as the sum of partial fractions.

Solution by equating coefficients:

$$\frac{2x^2 + 3}{(x^2 + 1)^2} = \frac{Ax + B}{x^2 + 1} + \frac{Cx + D}{(x^2 + 1)^2}$$

Thus, $2x^2 + 3 = (Ax + B)(x^2 + 1) + (Cx + D)$ or, equivalently, $2x^2 + 3 = Ax^3 + Bx^2 + (A + C)x + (B + D)$

Equating coefficients we get the system

$$
\begin{array}{rcl}
A & & = 0 \\
& B & = 2 \\
A & + C & = 0 \\
& B & + D = 3
\end{array}
$$

and $A = 0$, $B = 2$, $C = 0$, and $D = 1$

Hence $\dfrac{2x^2 + 3}{(x^2 + 1)^2} = \dfrac{2}{x^2 + 1} + \dfrac{1}{(x^2 + 1)^2}$

3.8 (B)   EXERCISES

Express each rational fraction in terms of partial fractions:

1. $\dfrac{1}{x^2 - 4}$

2. $\dfrac{x + 2}{x^3 + x^2 - 6x}$

3. $\dfrac{2x^2 + 3}{x(x - 1)^2}$

4. $\dfrac{3x^2 - 10x + 14}{(x + 1)(x - 2)^3}$

5. $\dfrac{1}{x(x^2 + x + 1)}$

6. $\dfrac{x^3 + x^2 + x + 2}{(x^2 + 1)(x^2 + 2)}$

7. $\dfrac{2x^2 + 1}{(x^2 - x + 1)^2}$

8. $\dfrac{x^4 - x^3 + 2x^2 - x + 2}{(x - 1)(x^2 + 2)^2}$

9. $\dfrac{x^3 + 5x^2 + 2x - 4}{x(x^2 + 4)^2}$

10. $\dfrac{7x^3 - 29x^2 + 46x - 55}{(x - 3)^2(x^2 + 2)}$

SUMMARY

Polynomial Functions

A **polynomial** is an expression of the form $a_n x^n + \cdots + a_1 x + a_0$, where n is a nonnegative integer and $a_n, \ldots, a_1, a_0$ are real numbers called **coefficients.** Two polynomials, $a_n x^n + \cdots + a_1 x + a_0$ and $b_m x^m + \cdots + b_1 x + b_0$, are **equal** if and only if they have the same degree and the same coefficients.

If p is a polynomial function defined by $p(x)$, then any value of x which makes $p(x) = 0$ is called a **root (zero)** of p.

Remainder Theorem. If p is a polynomial function defined by $p(x)$, and $p(x)$ is divided by $x - a$, then the remainder is given by $p(a)$.

Factor Theorem.   Let p be a polynomial function defined by $p(x)$. Then $x - a$ is a factor of $p(x)$ if and only if $p(a) = 0$.

Linear Functions

A function f defined by the relation $f(x) = mx + b$, where m and b are constants, is called a **linear function.** Its graph is a straight line. If (x_1, y_1) and (x_2, y_2)

are two distinct points on a line not parallel to the y-axis, then the **slope** of the line is

$$m = \frac{\Delta y}{\Delta x} = \frac{y_2 - y_1}{x_2 - x_1} = \frac{\text{change in } y}{\text{change in } x}$$

EQUATION OF LINES

General Equation. $Ax + By + C = 0$, A, B, and C constants and A and B not both zero.

Line Parallel to y-Axis. $x = a$, a is constant.

Line Parallel to x-Axis. $y = b$, b is constant.

Two-Point Form. $y - y_1 = \frac{y_2 - y_1}{x_2 - x_1} \cdot (x - x_1)$.

Point-Slope Form. $y - y_1 = m(x - x_1)$, m = slope.

Slope-Intercept. $y = mx + b$, b = y-intercept. Let L_1 and L_2 be two nonvertical lines with slopes m_1 and m_2, respectively.

(1) L_1 is **parallel** to L_2 if and only if $m_1 = m_2$.

(2) L_1 is **perpendicular** to L_2 if and only if $m_1 = \frac{-1}{m_2}$.

Systems of Linear Equations

TWO LINEAR EQUATIONS IN TWO VARIABLES

The system $\begin{cases} a_1 x + b_1 y = c_1 \\ a_2 x + b_2 y = c_2 \end{cases}$ results in one of three possible cases:

(1) If the lines intersect in exactly one point, then the system is **consistent** and the equations are **independent.** The system has a **unique solution** given by the point of intersection.

(2) If the lines are parallel, the system is **inconsistent** and the equations are **independent.** Since there are no points of intersection, there is **no solution** to the system.

(3) If the lines coincide, the system is **consistent** and the equations are **dependent.** The solutions are all the points on the line which represents both equations. Therefore, there are an **infinite number of solutions.**

GENERAL SYSTEMS OF LINEAR EQUATIONS

Theorem. Given a system of linear equations, each of the following operations on the equations of the system gives an **equivalent** system:

(1) Interchange any two equations.

(2) Multiply any equation by a nonzero real number.

(3) Multiply any equation by a nonzero real number and add the resulting equation to another equation.

A **matrix** is a rectangular array of numbers arranged in rows and columns.

Examples:

$$\begin{bmatrix} a_{11} & a_{12} \\ a_{21} & a_{22} \end{bmatrix} \quad \text{and} \quad \begin{bmatrix} a_{11} & a_{12} & a_{13} \\ a_{21} & a_{22} & a_{23} \\ a_{31} & a_{32} & a_{33} \end{bmatrix}$$

Theorem. **Row-equivalent** matrices are obtained by the following row operations:

(1) Two rows are interchanged.
(2) Any row is multiplied by a nonzero number.
(3) Multiply any row by a nonzero number and add the resulting row to another row.

Determinants

SECOND-ORDER DETERMINANTS

$$\begin{vmatrix} a_1 & b_1 \\ a_2 & b_2 \end{vmatrix} = a_1 b_2 - a_2 b_1$$

Cramer's Rule for two linear equations in two variables:

Given the system $\begin{cases} a_1 x + b_1 y = c_1 \\ a_2 x + b_2 y = c_2 \end{cases}$, the solution is

$$x = \frac{D_x}{D} = \frac{\begin{vmatrix} c_1 & b_1 \\ c_2 & b_2 \end{vmatrix}}{\begin{vmatrix} a_1 & b_1 \\ a_2 & b_2 \end{vmatrix}} \text{ and } y = \frac{D_y}{D} = \frac{\begin{vmatrix} a_1 & c_1 \\ a_2 & c_2 \end{vmatrix}}{\begin{vmatrix} a_1 & b_1 \\ a_2 & b_2 \end{vmatrix}}, \text{ provided } D \neq 0.$$

PROPERTIES OF DETERMINANTS

(1) If each element in a row (or column) is zero, then the determinant is zero.
(2) If each of the elements in a row (or column) of a determinant is multiplied by a real number k, then the value of the determinant is multiplied by k.
(3) If two rows (or columns) of a determinant are identical, the value of the determinant is zero.
(4) Interchanging any two rows (or columns) reverses the sign of the determinant.

Quadratic Functions

A function f defined by $f(x) = ax^2 + bx + c$, with a, b, and c constants and $a \neq 0$, is called a **quadratic function.**
Quadratic Formula. Given $ax^2 + bx + c = 0$, then

$$x = \frac{-b \pm \sqrt{b^2 - 4ac}}{2a}$$

Discriminant. $b^2 - 4ac$
Nature of Roots

$$\begin{cases} b^2 - 4ac > 0; \text{ two real roots} \\ b^2 - 4ac = 0; \text{ one real root} \\ b^2 - 4ac < 0; \text{ no real roots} \end{cases}$$

Theorem. The quadratic function f, defined by $f(x) = ax^2 + bx + c$, is
(i) a parabola with **maximum** at the vertex if $a < 0$.
(ii) a parabola with **minimum** at the vertex if $a > 0$.

The **vertex** is at the point $\left(\dfrac{-b}{2a}, \dfrac{4ac - b^2}{4a} \right)$.

Rational Roots

Theorem: Rational Roots. Given $p(x) = a_n x^n + \cdots + a_1 x + a_0$, $a_n, \ldots,$
a_1, a_0 integers and $a_n \neq 0$, if $\dfrac{b}{c}$ is a root of $p(x)$, then b is a factor of a_0 and c is a
factor of a_n.

Complex Numbers and Roots

A **complex number** is a number of the form $a + bi$, where $i = \sqrt{-1}$ and a
and b are real numbers; a is the **real part** and b is the **imaginary part**. The **conjugate** is $a - bi$.

THE FUNDAMENTAL THEOREM OF ALGEBRA

Every polynomial of degree $n \geq 1$, with real or complex coefficients, has at
least one real or complex root.

FACTORIZATION THEOREM

If p is a polynomial function defined by $p(x) = a_n x^n + \cdots + a_1 x + a_0$, then
$p(x)$ can be written as $p(x) = a_n(x - r_1)(x - r_2) \cdots (x - r_n)$.

THE COMPLEX CONJUGATES THEOREM

If the complex number $z = a + bi$ is a root of a polynomial $p(x)$ of degree
$n \geq 1$, with **real coefficients,** then its conjugate $\bar{z} = a - bi$ is also a root.

OPERATIONS WITH COMPLEX NUMBERS

Addition. $(a + bi) + (c + di) = (a + c) + (b + d)i$
Subtraction. $(a + bi) - (c + di) = (a - c) + (b - d)i$
Multiplication. $(a + bi)(c + di) = (ac - bd) + (ad + bc)i$
Division. $\dfrac{(a + bi)}{(c + di)} = \dfrac{(a + bi)(c - di)}{(c + di)(c - di)} = \dfrac{(a + bi)(c - di)}{c^2 + d^2}$
Magnitude. $|z| = |a + bi| = \sqrt{a^2 + b^2}$

Rational Function

A function f defined by $f(x) = \dfrac{g(x)}{h(x)}$, where $g(x)$, $h(x)$ are polynomials and
$h(x) \neq 0$, is called a **rational function.**

Rule for Vertical Asymptotes. If f is a rational function defined by $f(x) = \dfrac{g(x)}{h(x)}$, where $\dfrac{g(x)}{h(x)}$ have no common factors, then the line $x = a$ is a vertical
asymptote of f if $h(a) = 0$ and $g(a) \neq 0$.

Rule for Horizontal Asymptotes. Let $f(x) = \dfrac{g(x)}{h(x)}$ be a rational function, where $g(x) = a_n x^n + \cdots + a_1 x + a_0$, $a_n \neq 0$, and $h(x) = b_m x^m + \cdots + b_1 x + b_0$, $b_m \neq 0$.

(i) If $m > n$, then the x-axis is a horizontal asymptote of f.

(ii) If $m = n$, then the line $y = \dfrac{a_n}{b_m}$ is a horizontal asymptote of f.

(iii) If $m < n$, then the rational function f has no horizontal asymptote.

CHAPTER 3 EXERCISES

1. Given $p(x) = x^7 + x^6 + x^5 + x^4 + x^3 + x^2 + x + 1$. Show, using synthetic division, that $x + 1$ is a factor. Find the quotient that results when $p(x)$ is divided by $x + 1$.

2. Using the Remainder Theorem, find $p(5)$ where

$$p(x) = x^5 - 3x^3 + x^2 + 4x - 6$$

3. Determine the equation of the line satisfying the following conditions:
   (a) passing through $(7, 1)$ and $(-2, 0)$.
   (b) parallel to x-axis and passing through $(5, -3)$.
   (c) slope $= \dfrac{1}{2}$ and y-intercept $= -4$.
   (d) parallel to y-axis and passing through $(4, 7)$.
   (e) passing through origin and slope $= -\dfrac{3}{5}$.

4. (a) Find the equation of the line perpendicular to $-4x - y + 1 = 0$ and passing through the point $(-2, 1)$.
   (b) Find the line parallel to $2x - 3y + 6 = 0$ and passing through the point $(3, -4)$.

5. Using Cramer's Rule, solve the following system:

$$\begin{cases} 2x_1 - x_2 + x_3 = 1 \\ 3x_1 + 2x_3 = -1 \\ 4x_1 + x_2 + 2x_3 = 2 \end{cases}$$

6. Evaluate the following determinants:

   (a) $\begin{vmatrix} 3 & 0 & -6 \\ 4 & 1 & -8 \\ -1 & 3 & 2 \end{vmatrix}$

   (b) $\begin{vmatrix} 0 & 0 & 0 \\ -7 & 5 & 1 \\ 3 & -2 & 4 \end{vmatrix}$

   (c) $\begin{vmatrix} -4 & 0 & 0 \\ 0 & 3 & 0 \\ 0 & 0 & -2 \end{vmatrix}$

7. Solve the following quadratic equations using the Quadratic Formula:
   (a) $2x^2 - 3x - 4 = 0$
   (b) $x^2 - x + \dfrac{1}{4} = 0$
   (c) $3x^2 - 4x + 2 = 0$

8. Evaluate the discriminant and determine the nature of the roots for each of the following:
   (a) $p(x) = 9x^2 + 6x + 1$
   (b) $p(x) = x^2 + x + 5$
   (c) $p(x) = 4x^2 - 7$

9. Graph the quadratic function f defined by $f(x) = 2x^2 + 4x - 1$. Find the vertex and determine if it is a maximum or minimum.

10. Find all the rational roots of the following polynomials:
    (a) $f(x) = x^3 - x^2 + 3x - 27$
    (b) $f(x) = x^4 - 2x^3 + 2x^2 - 10x + 25$
    (c) $f(x) = 10x^4 - 2x^3 + 10x^2 - 22x + 4$

11. Find a polynomial with real coefficients that has roots $4i$, 7, and -5.

12. Perform the operations indicated and simplify:
    (a) $(7 + 4i) + (6 - i) - (5 + 2i)$
    (b) $(-1 + 3i)(-1 - 3i)$
    (c) $\dfrac{\sqrt{5} - 2i}{1 + 6i}$
    (d) $|3 + 5i|$

13. Determine the vertical and horizontal asymptotes and graph the following rational expressions:
    (a) $f(x) = \dfrac{3}{x + 2}$
    (b) $f(x) = \dfrac{x}{x^2 + 1}$
    (c) $f(x) = \dfrac{2x^2 + 8x + 6}{x + 3}$
    (d) $f(x) = \dfrac{2x^2}{x^2 + x - 2}$

14. Express each rational fraction in terms of partial fractions:
    (a) $\dfrac{1}{x^2 - 9}$ (b) $\dfrac{x}{(x - 1)^3}$

15. Decompose the following rational fractions into partial fractions:
    (a) $\dfrac{x^2 + x + 1}{(2x + 1)(x^2 + 1)}$
    (b) $\dfrac{2x^2 + 1}{(x^2 - x + 1)^2}$

CHAPTER 4

EXPONENTIAL AND LOGARITHMIC FUNCTIONS

INTRODUCTION

In the previous chapters we studied algebraic functions—that is, functions which can be obtained from the identity function and the constant function by means of the algebraic operations of addition, subtraction, multiplication, division, and roots. In this chapter, we introduce functions which are not algebraic, called **transcendental functions.** Examples of such functions are the exponential and logarithmic functions. In later chapters we shall discuss other transcendental functions such as the trigonometric functions.

We begin this chapter by reviewing the properties of exponents. Next, we study the exponential function and its applications. Exponential functions are used extensively in studying population growth and decay, compound interest, and in many other fields. We then answer the question: "Does the exponential function have an inverse?" We shall see that the inverse of each exponential function is a logarithmic function and, also, that the inverse of each logarithmic function is an exponential function. We conclude this chapter with applications of logarithmic functions.

4.1 EXPONENTS

Recall that exponents can be thought of as a short way of writing repeated multiplication of the same factor. For example, $3^4 = 3 \cdot 3 \cdot 3 \cdot 3$, $a^5 = a \cdot a \cdot a \cdot a \cdot a$, and $(2x - 3)^3 = (2x - 3)(2x - 3)(2x - 3)$.

> **Definition 1:** If a is a real number and n is a positive integer, then a^n means the product obtained by using a as a factor n times. That is,
>
> $$a^n = \underbrace{a \cdot a \cdot a \ldots a}_{n \text{ factors}},$$
>
> where a is called the **base** and n is called the **exponent** of a^n.

Before stating the laws of integral exponents, we give two additional definitions that define zero and negative exponents.

Definition 2: $a^0 = 1$, provided $a \neq 0$.

EXAMPLE 1

$7^0 = 1$

EXAMPLE 2

$(x + 3y + 7)^0 = 1$, provided $x + 3y + 7 \neq 0$.

EXAMPLE 3

$8x^0 = 8(1) = 8$

Definition 3: $a^{-n} = \dfrac{1}{a^n}$, provided $a \neq 0$.

EXAMPLE 4

$5^{-2} = \dfrac{1}{5^2} = \dfrac{1}{25}$

EXAMPLE 5

$(x + y)^{-3} = \dfrac{1}{(x + y)^3}$

EXAMPLE 6

$-14x^{-2}y = -14 \cdot \dfrac{1}{x^2} \cdot y = \dfrac{-14y}{x^2}$, provided $x \neq 0$.

The following theorem gives the laws for integral exponents.

Theorem 1: If a and b are real numbers, and m and n are integers, then
(1) $a^m \cdot a^n = a^{m+n}$
(2) $(a^m)^n = a^{mn}$
(3) $\dfrac{a^m}{a^n} = a^{m-n}$
(4) $(ab)^m = a^m b^m$
(5) $\left(\dfrac{a}{b}\right)^m = \dfrac{a^m}{b^m}$, $b \neq 0$

EXAMPLE 7

$$4^2 \cdot 4^3 = (4 \cdot 4)(4 \cdot 4 \cdot 4) = 4^{2+3} = 4^5$$

EXAMPLE 8

$$(-5^3)^2 = (-5^3)(-5^3) = (-5)(-5)(-5)(-5)(-5)(-5) = (-5)^{3 \cdot 2} = (-5)^6$$

EXAMPLE 9

$$(6x^2y)^2 = (6x^2y)(6x^2y) = 6 \cdot 6 \cdot x^2 \cdot x^2 \cdot y \cdot y = 6^2(x^2)^2y^2 = 36x^4y^2$$

EXAMPLE 10

$$\frac{10^4}{10^2} = 10^{4-2} = 10^2$$

EXAMPLE 11

$$\frac{x^3}{x^7} = x^{3-7} = x^{-4} = \frac{1}{x^4}$$

EXAMPLE 12

$$\left(\frac{7x^2}{3y}\right)^3 = \frac{(7x^2)^3}{(3y)^3} = \frac{7^3(x^2)^3}{3^3y^3} = \frac{343x^6}{27y^3}$$

EXAMPLE 13

$$\frac{6x}{5x^{-2}y^{-1}} = \frac{6}{5} \cdot \frac{x}{x^{-2}} \cdot \frac{1}{y^{-1}} = \frac{6}{5}x^3y$$

EXAMPLE 14

$$(a^{-2} + b^{-2})^{-2} = \left(\frac{1}{a^2} + \frac{1}{b^2}\right)^{-2} = \left(\frac{b^2 + a^2}{a^2b^2}\right)^{-2} = \frac{1}{\left(\frac{b^2 + a^2}{a^2b^2}\right)^2}$$

$$= \frac{1}{\frac{(b^2 + a^2)^2}{(a^2b^2)^2}} = \frac{1}{\frac{(b^4 + 2a^2b^2 + a^4)}{a^4b^4}}$$

$$= \frac{a^4b^4}{b^4 + 2a^2b^2 + a^4}$$

The laws of exponents can be extended to include rational exponents, but first we need the following definition.

> **Definition 4:** If $a > 0$ and n is a positive integer, then
>
> $$a^{1/n} = \sqrt[n]{a}$$
>
> and $a^{1/n}$ is called the **principal nth root** of a.

EXAMPLE 15

$$4^{1/2} = \sqrt[2]{4} = 2$$

EXAMPLE 16

$$(8)^{1/3} = \sqrt[3]{8} = 2$$

EXAMPLE 17

$$y^{1/5} = \sqrt[5]{y}$$

EXAMPLE 18

$$125^{1/3} = \sqrt[3]{125} = 5$$

Note that if $a < 0$ and n is even, then $a^{1/n}$ is not defined in the real number system.

EXAMPLE 19

$(-9)^{1/2} = \sqrt[2]{-9}$ is not defined in the real number system because there does not exist a real number x such that $x^2 = -9$.

If $a > 0$ and n is odd, then $(-a)^{1/n} = \sqrt[n]{-a} = -\sqrt[n]{a}$

EXAMPLE 20

$$(-8)^{1/3} = \sqrt[3]{-8} = -\sqrt[3]{8} = -2$$

EXAMPLE 21

$$(-1)^{1/7} = \sqrt[7]{-1} = -\sqrt[7]{1} = -1$$

EXAMPLE 22

$$(-64)^{1/3} = \sqrt[3]{-64} = -\sqrt[3]{64} = -4$$

Definition 5: If $a > 0$ and m is any integer, then

$$a^{m/n} = (a^{1/n})^m = (\sqrt[n]{a})^m$$

EXAMPLE 23

$$27^{2/3} = (27^{1/3})^2 = (\sqrt[3]{27})^2 = 3^2 = 9$$

EXAMPLE 24

$$(32)^{3/5} = (32^{1/5})^3 = (\sqrt[5]{32})^3 = 2^3 = 8$$

We see that $a^{m/n} = (a^{1/n})^m$ can also be written as $a^{m/n} = (a^m)^{1/n}$. However, it is usually easier to use the form in Definition 5 when evaluating rational exponents. The laws for rational exponents are similar to those for integral exponents.

Theorem 2: If a and b are positive real numbers, and p and q are rational numbers, then

(1) $a^p \cdot a^q = a^{p+q}$

(2) $(a^p)^q = a^{pq}$

(3) $\dfrac{a^p}{a^q} = a^{p-q}$

(4) $(ab)^p = a^p b^p$

(5) $\left(\dfrac{a}{b}\right)^p = \dfrac{a^p}{b^p}, b \neq 0$

EXAMPLE 25

$$(5)^{1/3}(5)^{1/4} = 5^{1/3+1/4} = 5^{7/12}$$

EXAMPLE 26

$$(\sqrt[3]{7})^{1/5} = (7^{1/3})^{1/5} = 7^{1/3 \cdot 1/5} = 7^{1/15}$$

EXAMPLE 27

$$\frac{x^{1/2}}{x^{1/3}} = x^{1/2-1/3} = x^{1/6}$$

EXAMPLE 28

$$(9x^2)^{1/3} = 9^{1/3}(x^2)^{1/3} = 9^{1/3}x^{2/3}$$

EXAMPLE 29

$$\left(\frac{4x^3}{25y^6}\right)^{1/2} = \frac{(4x^3)^{1/2}}{(25y^6)^{1/2}} = \frac{4^{1/2}(x^3)^{1/2}}{25^{1/2}(y^6)^{1/2}} = \frac{2x^{3/2}}{5y^3}$$

EXAMPLE 30

$$\left(\frac{a^{n^2-1}}{a^{n+1}}\right)^{1/n+1} = \frac{a^{n^2-1/n+1}}{a^{n+1/n+1}} = \frac{a^{n-1}}{a} = a^{n-2}$$

EXAMPLE 31

$$\sqrt[3]{\sqrt[4]{20}} = \sqrt[3]{20^{1/4}} = (20^{1/4})^{1/3} = 20^{1/12}$$

PROBLEM 1

For what value of x does $3^x = 27$?

Solution: Since $27 = 3^3$, we have $3^x = 3^3$ or $x = 3$.

PROBLEM 2

For what value of x does $32^x = 64$?

Solution: Since $32 = 2^5$, we have $32^x = (2^5)^x = 2^{5x}$. Also, $64 = 2^6$. Thus $32^x = 64$ or, equivalently, $2^{5x} = 2^6$, and $5x = 6$, which implies that $x = \dfrac{6}{5}$.

Problem 2 illustrates the following property of exponents.

> If $a > 0$, $a \neq 1$, then $a^x = a^y$ if and only if $x = y$.

PROBLEM 3

Solve $2^{2x+3} - 33(2^x) + 4 = 0$.

Solution: We write the equation in the form $2^3 \cdot 2^{2x} - 33(2^x) + 4 = 0$. Let $y = 2^x$; then $2^{2x} = (2^x)^2 = y^2$ and the equation now becomes a quadratic equation in y.

$$(2^3)y^2 - 33y + 4 = 0; \text{ or } 8y^2 - 33y + 4 = 0$$

Solving for y we get $y = 4$ and $\dfrac{1}{8}$. Thus, for $y = 4$ we have $2^x = 4$ or $2^x = 2^2$, and $x = 2$. Finally, for $y = \dfrac{1}{8}$ we get $2^x = \dfrac{1}{8}$ or $2^x = 2^{-3}$, and $x = -3$. Therefore $x - 2$ or $x = -3$.

Remark: It can be shown that the properties of exponents can be extended to include irrational numbers. In such a case we must have $a > 0$ and $b > 0$.

Theorem 3: If $a > 0$ and $b > 0$ and x, y are any real numbers, then

(1) $a^x \cdot a^y = a^{x+y}$ (2) $(a^x)^y = a^{xy}$ (3) $\dfrac{a^x}{a^y} = a^{x-y}$

(4) $(ab)^x = a^x b^x$ (5) $\left(\dfrac{a}{b}\right)^x = \dfrac{a^x}{b^x}$

EXAMPLE 32

$$4^\pi \cdot 4^2 = 4^{2+\pi}$$

EXAMPLE 33

$$(6^{\sqrt{2}})^3 = 6^{3\sqrt{2}}$$

EXAMPLE 34

$$\frac{5^2}{5^{\sqrt{3}}} = 5^{2-\sqrt{3}}$$

EXAMPLE 35

$$(4x^2 y)^\pi = 4^\pi x^{2\pi} y^\pi$$

EXAMPLE 36

$$\left[\frac{(x+y)}{\pi}\right]^{\sqrt{5}} = \frac{(x+y)^{\sqrt{5}}}{\pi^{\sqrt{5}}}$$

4.1 EXERCISES

1. Simplify each expression:
   (a) $(-7)^2$
   (b) $(4)^{-2}(5)^3$
   (c) $a^2 \cdot a^3$
   (d) $x^{-2} \cdot x^2 \cdot x^3$
   (e) $(17^3)^2$
   (f) $(abc^2)^5$
   (g) $\dfrac{a^2 b^4}{ab^2}$
   (h) $\dfrac{10^{14}}{10^6}$
   (i) $\left(\dfrac{-2a}{3b^2}\right)^3$
   (j) $\left(\dfrac{-3y}{2x^2}\right)^4$
   (k) $(7x)(-7x)^3$
   (l) $(xy^2)^3(-3y^2)^2$
   (m) $a^{m+1} \cdot a^{m-1}$
   (n) $\dfrac{10^n \cdot 10^{n+1}}{10^{n-1}}$
   (o) $\dfrac{(y^{n+1} \cdot y^{2n-1})^2}{y^{3n}}$
   (p) $\dfrac{(a^{m+1})^m}{a^m}$

2. Simplify the following expressions and express answers with positive exponents:
   (a) $4 \cdot 3^{-5}$
   (b) $3^2 + 3^{-2}$
   (c) $\dfrac{1}{5^{-2}}$
   (d) $\dfrac{4^0}{2^{-3}}$
   (e) $\left(\dfrac{4}{5}\right)^{-1}$
   (f) $\dfrac{4^{-2}}{7^{-3}}$
   (g) $8^{-3} - 6^0$
   (h) $7(25 + 46)^0$
   (i) $a^2 b^{-3}$
   (j) $\dfrac{a^2}{b^{-3}}$
   (k) $\dfrac{(ab^2)^3}{(a^2 b)^2}$
   (l) $(a^{-3} b^2)^0$
   (m) $\dfrac{7^{-1} a^0 y^{-2}}{(3ab)^{-4}}$
   (n) $\left(\dfrac{a^{-1} b^2}{4x^0 y^{-3}}\right)^{-2}$
   (o) $\left[\left(\dfrac{xy^{-2} z^{-1}}{y^{-2} x}\right)^{-1}\right]^3$
   (p) $x^{-2} + y^{-2}$
   (q) $\dfrac{a^{-1} + b^{-1}}{(ab)^{-1}}$
   (r) $\dfrac{x^{-1} + y^{-1}}{x^{-1} - y^{-1}}$

3. Simplify:
   (a) $(64)^{2/3}$
   (b) $(-32)^{3/5}$
   (c) $(-27)^{2/3}$
   (d) $(125)^{2/3}$
   (e) $27^{-1/3}$
   (f) $(-64)^{-2/3}$
   (g) $a^{1/3}a^{1/2}$
   (h) $\dfrac{x^{2/3}}{x^{1/3}}$
   (i) $(a^7)^{3/7}$
   (j) $(x^{2/3}y)^{1/2}$
   (k) $\left(\dfrac{a^9}{b^6}\right)^{2/3}$
   (l) $a^{1/2}(a^{1/3} + a^{-1/2})$
   (m) $y^{-1/5}(y^5 + y^0)$
   (n) $\left(\dfrac{a^n}{b}\right)^{1/2}\left(\dfrac{a}{b^{4n}}\right)^2$
   (o) $\left[\left(\dfrac{a^{2m}}{y^{3m}}\right)^{1/12}\right]^{1/m}$

4. Solve for x:
   (a) $10^x = \dfrac{1}{1000}$
   (b) $2^x = 32$
   (c) $4^x = 8$
   (d) $\left(\dfrac{1}{2}\right)^x = 16$
   (e) $27^x = 9$
   (f) $\left(\dfrac{9}{4}\right)^x = \dfrac{3}{2}$

5. Solve for x:
   (a) $2^{2x+1} - 3(2^x) + 1 = 0$
   (b) $3^{2x+1} - 28(3^x) + 9 = 0$

4.2   THE EXPONENTIAL FUNCTION

Now that we understand that there corresponds a real number to every expression of the form a^x, where $a > 0$ and x is any real number, we can examine more closely the idea of an exponential function.

EXAMPLE 1

Consider the function defined by $f(x) = 2^x$. We sketch the graph of $f(x) = 2^x$ (see Table 2).

TABLE 2

x	−3	−2	−1	0	1	2	3
$f(x) = 2^x$	$\dfrac{1}{8}$	$\dfrac{1}{4}$	$\dfrac{1}{2}$	1	2	4	8

Note that the line $y = 0$ is a horizontal asymptote (see Fig. 106). Note

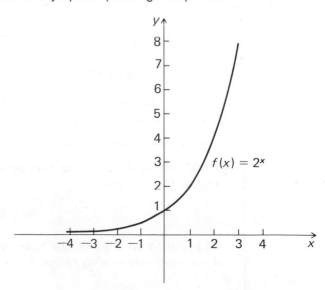

Figure 106

that the function $f(x) = 2^x$ is one-to-one and therefore must have an inverse. Also, $f(x) = 2^x$ is an increasing function and passes through the point (0, 1).

EXAMPLE 2

Consider $g(x) = \left(\frac{1}{2}\right)^x$. We sketch the graph of $g(x)$ (see Table 3).

TABLE 3

x	-3	-2	-1	0	1	2	3
$g(x) = \left(\frac{1}{2}\right)^x$	8	4	2	1	$\frac{1}{2}$	$\frac{1}{4}$	$\frac{1}{8}$

Note that as x increases, $g(x) = \left(\frac{1}{2}\right)^x$ decreases and that it passes through the point (0, 1). Also, $g(x) = \left(\frac{1}{2}\right)^x = \frac{1}{2^x} = 2^{-x}$. Thus, the graph of $g(x)$ is the graph of $f(x) = 2^x$ reflected in the y-axis (see Fig. 107). As with $f(x) = 2^x$,

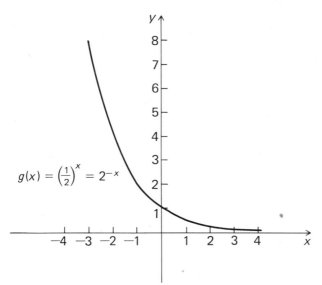

Figure 107

$g(x) = \left(\frac{1}{2}\right)^x$ is one-to-one and therefore must have an inverse.

The previous examples illustrate the concept of an exponential function. We now define an exponential function.

Definition: Let a be any real number such that $a > 0$. Then the function f is defined by

$$f(x) = a^x, \ a \neq 1$$

and is called an **exponential function** with base a. Its domain is the set of all real numbers and its range is the set of positive real numbers provided $a \neq 1$.

PROBLEM 1

Graph $h(x) = 2^{x-1} - 1$.

Solution: Using the properties of the graph of a function that we discussed in Chapter 2, we see that the graph of $h(x)$ is the graph of $f(x) = 2^x$ shifted to the right one unit and down one unit (see Fig. 108).

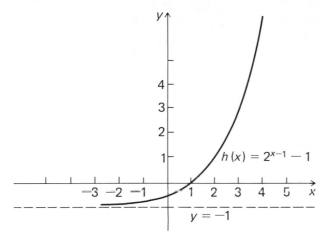

Figure 108

PROBLEM 2

Graph $f(x) = 1^x$.

Solution: For any finite value of x, $f(x) = 1^x$ is equal to 1 (see Fig. 109).

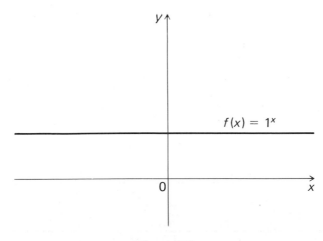

Figure 109

Note that $f(x) = 1^x$ is not one-to-one, it fails the horizontal line test, and thus it has no inverse.

The properties of exponential functions can be summarized in the following theorem.

Theorem 4: Given the exponential function defined by $f(x) = a^x$, where $a > 0$, then
(1) $f(x) = a^x$ is an increasing function if $a > 1$.
(2) $f(x) = a^x$ is a decreasing function if $a < 1$.
    The graph always passes through the point (0, 1).

PROBLEM 3

Graph $f(x) = 3^{-x^2}$.

Solution: We first note that $f(x) = 3^{-x^2} = \dfrac{1}{3^{x^2}}$ is an even function, and thus is symmetric with respect to the y-axis. Assigning values to x we have Table 4. The graph is given in Figure 110.

TABLE 4

x	0	1	2
$f(x) = 3^{-x^2}$	1	$\dfrac{1}{3}$	$\dfrac{1}{81}$

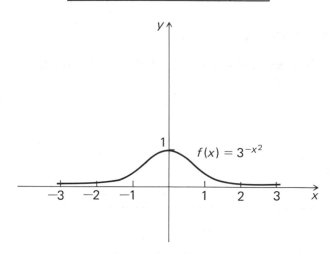

Figure 110

Theorem 5: If f is an exponential function, and x and y are real numbers, then

$$f(x + y) = f(x) \cdot f(y)$$

Proof: If *f* is an exponential function, then it is defined by $f(x) = a^x$, $a > 0$. Then $f(x + y) = a^{x+y} = a^x \cdot a^y = f(x) \cdot f(y)$.

EXAMPLE 3

Let *f* be the function defined by $f(x) = 3^x \cdot 3^{2x}$. Then $f(x + y) = 3^{x+y}$. $3^{2(x+y)} = 3^x \cdot 3^y \cdot 3^{2x} \cdot 3^{2y} = (3^x \cdot 3^{2x})(3^y \cdot 3^{2y}) = f(x)f(y)$.

PROBLEM 4

Determine whether $f(x) = 4^x + 4^{-x}$ defines an exponential function.

Solution: $f(x + y) = 4^{x+y} + 4^{-(x+y)} = 4^x 4^y + 4^{-x} \cdot 4^{-y}$. However, $f(x) \cdot f(y) = (4^x + 4^{-x})(4^y + 4^{-y}) = 4^x 4^y + 4^{-x} 4^y + 4^x 4^{-y} + 4^{-x} 4^{-y}$. Thus, $f(x + y) \neq f(x)f(y)$ and *f* is *not* an exponential function, even though it is the sum of two exponential functions.

Euler's Number *e*

One of the most important exponential functions is the function defined by $f(x) = e^x$, where *e* is Euler's number. The number *e* is an irrational number whose infinite decimal representation can be approximated by $e \approx 2.71828182845904.\ldots$. This number was used by Leonard Euler (1707–1783), a Swiss mathematician, in a study of population growth. A definition of the number *e* is given by

$$e = \lim_{n \to \infty} \left(1 + \frac{1}{n}\right)^n$$

The concept of limit can be illustrated by substituting values for *n*:

$$n - 1, \left(1 + \frac{1}{1}\right)^1 = (2)^1 = 2$$

$$n = 2, \left(1 + \frac{1}{2}\right)^2 = \left(\frac{3}{2}\right)^2 = \frac{9}{4} = 2.25 \ldots$$

$$n = 3, \left(1 + \frac{1}{3}\right)^3 = \left(\frac{4}{3}\right)^3 = \frac{64}{27} \approx 2.37 \ldots$$

$$n = 4, \left(1 + \frac{1}{4}\right)^4 = \left(\frac{5}{4}\right)^4 = \frac{625}{256} \approx 2.4414 \ldots$$

$$\vdots$$

$$n = 100, \left(1 + \frac{1}{100}\right)^{100} \approx 2.70 \ldots$$

Continuing this process we get an infinite nonrepeating decimal, which represents an irrational number. Therefore,

$$\lim_{n \to \infty} \left(1 + \frac{1}{n}\right)^n = e \approx 2.71828182845904 \ldots$$

The exponential function defined by $f(x) = e^x$ obeys all the properties given above. That is,

(1) $e^0 = 1$

(2) $e^x \cdot e^y = e^{x+y}$

(3) $e^{-x} = \dfrac{1}{e^x}$

(4) $\dfrac{e^x}{e^y} = e^{x-y}$

(5) $e^x > 0$ for all values of x

(6) e^x is an increasing function

The graph of $f(x) = e^x$ is given in Figure 111.

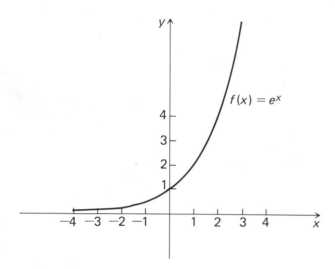

Figure 111

4.2 EXERCISES

1. Given that the following points lie on the graph of an exponential function, determine the bases of the exponential functions.

   (a) $\left(2, \dfrac{1}{4}\right)$ (b) $(3, 27)$

   (c) $\left(-2, \dfrac{1}{16}\right)$ (d) $(-3, 64)$

   (e) $\left(2, \dfrac{1}{49}\right)$ (f) $(0, 1)$

2. For each of the following functions indicate the domain and range, whether the function is increasing or decreasing, and whether the function has an inverse.

   (a) $f(x) = 4^x$ (b) $f(x) = \left(\dfrac{1}{3}\right)^x$

   (c) $f(x) = 2^{x+1}$ (d) $f(x) = (\sqrt{2})^x$

   (e) $f(x) = 1^x$ (f) $f(x) = 3^{-x}$

3. Graph each of the following functions:

   (a) $f(x) = 3^x$ (b) $f(x) = \left(\dfrac{1}{3}\right)^x$

   (c) $f(x) = 4^x$ (d) $f(x) = \left(\dfrac{1}{4}\right)^x$

   (e) $f(x) = 5^x$ (f) $f(x) = \left(\dfrac{1}{5}\right)^x$

4. Graph:

   (a) $f(x) = 3^{x+1} - 1$

   (b) $f(x) = \left(\dfrac{1}{2}\right)^{x-1} + 1$

   (c) $f(x) = 2^{-x^2}$

5. Use Theorem 5 to determine whether the following are exponential functions.
   (a) $f(x) = 3^{2x+1}$
   (b) $f(x) = 2^x \cdot 2^{3x}$
   (c) $f(x) = 5^x - 5^{-x}$
   (d) $f(x) = 7^{2x}$

6. Why are exponential functions not defined for $a \leq 0$?

7. Determine whether an exponential function can be an even function or an odd function.

8. Let $f(x) = b \cdot 10^{nx}$, and let $f(0) = 7$ and $f(1) = 700$. Determine the values of b and n.

9. Graph:
   (a) $f(x) = e^{-x}$
   (b) $f(x) = -e^{-x}$
   (c) $f(x) = -e^x$
   (d) $f(x) = e^{-x^2}$

10. Let $f(x) = a \cdot e^{kx}$. Determine $f(x + y)$ and $f(x - y)$ in terms of $f(x)$ and $f(y)$.

4.3 APPLICATIONS OF EXPONENTIAL FUNCTIONS

Exponential functions are used in many fields. They occur, for instance, in the natural growth and decay of populations, cell division, compound interest, and in other areas. In this section we give just a few examples of these applications.

Business

Suppose we invest a certain amount of money, p, which is compounded annually at a rate of r per cent. Also, suppose that we leave the original principal p and the interest $p \cdot r$ in the account at the end of the year. The amount left on deposit for the second year is

$$p + p \cdot r = p(1 + r)$$

If we continue to leave the interest, then at the end of two years the new principal is

$$p(1 + r) + p(1 + r) \cdot r = p(1 + r)(1 + r) = p(1 + r)^2$$

If the money is left for t years, then the original amount has been compounded to

$$A = p(1 + r)^t \tag{I}$$

PROBLEM 1

Suppose $100 is invested at 8% annual interest. If the money is left on deposit for ten years, how much money will be in the account at the end of that time?

Solution: We have $p = 100$, $r = 8\% = .08$, and $t = 10$. The compound interest formula gives

$$A = 100(1 + .08)^{10}$$

In a later section we will show how we can compute the amount A more precisely.

Usually an institution will compound interest more often than annually. If the interest is compounded n times per year, then the rate of interest is $\frac{r}{n}$ per period. In a span of t years, the number of periods will be $n \cdot t$. Therefore, the formula which expresses the amount A which the original principal p will be after t years at an interest rate of r per cent compounded n times each year is

$$A = p \left(1 + \frac{r}{n}\right)^{nt} \tag{II}$$

PROBLEM 2

Suppose $1000 is invested at 5% interest, compounded six times a year. What is the balance after ten years?

Solution: For this problem we have $p = 1000$, $r = 5\% = .05$, $n = 6$, and $t = 10$. The formula gives the balance at the end of ten years:

$$A = 1000 \left(1 + \frac{.05}{6}\right)^{6 \cdot 10} \approx 1000(1 + .008)^{60}$$

What would the balance be if the principal were compounded *continuously*? This would occur if the number of times the interest is compounded increases—that is, n becomes larger and larger. Recall that

$$\lim_{n \to \infty} \left(1 + \frac{1}{n}\right)^n = e, \text{ or equivalently, } \lim_{k \to 0} (1 + k)^{1/k} = e$$

To derive a formula for continuous interest we rewrite formula (II) by setting $k = \frac{r}{n}$, which gives $n = \frac{r}{k}$. Substituting into (II) gives

$$A = p \left(1 + \frac{r}{n}\right)^{nt} = p(1 + k)^{rt/k} = p[(1 + k)^{1/k}]^{rt}$$

It is easy to see that as n gets larger and larger, $k = \frac{r}{n}$ gets smaller and smaller for a fixed r. That is, n approaching ∞ is equivalent to k approaching zero. Therefore,

$$A = p \left[\lim_{k \to 0} (1 + k)^{1/k} \right]^{rt} = pe^{rt}$$

If p dollars are deposited at a rate r and compounded continuously, then the balance after t years is

$$A = pe^{rt} \qquad \text{(III)}$$

PROBLEM 3

What is the balance after five years if \$250 is compounded continuously at a rate of 7%.

Solution: We have $t = 5$, $p = 250$, and $r = 7\% = .07$. Using formula (III) we get

$$A = 250 \cdot e^{(.07)(5)} = 250e^{.35} \qquad 1.41906\cdot$$

Population Growth and Decay

Many laws of growth and decay which occur in biology and chemistry can be described using an exponential function of the form $N(t) = N_0 e^{kt}$, where t is the time, k is called the **constant of proportionality**, and N_0 is the amount of substance present at the initial time. Note that (see Fig. 112):

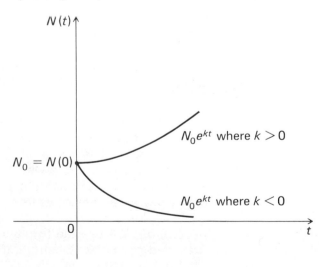

Figure 112

(1) If $k > 0$, N is an increasing function and we have the **Law of Natural Growth.**

(2) If $k < 0$, N is a decreasing function and we have the **Law of Natural Decay.**

Also, when $t = 0$, we have $N(0) = N_0 e^{k \cdot 0} = N_0 e^0 = N_0$. Thus, N_0 is the **initial value.**

PROBLEM 4

The size of a population of bacteria at time t in hours is given by $N(t) = N_0 e^{.5t}$. The initial population is 10^3. What is the population after 6 hours?

Solution: We have $N_0 = 10^3$ and $t = 6$. Therefore, the population of the bacteria will be

$$N = 10^3 \cdot e^{(.5)(6)} = 10^3 \cdot e^3 \approx 20{,}086$$

Radioactive material will decay at a rate that is proportional to the amount present at a given time. The amount $N(t)$ which is present after a time interval t is given by the formula

$$N(t) = N_0 \left(\frac{1}{2}\right)^{t/h} = N_0 \cdot 2^{-t/h} \qquad \textbf{(IV)}$$

where N_0 is the initial amount, and h is the **half-life** of the radioactive substance. The **half-life** of a substance is the time required for one half of the initial amount to decay.

PROBLEM 5

One technique of determining the age of an object which was once living is that of carbon-14 dating. While the object is alive the amount of carbon-14 present is constant; however, once the object is dead the carbon-14 decays at a rate which is a function of its half-life, 6000 years. Suppose the initial amount of carbon-14 present is 10^4 atoms. How many atoms will be left after 40 years?

Solution: We use the formula $N(t) = N = N_0 \cdot 2^{-t/h}$, where $N_0 = 10^4$, $t = 40$, and $h = 6000$. Substituting into the formula we obtain

$$N = 10^4 \cdot 2^{-40/6000} = 10^4 \cdot 2^{-1/150}$$

PROBLEM 6

The absorption of a beam of light, when it passes through a medium, is proportional to the intensity of the beam. The formula governing the intensity is $I = I_0 e^{-rt}$, where I_0 is the initial intensity of the beam, r is the absorption coefficient of the medium, and t is the thickness, in centimeters, of the medium. The intensity I is measured in lumens. Suppose $I_0 = 400$ lumens, the medium has a thickness of 5 centimeters, and an absorption coefficient of .4. Find the intensity.

Solution: We have $I_0 = 400$, $r = .4$, and $t = 5$. Therefore, $I = 400e^{-(.4)(5)} = 400e^{-2.0} \approx 54.13$.

4.3 EXERCISES

1. If a person opens a savings account with $1000 at an interest rate of 7%, then how much will there be in the savings account after 20 years if the interest is compounded:
   (a) daily   (b) monthly
   (c) quarterly   (d) semiannually
   (e) annually   (f) continuously

2. An element decays according to the law $N = N_0 2^{-.03t}$. How much of the element will be left after 25 years if initially there was 10 milligrams of the element present?

3. The size of a population of bacteria at time t is given by $N(t) = N_0 5^t$. Find the number of bacteria at the end of 13 hours if we started with 2500 bacteria.

4. Suppose an object has 10^3 atoms of carbon-14. How many atoms will remain after 50 years?

5. Strontium-90 has a half-life of 25 years. If we start with 10 milligrams of strontium-90, how much will remain after 15 years?

6. Voltage in an electrical circuit decays according to the formula $I = I_0 e^{-.3t}$, where t is measured in seconds and I_0 is the current at the time the voltage is turned off. If the current is 10 amperes, then what will the voltage be at the end of 30 seconds?

7. Find the intensity of a light beam passing through a medium which has thickness of 2 centimeters, absorption coefficient of .01, and initial intensity of 100 lumens.

8. The concentration of a drug in the body fluid is a function of the time elapsed since the drug was administered. The concentration is given by $C(t) = C_0 e^{-t}$, where C_0 is the initial dose. Determine the concentration of a drug after 4 hours, if the initial dose was 50 cc.

9. *Newton's Cooling Law* states that the temperature of a cooling object is a function of the initial temperature of the object and the temperature of the medium surrounding the object. The formula which gives the temperature is $T_c = T_m + (T_0 - T_m)e^{-ct}$, where T_0 is the temperature of the object, T_m is the temperature of the medium, c is a constant which depends on the object, and t is measured in minutes later. Suppose that for a heated iron bar $c = .01$ and the initial temperature of the bar is 200°F. If the bar is cooling in air with a temperature of 60°F, then find the temperature of the bar 10 minutes later.

10. The *reliability R* of an electrical component during t hours of use is given by $R = e^{-.02t}$. Determine the reliability during the first 15 hours of use.

4.4 LOGARITHMIC FUNCTION

Consider the function f defined by $f(x) = 2^x$ and whose graph is shown in Figure 113. We see that $f(x) = 2^x$ is a strictly increasing function and, therefore,

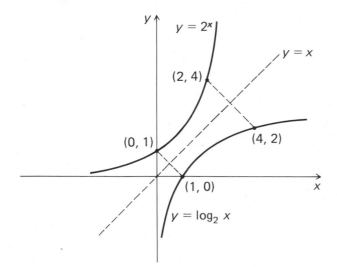

Figure 113

is one-to-one. Thus, the function f has an inverse. Recall that the graph of the inverse of a function is the reflection of the graph of the function about the line with equation $y = x$. The inverse of f is the function f^{-1} defined by $y = \log_2 x$, which is read "y is the logarithm of x to the base 2." This inverse function, $\log_2$, is an example of what is called a **logarithmic function** and is graphed in Figure 113.

Generally, the exponential function f defined by $f(x) = b^x$, $b \neq 1$, is either a strictly increasing or strictly decreasing function. We know that such functions are one-to-one and, therefore, have an inverse. This inverse function is defined as follows.

Definition: Let f be the exponential function defined by $f(x) = b^x$, where $b \neq 1$. The inverse function f^{-1} defined by $f^{-1}(x) = \log_b x$ is called the **logarithmic function with base b.**

The domain of an exponential function is all the real numbers, and the range is the positive real numbers. Since the logarithmic function is the inverse function of an exponential function, with base not equal to one, the domain and range must be interchanged. Thus, the **domain** of the logarithmic function is the range of the exponential function and the **range** of the logarithmic function is the domain of the exponential function. That is, the domain of $f^{-1}(x) = \log_b x$ is all positive reals and the range is all real numbers. If we use set notation, then the exponential function f defined by $f(x) = b^x$ can be written $f = \{(x, y) | y = b^x\}$. We can now give an equivalent definition for the logarithmic function.

Definition: If b is any positive real number, where $b \neq 1$, then

$$y = \log_b x \text{ if and only if } x = b^y$$

The graph of the logarithmic function is given in Figure 114.

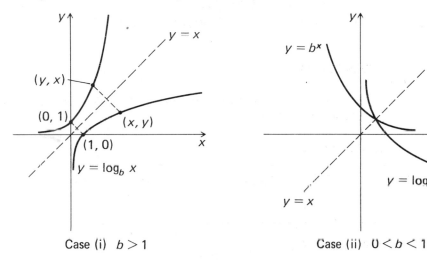

Figure 114   Case (i) $b > 1$ Case (ii) $0 < b < 1$

Logarithmic statements can be written in exponential form.

Example	Logarithmic Equation		Exponential Equation
	$y = \log_b x$	is equivalent to	$x = b^y$
1	$2 = \log_4 16$	is equivalent to	$16 = 4^2$
2	$3 = \log_5 125$	is equivalent to	$125 = 5^3$
3	$2 = \log_{1/3} \dfrac{1}{9}$	is equivalent to	$\dfrac{1}{9} = \left(\dfrac{1}{3}\right)^2$
4	$-3 = \log_2 \dfrac{1}{8}$	is equivalent to	$\dfrac{1}{8} = 2^{-3}$
5	$0 = \log_7 1$	is equivalent to	$1 = 7^0$

Also, exponential statements may be written in logarithmic form.

Example	Exponential Equation		Logarithmic Equation
	$b^y = x$	is equivalent to	$\log_b x = y$
6	$10^2 = 100$	is equivalent to	$\log_{10} 100 = 2$
7	$3^4 = 81$	is equivalent to	$\log_3 81 = 4$
8	$6^0 = 1$	is equivalent to	$\log_6 1 = 0$
9	$16^{-1/2} = \dfrac{1}{4}$	is equivalent to	$\log_{16} \dfrac{1}{4} = -\dfrac{1}{2}$
10	$5^{-1} = \dfrac{1}{5}$	is equivalent to	$\log_5 \dfrac{1}{5} = -1$

Recall that the composition of a function f and its inverse function f^{-1} acts the same as the identity function. That is, for all x in the domain of f we have $(f \cdot f^{-1})(x) = f[f^{-1}(x)] = x$ and $(f^{-1} \cdot f)(x) = f^{-1}[f(x)] = x$. We know that the exponential and logarithmic functions are inverse functions of each other. Therefore, for $y = b^x$ we have $\log_b y = \log_b b^x = x$, and for $y = \log_b x$ we have $b^y = b^{\log_b x} = x$. Thus, we have the relations

$$\log_b b^x = x \text{ and } b^{\log_b x} = x$$

EXAMPLE 11

$\log_7 7^2 = 2$ because this logarithmic statement is equivalent to the exponential statement $7^2 = 7^2$.

EXAMPLE 12

$10^{\log_{10} 7} = 7$ since this is equivalent to $\log_{10} 7 = \log_{10} 7$.

PROBLEM 1

Solve $\log_3 x = 4$ for x.

Solution: $\log_3 x = 4$ is equivalent to $x = 3^4 = 81$.

PROBLEM 2

If $\log_b \left(\dfrac{1}{8}\right) = -3$, solve for b.

Solution: $\log_b \left(\dfrac{1}{8}\right) = -3$ is equivalent to $\dfrac{1}{8} = b^{-3}$. Therefore, $b = 2$.

PROBLEM 3

If $\log_b 0.01 = -2$, solve for b.

Solution: $\log_b 0.01 = -2$ is equivalent to $0.01 = b^{-2}$. However, $0.01 = 10^{-2}$. Therefore, $10^{-2} = b^{-2}$ and $b = 10$.

PROBLEM 4

Simplify $\log_3 (\log_5 125)$.

Solution: Since $\log_5 125 = 3$, we have $\log_3 (\log_5 125) = \log_3 (3) = 1$.

PROBLEM 5

What is the $\log_b b$?

Solution: Let $x = \log_b b$. Then $x = \log_b b$ is equivalent to $b^x = b$. That is, $b^x = b^1$ and $x = 1$. Thus, $\log_b b = 1$.

4.4 EXERCISES

1. Express each of the following exponential statements in an equivalent logarithmic statement:
   (a) $3^2 = 9$
   (b) $5^2 = 25$
   (c) $4^3 = 64$
   (d) $3^3 = 27$
   (e) $\left(\frac{1}{2}\right)^3 = \frac{1}{8}$
   (f) $\left(\frac{1}{4}\right)^2 = \frac{1}{16}$
   (g) $27^{-1/3} = \frac{1}{3}$
   (h) $32^{-1/5} = \frac{1}{2}$
   (i) $10^3 = 1000$
   (j) $10^{-1} = .1$
   (k) $13^0 = 1$
   (l) $49^{1/2} = 7$
   (m) $e^0 = 1$
   (n) $8^{2/3} = 4$
   (o) $\left(\frac{3}{5}\right)^2 = \frac{9}{25}$
   (p) $b^x = 6$

2. Express each of the following logarithmic statements in an equivalent exponential statement:
   (a) $\log_2 81 = 9$
   (b) $\log_4 \left(\frac{1}{16}\right) = -2$
   (c) $\log_{11} 1 = 0$
   (d) $\log_{1/6} 36 = -2$
   (e) $\log_{10} 0.0001 = -4$
   (f) $\log_e (e^3) = 3$
   (g) $\log_4 256 = 4$
   (h) $\log_2 \sqrt{16} = 2$

3. Solve the following equations for x:
   (a) $\log_2 x = 5$
   (b) $\log_e 1 = x$
   (c) $\log_x 16 = \dfrac{1}{2}$
   (d) $\log_9 \left(\dfrac{1}{3}\right) = x$
   (e) $\log_4 x = -3$
   (f) $\log_x 0.01 = -2$
   (g) $\log_x 10 = \dfrac{1}{2}$
   (h) $\log_{10} x = -1$
   (i) $\log_{1/2} x = -4$

4. Simplify each of the following expressions:
   (a) $\log_4 (\log_9 3)$
   (b) $\log_x (\log_x x)$
   (c) $\log_x (\log_a a^x)$

4.5 PROPERTIES OF LOGARITHMIC FUNCTIONS

The definition of a logarithm involves exponents and many of the rules of exponents yield similar rules for logarithms. These properties allow us to solve complicated problems by transforming multiplication problems into addition problems, division problems into subtraction problems, and exponential problems into multiplication problems. The following theorems give the basic properties of the logarithmic function.

Theorem 6: Let b, M, and N be any positive real numbers and $b \neq 1$. Then

$$\log_b (MN) = \log_b M + \log_b N$$

That is, provided the base remains the same, the logarithm of the product is the sum of the logarithms of the factors.

Proof: Let $x = \log_b M$, $y = \log_b N$, and $z = \log_b MN$. We want to establish a relation between x, y, and z. By the definition of logarithms $x = \log_b M$ if and only if $M = b^x$, $y = \log_b N$ if and only if $N = b^y$, and $z = \log_b MN$ if and only if $MN = b^z$. However, we have $M \cdot N = b^x \cdot b^y = b^{x+y}$. Thus, $M \cdot N = b^z = b^{x+y}$ and therefore we must have $z = x + y$. That is,

$$\log_b MN = \log_b M + \log_b N$$

EXAMPLE 1

$$\log_5 (18) = \log_5 (6 \cdot 3) = \log_5 6 + \log_5 3$$

EXAMPLE 2

$$\log_2 128 = \log_2 (4 \cdot 32) = \log_2 4 + \log_2 32 = 2 + 5 = 7$$

PROBLEM 1

Given $\log_b 7 = x$ and $\log_b 3 = y$, express $\log_b 21$, $\log_b 49$, and $\log_b 27$ in terms of x and y.

Solution: (a) $\log_b 21 = \log_b (7 \cdot 3) = \log_b 7 + \log_b 3 = x + y$
(b) $\log_b 49 = \log_b (7 \cdot 7) = \log_b 7 + \log_b 7 = x + x = 2x$
(c) $\log_b 27 = \log_b (9 \cdot 3) = \log_b 9 + \log_b 3 = \log_b (3 \cdot 3) + \log_b 3 = \log_b 3 + \log_b 3 + \log_b 3 = 3 \log_b 3 = 3y$

Problem 1 (c) illustrates the fact that Theorem 6 can be generalized to more than two factors.

Theorem 7: If $b, M_1, M_2, \ldots, M_n$ are positive real numbers and $b \neq 1$, then $\log_b (M_1 \cdot M_2 \cdots M_n) = \log_b M_1 + \log_b M_2 + \cdots + \log_b M_n$

EXAMPLE 3

$$\log_b (3xy) = \log_b 3 + \log_b x + \log_b y$$

EXAMPLE 4

$$\underbrace{\log_b 4 + \log_b M}_{} + \underbrace{\log_b 5 + \log_b N}_{}$$
$$= \log_b (4 \cdot M) + \log_b (5 \cdot N)$$
$$= \log_b (4 \cdot M \cdot 5 \cdot N) = \log_b 20MN$$

Note: The reader should be careful and recognize that

$$\log_b (M + N) \neq \log_b M + \log_b N$$

We have shown that $\log_b M + \log_b N = \log_b MN$ and, if the above expression were true, then $\log_b (M + N) = \log_b MN$ and $M + N = MN$ for any positive real numbers, and this is definitely not true.

Theorem 8: Let b, M, and N be any positive real numbers and $b \neq 1$. Then

$$\log_b \left(\frac{M}{N} \right) = \log_b M - \log_b N$$

That is, the logarithm of a quotient is the logarithm of the numerator minus the logarithm of the denominator, and the base remains the same.

Proof: Let $x = \log_b M$, $y = \log_b N$, and $z = \log_b \left(\dfrac{M}{N}\right)$. Then we have $M = b^x$, $N = b^y$, and $\left(\dfrac{M}{N}\right) = b^z$. Since $M = b^x$ and $N = b^y$, we also have $\left(\dfrac{M}{N}\right) = \dfrac{b^x}{b^y} = b^{x-y}$. Thus, $b^z = b^{x-y}$ and $z = x - y$. Therefore, $\log_b \left(\dfrac{M}{N}\right) = \log_b M - \log_b N$.

EXAMPLE 5

$$\log_2 \left(\frac{3}{5}\right) = \log_2 3 - \log_2 5$$

EXAMPLE 6

$$\log_b \left(\frac{xy}{z}\right) = \log_b (xy) - \log_b z = \log_b x + \log_b y - \log_b z$$

EXAMPLE 7

$$\log_b \left(\frac{M}{NP}\right) = \log_b M - \log_b (NP) = \log_b M - \{\log_b N + \log_b P\}$$
$$= \log_b M - \log_b N - \log_b P$$

PROBLEM 2

Express $\log_b x - \log_b 3 + \log_b 4$ as a single logarithm.

Solution: We arrange the expression so that all the positive logarithms are together and do the same for the negative logarithms.

$$\log_b x + \log_b 4 - \log_b 3 = \log_b 4x - \log_b 3 = \log_b \left(\frac{4x}{3}\right)$$

PROBLEM 3

Simplify $\log_2 M + \log_7 N - \log_2 P - \log_2 Q + \log_7 R - \log_7 S$.

Solution: Note that not all of bases are the same, and that the theorems apply only for logarithms with the same base. In order to simplify the expression we gather those logarithms with like bases before using the theorems.

$$\log_2 M - \log_2 P - \log_2 Q + \log_7 N + \log_7 R - \log_7 S$$
$$= \log_2 M - (\log_2 P + \log_2 Q) + (\log_7 N + \log_7 R) - \log_7 S$$
$$= \log_2 M - \log_2 (PQ) + \log_7 (NR) - \log_7 S$$
$$= \log_2 \left(\frac{M}{PQ}\right) + \log_7 \left(\frac{NR}{S}\right)$$

Note: Again, the reader should be careful to recognize that

$$\frac{\log_b M}{\log_b N} \neq \log_b M - \log_b N$$

Theorem 9: Let p be any real number and b any real positive number such that $b \neq 1$. Then

$$\log_b M^p = p \cdot \log_b M$$

That is, the logarithm of the pth power of a number is p times the logarithm of the number.

Proof: Let $x = \log_b M$. Then $M = b^x$ and $M^p = (b^x)^p = b^{xp} = b^{px}$. Thus we have $M^p = b^{px}$ and the definition of a logarithm gives $\log_b M^p = px$. Hence, $\log_b M^p = p \log_b M$.

EXAMPLE 8

$$\log_2 17^3 = 3 \log_2 17$$

EXAMPLE 9

$$\log_3 \sqrt{11} = \log_3 11^{1/2} = \frac{1}{2} \log_3 11$$

EXAMPLE 10

$$\log_b \left(\frac{M}{N}\right)^{2/3} = \frac{2}{3} \log_b \left(\frac{M}{N}\right) = \frac{2}{3} \{\log_b M - \log_b N\}$$

EXAMPLE 11

$$\log_4 (3xy^2) = \log_4 3 + \log_4 x + \log_4 y^2 \text{ by Theorem 7}$$
$$= \log_4 3 + \log_4 x + 2 \log_4 y \text{ by Theorem 9}$$

PROBLEM 4

Express $\frac{1}{2} \left\{ \log_b 3 - 5 \log_b x + \frac{3}{2} \log_b 4 \right\}$ as a single logarithm.

Solution: $\frac{1}{2} \left\{ \log_b 3 - 5 \log_b x + \frac{3}{2} \log_b 4 \right\}$

$$= \frac{1}{2} \{\log_b 3 - \log_b x^5 + \log_b 4^{3/2}\}$$

$$= \frac{1}{2}\{\log_b 3 - \log_b x^5 + \log_b 8\} = \frac{1}{2}\left\{\log_b\left(\frac{3\cdot 8}{x^5}\right)\right\}$$

$$= \log_b\left(\frac{24}{x^5}\right)^{1/2} = \log_b\sqrt{\frac{24}{x^5}}$$

PROBLEM 5

Use Theorems 6 to 9 to simplify:

$$\log_5\sqrt{\frac{2z(x+y)^3}{7p^4 s^{1/3}}}$$

Solution: $\log_5\sqrt{\dfrac{2z(x+y)^3}{7p^4 s^{1/3}}} = \log_5\left(\dfrac{2z(x+y)^3}{7p^4 s^{1/3}}\right)^{1/2}$

$$= \frac{1}{2}\log_5\left(\frac{2z(x+y)^3}{7p^4 s^{1/3}}\right) \text{ by Theorem 9}$$

$$= \frac{1}{2}\{\log_5[2z(x+y)^3] - \log_5[7p^4 s^{1/3}]\}$$

$$= \frac{1}{2}\{\log_5 2 + \log_5 z + \log_5(x+y)^3$$

$$- [\log_5 7 + \log_5 p^4 + \log_5 s^{1/3}]\}$$

$$= \frac{1}{2}\{\log_5 2 + \log_5 z + 3\log_5(x+y)$$

$$- \log_5 7 - 4\log_5 p - \frac{1}{3}\log_5 s\}$$

PROBLEM 6

Solve the logarithmic equation

$$\log_{10}(x - 21) + \log_{10} x = 2, \text{ for } x.$$

Solution: Using Theorem 6 we obtain $\log_{10}[x(x - 21)] = 2$, or $\log_{10}(x^2 - 21x) = 2$. Now we convert this logarithmic equation into its exponential form. Thus, $\log_{10}(x^2 - 21x) = 2$ is equivalent to $x^2 - 21x = 10^2$, or,

$$x^2 - 21x - 100 = 0$$
$$(x - 25)(x + 4) = 0$$
$$x = 25 \text{ or } x = -4$$

Recall that the logarithmic function has the positive real numbers as its domain. Therefore, $x = -4$ does not belong to the domain and is not a solution to the logarithmic equation. Thus, the solution is $x = 25$.

Check: $\log_{10}(25 - 21) + \log_{10} 25 = \log_{10} 4 + \log_{10} 25 = \log_{10}(4\cdot 25) = \log_{10} 100 = 2.$

PROBLEM 7

Solve the exponential equation $4^{3x+5} = 7$.

Solution: To solve the exponential equation we use the fact that:

If $M = N$ and $M > 0$, $N > 0$, then $\log_b M = \log_b N$.

Thus,

$$4^{3x+5} = 7$$
$$\log_4 4^{3x+5} = \log_4 7$$
$$(3x + 5) \log_4 4 = \log_4 7$$
$$3x + 5 = \log_4 7, \text{ since } \log_4 4 = 1$$
$$x = \frac{\log_4 7 - 5}{3}$$

PROBLEM 8

Solve for x if $\log_b x = \frac{2}{3} \log_b 8 - \frac{1}{2} \log_b 25 + 3 \log_b 3$.

Solution:

$$\log_b x = \frac{2}{3} \log_b 8 - \frac{1}{2} \log_b 25 + 3 \log_b 3$$

$$\log_b x = \log_b 8^{2/3} - \log_b 25^{1/2} + \log_b 3^3$$
$$\log_b x = \log_b 4 - \log_b 5 + \log_b 27$$

$$\log_b x = \log_b \left(\frac{27 \cdot 4}{5} \right) = \log_b \left(\frac{108}{5} \right)$$

Thus

$$x = \frac{108}{5}$$

This problem makes use of the fact that:

If $\log_b M = \log_b N$, then $M = N$.

4.5 EXERCISES

1. Use Theorems 6 to 9 to simplify the following expressions:

   (a) $\log_b \sqrt{xy}$

   (b) $\log_b 3r^2$

   (c) $\log_b \sqrt[3]{ac^2}$

   (d) $\log_b \left(\dfrac{r^2 s}{t^3} \right)$

   (e) $\log_b u^{-4/5}$

   (f) $\log_b 2\pi \sqrt{\dfrac{l}{g}}$

   (g) $\log_b \sqrt[3]{(x + y)^2 (x - y)^2}$

   (h) $\log_b (\sqrt[4]{x} \cdot \sqrt[3]{y})$

   (i) $\log_b \sqrt[6]{\dfrac{x^{1/3} z^5}{y^2}}$

2. Express each of the following as a single logarithm:

   (a) $\log_b x + \log_b y - \log_b z$

   (b) $\dfrac{1}{4} \log_b x - \dfrac{2}{3} \log_b y$

   (c) $4 \log_b m - \log_b n - 3 \log_b t$

   (d) $\dfrac{1}{3} \{ \log_b A + 3 \log_b C - 2 \log_b D \}$

   (e) $- \log_b M$

   (f) $\log_b (x^2 - 9) - \log_b (x + 3)$

3. Solve the following logarithmic equations:

   (a) $\log_{10} x + \log_{10} 4 = 2$

   (b) $\log_{10} x + \log_{10} (x - 3) = 1$

   (c) $\log_6 (x - 9) + \log_6 x = 2$

4. Solve for x:

   (a) $\log_b x = \dfrac{1}{2} \log_b 36 - 2 \log_b 3$
   $$+ \dfrac{3}{2} \log_b 16$$

   (b) $\log_b x = \log_b 30 + \dfrac{1}{5} \log_b 32$
   $$- \dfrac{2}{3} \log_b 8$$

5. Find x such that:

   (a) $2^x = 5$ (b) $5^{x+1} = 13$

   (c) $4^{-x} = 10$ (d) $3^{x^2} = 7$

   (e) $5^{1-x} = 15$ (f) $e^{x-2} = 4$

4.6 COMPUTATION OF LOGARITHMS

In this section we show how logarithms are used as an aid to computation. Since our number system has 10 as its base, we shall use logarithms to the base 10. A logarithm to the base 10 is called a **common logarithm.** Also, we shall need to compute logarithms to the base e. A logarithm to the base e is called a **natural logarithm.** Natural logarithms are used in such fields as biology, chemistry, business, and physics. Therefore, most of our computations will be with common and natural logarithms. Since the numbers 10 and $e \approx 2.7183$ are used most frequently, we have special notation. By convention, we do not write the base 10, so $\log N$ is written instead of $\log_{10} N$. Logarithms to the base e, $\log_e N$, are written $\ln N$.

It may be necessary to compute logarithms in a base other than 10 or e, or to change from one base to another. Later in this section we shall state a theorem which will demonstrate how this change of bases can be achieved.

We saw in Section 4.4 that $10^{\log_{10} N} = N$, for $N > 0$. That is, log N is the exponent of 10 such that the result is N. In this section we are interested in finding log N. When N is an integral power of 10, log N is easily determined. Consider the following list of integral powers of 10 and their corresponding logarithmic form:

$$\log_{10} 100 = 2 \text{ is equivalent to } 10^2 = 100$$

$$\log_{10} 10 = 1 \text{ is equivalent to } 10^1 = 10$$

$$\log_{10} 1 = 0 \text{ is equivalent to } 10^0 = 1$$

$$\log_{10} 0.1 = -1 \text{ is equivalent to } 10^{-1} = 0.1$$

$$\log_{10} 0.01 = -2 \text{ is equivalent to } 10^{-2} = 0.01$$

Note that $\log_{10} 100 = \log_{10} 10^2 = 2$; $\log_{10} 0.1 = \log_{10} 10^{-1} = -1$. That is, when N is an integral power of 10, the logarithm of N is the exponent of 10. However, most values of N are not integral powers of 10. Therefore, we must use a table to compute log N. We shall show that if we can find log N for $1 \leq N < 10$, then log N can be determined for any $N > 0$. Consider the partial table showing some values of log N for $1 \leq N < 10$ (Fig. 115). The common logarithm table, Table A, is given in Appendix III.

N	0	1	2	3	4	5	6	7	8	9
4.8	.6812	.6821	.6830	.6839	.6848	.6857	.6866	.6875	.6884	.6893
4.9	.6902	.6911	.6920	(.6928)	.6937	.6946	.6955	.6964	.6972	.6981
5.0	.6990	.6998	.7007	.7016	.7024	.7033	.7042	.7050	.7059	.7067
5.1	.7076	.7084	.7093	.7101	.7110	.7118	.7126	.7135	.7143	.7152
5.2	.7160	.7168	.7177	.7185	.7193	.7202	.7210	(7218)	.7226	.7235
5.3	.7243	.7251	.7259	.7267	.7275	.7284	.7292	.7300	.7308	.7316

Figure 115

PROBLEM 1

Find log 5.27.

Solution: We locate the row containing 5.2 and then move to the column headed 7. Thus log 5.27 = 0.7218.

PROBLEM 2

Find log 4.93.

Solution: We locate the intersection of the row containing 4.9 under N and the column containing 3. Therefore, log 4.93 = 0.6928.

Note that even though we have used an equals sign, the logarithms we found were only approximations. The reason for this is that most logarithms are irrational and cannot be represented by a terminating decimal.

Until now, we have limited ourselves to finding $\log N$, where $1 \leq N < 10$ (and integral powers of 10). How do we compute $\log N$ when $0 < N < 1$ or $N > 10$? We shall show that this can be achieved by representing the number N in **scientific notation**—that is, as the product of a number between 1 and 10 and a power of 10.

Definition: Any positive number N can be represented as $N = A \cdot 10^c$, where $1 \leq A < 10$ and c is an integer. N is said to be written in **scientific notation.**

PROBLEM 3

Write $N = 732$ in scientific notation.

Solution: $N = 732 = (7.32)(100) = (7.32) \cdot 10^2$

PROBLEM 4

Express $N = 0.0034$ in scientific notation.

Solution: $N = 0.0034 = (3.4) \cdot 10^{-3}$

PROBLEM 5

Write $N = 2.73$ using scientific notation.

Solution: $N = 2.73 = (2.73) \cdot 1 = (2.73) \cdot 10^0$

Theorem 10: If N is any positive real number written in its scientific notation, $N = A \cdot 10^c$, where $1 \leq A < 10$ and c is an integer, then

$$\log N = \log(A \cdot 10^c) = c + \log A$$

Proof:
$$
\begin{aligned}
\log(A \cdot 10^c) &= \log A + \log 10^c && \text{by Theorem 6} \\
&= \log A + c \log 10 && \text{by Theorem 9} \\
&= \log A + c(1) && \text{since } \log 10 = 1 \\
&= \log A + c \\
&= c + \log A
\end{aligned}
$$

PROBLEM 6

Find log 527.

Solution: Writing 527 in scientific notation we get $(5.27) \cdot 10^2$. Thus, log 527 = log$[(5.27) \cdot 10^2]$ = 2 + log 5.27 = 2 + 0.7218. Note that $c = 2$ and $A = 5.27$ and the log 5.27 was found in Problem 1.

PROBLEM 7

Find log 0.0485.

Solution: log 0.0485 = log$[(4.85) \cdot 10^{-2}]$ = -2 + 0.6857

This technique of writing the logarithm of any positive real number as an integral part, c, and a nonnegative decimal fraction, log A, leads to the following definition:

Definition: When writing log N = log$(A \cdot 10^c)$ = c + log A, the integer c is called the **characteristic** of log N and the decimal fraction log A is called the **mantissa** of log N.

EXAMPLE 1

In Problem 6, log 527 has characteristic $c = 2$ and mantissa log A = log 5.27 = 0.7218.

EXAMPLE 2

In Problem 7, log 0.0485 has characteristic $c = -2$ and mantissa log A = log 4.85 = 0.6857.

EXAMPLE 3

log 2.73 = log$[(2.73) \cdot 10^0]$ = 0 + log 2.73 = 0 + 0.4362. Thus, characteristic $c = 0$ and the mantissa is 0.4362. The mantissa can be found in Table A of Appendix III.

PROBLEM 8

Find log 0.0005.

Solution: log 0.0005 = log$(5 \cdot 10^{-4})$ = -4 + log 5 = -4 + 0.6990. The characteristic $c = -4$ and the mantissa is 0.6990.

Note that the characteristic and the mantissa are written separately. This is done for computational reasons that will be explained later.

Until now, we have been given a positive real number N and found log N. However, suppose we are given log N and told to find N. We have solved a few such problems in previous sections. For example, if log $N = 2$ then $N = 10^2 = 100$. Suppose log $N = 1 + 0.7016$ and we want to find N. The procedure is basically the same as before, with the help of the logarithm table. If log $N = 1 + 0.7016$, then $N = 10^{1+0.7016}$. Using the properties of exponents we get

$$N = 10^{1+7016} = 10^1 \cdot 10^{0.7016} = 10 \cdot (5.03) = 50.3,$$

where $10^{0.7016}$ was obtained from the table given in Figure 115.

Definition: Given a number x. The number N, such that log $N = x$, is called the **antilogarithm** of x and is written as $N =$ antilog x.

EXAMPLE 4

As shown above, if log $N = 1 + 0.7016$, then the antilog $N = 50.3$.

PROBLEM 9

Find N if log $N = -2 + 0.6964$.

Solution: Using the table in Figure 115, we have log $4.97 = 0.6964$. By Theorem 10, we obtain

$$\begin{aligned} \log N &= -2 + 0.6964 \\ &= \log[(4.97) \cdot 10^{-2}] \\ &= \log .0497 \end{aligned}$$

Thus,

$$N = .0497$$

PROBLEM 10

Find N if log $N = 3.9117$.

Solution: First we write log $N = 3.9117$ in standard form—that is, as the sum of an integer and a nonnegative decimal fraction. log $N = 3.9117 = 3 + .9117$. Table A of Appendix III gives $.9117 = $ log 8.16. Thus, Theorem 10 gives

$$\begin{aligned} \log N &= 3 + .9117 \\ &= \log[(8.16) \cdot 10^3] \\ &= \log 8160 \\ N &= 8160 \end{aligned}$$

Linear Interpolation

So far we have computed logarithms and antilogarithms which we could find in Table A of Appendix III. What happens if we cannot find such values in the table? For example, we used Table A to find logarithms of N when N had three digits. However, what occurs if N has four or more significant digits? Suppose we want to find log 5.235. Similarly, what if we were looking for the antilogarithm and there were no value in the table corresponding to the mantissa? For example, find N if log N = 1.7206. These problems can be accomplished by using an approximation method called **linear interpolation.** We illustrate linear interpolation by finding log 5.235. We begin by looking at the graph of y = log N (see Fig. 116).

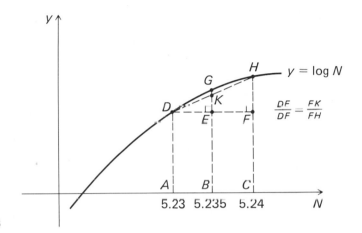

Figure 116

We shall find the value of log 5.235 by approximating the curve y = log N by the line DH. The distance BG = log 5.235, and we approximate BG using the distance DK. Therefore, an approximation of log 5.235 can be obtained by finding the distance BK. The distance AD = log 5.23 and the distance CH = log 5.24, and these values can be found in the table given in Figure 115: 0.7185 and 0.7193, respectively. Note that the triangles DEK and DFH are similar right triangles. Therefore, the sides of these triangles are proportional. That is, $\dfrac{DE}{DF} = \dfrac{EK}{FH}$.

We now use this information to linearly approximate log 5.235:

(a) Find the characteristic of 5.235. Writing 5.235 in scientific notation we have 5.235 = (5.235) · 10^0 and the characteristic c = 0.

(b)

$$
.010 \begin{bmatrix} 5.230 \\ 5.235 \\ 5.240 \end{bmatrix} .005 \qquad d \begin{bmatrix} .7185 \\ m \\ .7193 \end{bmatrix} .0008
$$

$$
\frac{.005}{.010} = \frac{d}{.0008} \quad \text{because} \quad \frac{DE}{DF} = \frac{EK}{FH}
$$

$$
d = \left(\frac{.005}{.010}\right)(.0008) = 0.0004
$$

(c) Thus, m = log 5.235 = log 5.230 + d = .7185 + .0004 = .7189

PROBLEM 11

Find log 0.004236.

Solution: (a) Since $0.004236 = (4.236) \cdot 10^{-3}$, the characteristic is -3.

(b)

$$
\begin{array}{ccc}
N & & \log N \\
\end{array}
$$

$$
.010\begin{bmatrix} 4.230 \\ 4.236 \\ 4.240 \end{bmatrix}.006 \qquad d\begin{bmatrix} .6263 \\ m \\ .6274 \end{bmatrix}.0011
$$

$$
\frac{.006}{.010} = \frac{d}{.0011} \text{ or } d = \frac{3}{5}(.0011) \approx .0007
$$

(c) Thus, $m = \log 4.236 = .6263 + .0007 = .6270$

PROBLEM 12

Find N if $\log N = 1.7206$.

Solution: (a) $\log N = 1.7206 = 1 + .7206$. Therefore, the characteristic is 1 and the mantissa is .7206.

(b) Find the antilog of the mantissa 0.7206. The table in Figure 115 gives

$$
\begin{array}{ccc}
N & & \log N \\
\end{array}
$$

$$
.010\begin{bmatrix} 5.250 \\ k \\ 5.260 \end{bmatrix}d \qquad .0004\begin{bmatrix} .7202 \\ .7206 \\ .7210 \end{bmatrix}.0008
$$

$$
\frac{d}{.010} = \frac{.0004}{.0008} \text{ or } d = \frac{1}{2}(.010) = .005
$$

(c) Thus, $k = 5.250 + 0.005 = 5.255$

(d) Finally, $\log N = 1 + \log 5.255$

$$
\begin{aligned}
&= \log[(5.255) \cdot 10^1] \\
&= \log 52.55 \\
N &= 52.55
\end{aligned}
$$

Logarithms Base e (Natural Logarithms)

In many scientific fields, the natural logarithms are more important than common logarithms. This follows from the fact that many natural phenomena are expressed by the exponential e^x and the natural logarithm $\ln x$ is the inverse of such expressions. Table B in Appendix III lists the values of $\ln N$, for $1 \leq N < 10$.

PROBLEM 13

Use natural logarithms to solve $e^{2t} = 5.2$ for t.

Solution: Taking the natural logarithm of both sides of the equation we obtain $\ln e^{2t} = \ln 5.2$, so that $2t \ln e = \ln 5.2$. However, $\ln e = 1$ and the equation becomes $t = \dfrac{\ln 5.2}{2}$. Using Table B,

$$t = \frac{1.6487}{2} = .8243$$

As mentioned above, Table B gives $\ln N$ for $1 \leq N < 10$. When computing $\ln N$ for $N \geq 10$ or $0 < N < 1$ we must be careful because the natural logarithms of integral powers of 10 are *not* integers. Once again, we use scientific notation to find the natural logarithm for any N.

To find $\ln N$, write N in scientific notation: $N = A \cdot 10^n$ where $1 \leq A < 10$. Then

$$\ln N = \ln (A \cdot 10^n) = \ln A + n \cdot \ln 10$$

Since $\ln 10 = 2.3026$, we have $\ln N = \ln A + n(2.3026)$.

EXAMPLE 5

(a) $\ln 3.26 = 1.1817$
(b) $\ln 32.6 = \ln (3.26 \cdot 10) = \ln 3.26 + \ln 10 = 1.1817 + 2.3026$
$= 3.4843$
(c) $\ln 3260 = \ln (3.26 \cdot 10^3) = \ln 3.26 + 3 \ln 10$
$= 1.1817 + 3(2.3026) = 1.1817 + 6.9078$
$= 8.0895$
(d) $\ln .0326 = \ln (3.26 \cdot 10^{-2}) = 1.1817 - 2(2.3026)$
$= 1.1817 - 4.6052 = -3.4235$

To find N, given $\ln N$, we do the following:
(1) If $0 \leq \ln N < 2.3026$, then we merely look in Table B.
(2) If $\ln N$ does not lie in the above interval, then we first express the given value $\ln N$ in the form $b + n(2.3026)$, where $0 < b < 2.3026$ and n is some integer. Thus, N is 10^n times the anti-ln of b.

PROBLEM 14

Find N if $\ln N = 2.2576$.

Solution: Note that $0 \le \ln N < 2.3026$. That is, $0 \le 2.2576 < 2.3026$ and we use Table B to find N. Thus, $N = 9.56$.

PROBLEM 15

Find N if $\ln N = 4.0092$.

Solution: We write 4.0092 in the form $b + n(2.3026)$. Thus $4.0092 = 1.7066 + 1(2.3026)$, and the anti-ln of 1.7066 is 5.51. Therefore, $N = 5.51 \cdot 10^1 = 55.1$.

PROBLEM 16

Find N if $\ln N = -2.4024$.

Solution: $\ln N = -2.4024 = 2.2028 - 2(2.3026)$. The anti-ln of 2.2028 is 9.05. Thus, $N = 9.05 \cdot 10^{-2} = .0905$.

Change of Bases

We now find a method that enables us to go from $\log_a N$ to $\log_b N$. Recall that for $a > 0$, $a \ne 1$, and $N > 0$,

$$N = a^{\log_a N}$$

However, we know that if $A = B$, where $A > 0$ and $B > 0$, then $\log_b A = \log_b B$. Applying this principle we obtain

$$\log_b N = \log_b a^{\log_a N}$$

and, by Theorem 9, Section 4.5,

$$\log_b N = \log_a N \cdot \log_b a$$

or

$$\log_a N = \frac{\log_b N}{\log_b a}$$

We can now state a theorem relating logarithms with different bases.

> **Theorem 11:** Let a and b be positive real numbers, and $a \neq 1$ and $b \neq 1$.
> Then
>
> $$\log_a N = \frac{\log_b N}{\log_b a}$$

PROBLEM 17

Express ln N in terms of the common logarithm.

Solution: $\ln N = \dfrac{\log N}{\log e} = \dfrac{\log N}{\log(2.718)} = \dfrac{\log N}{0.4343}$

$\qquad\qquad = \dfrac{1}{0.4343} \cdot \log N$

Thus,

$$\ln N = 2.303 \log N$$

PROBLEM 18

Find $\log_3 8.2$.

Solution: $\log_3 8.2 = \dfrac{\log 8.2}{\log 3} = \dfrac{0.9138}{0.4771} \approx 1.92$

PROBLEM 19

Find $\log_a b$ for $a > 0$, $b > 0$, and $a \neq 1$, $b \neq 1$, $\log_b a \neq 0$.

Solution: Theorem 11 gives $\log_a b = \dfrac{\log_b b}{\log_b a} = \dfrac{1}{\log_b a}$. Thus,

$$\log_a b = \frac{1}{\log_b a}$$

4.6 EXERCISES

1. Write the characteristic of the logarithm of each of the following numbers:
   (a) log 738
   (b) log 73.8
   (c) log 7.38
   (d) log .738
   (e) log $(13 \cdot 10^4)$
   (f) log $(9 \cdot 10^0)$
   (g) log $(5 \cdot 10^{-2})$

2. Find the logarithm of each of the following numbers:
   (a) log .0738
   (b) log 23.4
   (c) log 892
   (d) log .000316
   (e) log $(14 \cdot 10^7)$
   (f) log $(3 \cdot 10^2)$

3. Find N if:
    (a) $\log N = 0.9814$
    (b) $\log N = 1.8488$
    (c) $\log N = -1 + 0.7803$
    (d) $\log N = -2 + 0.7042$
    (e) $\log N = 3.4900$
    (f) $\log N = 0.0043$

5. Find N if:
    (a) $\log N = 3.5495$
    (b) $\log N = -1 + 0.9045$
    (c) $\log N = 0.8410$

7. Find N if:
    (a) $\ln N = 4.5218$ (b) $\ln N = -3.4999$

9. (a) Given $\log 9 = 0.9542$ find $\log_9 10$.
    (b) Show that $\log_4 5 = \dfrac{1}{2} \log_2 5$. (Hint:

    $\log_4 5 = \dfrac{1}{\log_5 4}$ and $4 = 2^2$.)

4. Find each logarithm using linear inter-
    polation.
    (a) $\log 13.57$
    (b) $\log 452.3$
    (c) $\log .2634$

6. Find each of the following natural loga-
    rithms using Table B, Appendix III:
    (a) $\ln 2.4$
    (b) $\ln 38.6$
    (c) $\ln (7.25e^3)$

8. Using Theorem 11, find:
    (a) $\log_2 7.36$
    (b) $\log_5 21$
    (c) $\log_{13} 437$

10. Show that $\ln (\ln N) = \ln (\log N) - \ln (\log e)$.

4.7 APPLICATIONS OF LOGARITHMIC FUNCTIONS

Logarithms can be used to help solve many problems in such areas as compound interest, biology, physics, and other fields. Despite the present trend
toward high-speed calculators, the techniques used in the following problems
give more insight into the properties of the logarithmic function.

PROBLEM 1

Using logarithms, compute $N = \sqrt{.7284}$.

Solution: Since $N = \sqrt{.7284} = (.7284)^{1/2}$, we have $\log N = \dfrac{1}{2} \log$

.7284. Using linear interpolation, we find $\log N = \dfrac{1}{2} (-1 + 0.8623)$.

Before multiplying by $\dfrac{1}{2}$, we write the characteristic -1 as $19 \cdot \cdot \cdot -20$.
This makes for easier calculations and, more importantly, the decimal
part will remain positive. The mantissa of any logarithm is always positive or zero. Therefore, $\log N = \dfrac{1}{2} (19.8623 - 20) = 9.9312 - 10 = -1 +$
0.9312. Finding the antilog we obtain $N = .8535$.

Since the logarithmic function is the inverse of the exponential function, we can use logarithms to evaluate more precisely the numbers we obtained in the examples of Section 4.3, Applications of Exponential Functions.

PROBLEM 2

Evaluate the compound interest in Problem 1 of Section 4.3. That is $A = 100(1 + .08)^{10}$.

Solution: If $A = 100(1.08)^{10}$, then $\log A = \log 100 + 10 \log (1.08) = 2 + 10(0.0334) = 2.3340$. Thus, $\log A = 2.3340$. Using antilogs, we find that $A = 215.8$.

PROBLEM 3

Using natural logarithms, evaluate the population of the bacteria in Problem 4 of Section 4.3. That is, $N = 10^3 \cdot e^3$.

Solution: If $N = 10^3 e^3$, then $\ln N = 3 \ln 10 + 3 \ln e$. However, $\ln 10 = 2.3026$ and $\ln e = 1$ gives $\ln N = 3(2.3026) + 3 = 9.9078$. Again, we use antilogarithms to find N. First, we write $\ln N = 9.9078$ in the form $\ln N = b + n(2.3026)$, where n is an integer and $0 \le b < 2.3026$. Therefore, $\ln N = 9.9078 = 0.6974 + 4(2.3026)$, and $n = 4$, $b = 0.6974$. Thus, $N = 10^4$ (anti-ln of 0.6974) $= 10^4 (2.0085) = 20,085$.

PROBLEM 4

The period T of a simple pendulum of length l is given by the formula $T = (2\pi) \sqrt{\dfrac{l}{g}}$, where g is the acceleration due to gravity. Find T, in seconds, if $l = 632.8$ cm and $g = 981$ cm/sec².

Solution: We have $2\pi = 2(3.1416) = 6.283$. Therefore,

$$T = (2\pi) \sqrt{\frac{l}{g}} = 6.283 \sqrt{\frac{632.8}{981}}$$

and

$$\log T = \log 6.283 + \frac{1}{2}(\log 632.8 - \log 981)$$

$$\log 6.283 \qquad\qquad = 0.7982$$

$$(+) \frac{1}{2}(\log 632.8) = \frac{1}{2}(2.8013) \quad = 1.4007$$

$$\overline{\qquad\qquad\qquad 2.1989}$$

$$(-) \frac{1}{2} (\log 981) = \frac{1}{2} (2.9917) \qquad = \underline{1.4959}$$

$$\log T = 0.7030$$

Thus, $T = 5.047$ seconds.

4.7 EXERCISES

1. Calculate each of the following using logarithms:
   (a) $\sqrt[3]{0.0214}$
   (b) $(7.284)^5$
   (c) $(1881)^{-4/5}$

2. Compute, using logarithms,
$$N = \frac{(96.15)\sqrt[4]{468}}{0.3132}.$$

3. Find the period of a pendulum of length 3.51 feet. Use $g = 32.2$ ft/sec².

4. Let the number of bacteria N present at a certain time be given by $N = N_0 e^{2t}$, where N_0 is the number of bacteria present when $t = 0$. If t is measured in hours, how long will it take for the number of bacteria to double?

5. The decay of a radioactive material is given by $A = A_0 e^{-kt}$, where A represents the amount present after t years, A_0 represents the initial amount, and k is a constant. Recall that the half-life is defined to be the time required for a given amount of material to decrease by one half. Find k if the half-life is 25 years.

SUMMARY

Exponents

If $a > 0$, $b > 0$, and x, y are any real numbers,
(i) $a^x \cdot a^y = a^{x+y}$
(ii) $(a^x)^y = a^{xy}$
(iii) $\dfrac{a^x}{a^y} = a^{x-y}$
(iv) $(ab)^x = a^x b^x$
(v) $\left(\dfrac{a}{b}\right)^x = \dfrac{a^x}{b^x}, b \neq 0$

Exponential Function

1. Let a be any real number such that $a > 0$. Then the function f defined by $f(x) = a^x$, $a \neq 1$, is called an **exponential function** with base a.
2. If $a > 1$, the exponential function is an **increasing function.** If $0 < a < 1$, then it is a **decreasing function.**

Logarithmic Function

1. Let f be the exponential function defined by $f(x) = b^x$, $b \neq 1$. The inverse function f^{-1} defined by $f^{-1}(x) = \log_b x$ is called the **logarithmic function** with base b.
2. If b is any positive real number, where $b \neq 1$, then

$$y = \log_b x \text{ if and only if } x = b^y$$

3. $\log_b b^x = x$ and $b^{\log_b x} = x$.

Properties of Logarithmic Functions

1. Let b, M, and N be any positive real numbers, and p any real number. Then
   (i) $\log_b (M \cdot N) = \log_b M + \log_b N$

   (ii) $\log_b \left(\dfrac{M}{N}\right) = \log_b M - \log_b N$

   (iii) $\log_b M^p = p \cdot \log_b M$
2. If $M = N$ and $M > 0$, $N > 0$, then $\log_b M = \log_b N$.
3. If $\log_b M = \log_b N$, then $M = N$.

Computation of Logarithms

1. LOGARITHMS BASE 10:
COMMON LOGARITHMS (log)

   If N is any positive real number written in its scientific notation, $N = A \cdot 10^c$, where $1 \leq A < 10$ and c is an integer, then $\log N = \log(A \cdot 10^c) = c + \log A$. The integer c is called the **characteristic** of $\log N$ and the decimal fraction $\log A$ is called the **mantissa** of $\log N$. Given $\log N$, a table can also be used to find N. N is called the **antilog** of $\log N$. Thus, $N = \text{antilog} (\log N)$.

2. LOGARITHMS BASE e:
NATURAL LOGARITHMS (ln)

   To find $\ln N$, write N in scientific notation: $N = A \cdot 10^n$ where $1 \leq A < 10$. Then

$$\ln N = \ln (A \cdot 10^n) = \ln A + n \ln 10 = \ln A + n(2.3026)$$

   To find N, given $\ln N$, do the following:
   (i) If $0 \leq \ln N < 2.3026$, look in Table B, Appendix III.
   (ii) If $\ln N$ does not lie in the above interval, we first express the given value $\ln N$ in the form $b + n(2.3026)$, where $0 < b < 2.3026$ and n is some integer. Thus, N is 10^n times the anti-ln of b.

Change of Bases

Let a and b be positive real numbers, and $a \neq 1$, $b \neq 1$. Then

$$\log_a N = \frac{\log_b N}{\log_b a}$$

CHAPTER 4 EXERCISES

1. Write the following in logarithmic form:
   (a) $3^4 = 81$ (b) $b^0 = 1$
   (c) $8^{2/3} = 4$ (d) $a^x = 16$
   (e) $4^{-2} = \dfrac{1}{16}$ (f) $5^y = x$

2. Write the following in exponential form:
   (a) $\log_2 8 = 3$
   (b) $\log_5 25 = 2$
   (c) $\log_b 1 = 0$

3. Solve each of the following exponential equations:
   (a) $\left(\dfrac{1}{3}\right)^x = 27$
   (b) $10^x = 7$
   (c) $(1.52)^x = 4$

4. Solve each of the following logarithmic equations:
   (a) $\log_5 x = -2$
   (b) $\log_x 49 = 2$
   (c) $\log_{25}\left(\dfrac{1}{5}\right) = x$

5. Write the following as a single logarithm:
   (a) $2 \log x - 3 \log y + \log z$
   (b) $\left(\dfrac{1}{2}\right) \log (x + y) - \left(\dfrac{1}{2}\right) \log (x - y)$

6. Express as the sum and/or difference of logarithms:
   (a) $\log \left(\sqrt[5]{\dfrac{x^2 y^{-2/3}}{(x + y)}} \right)$
   (b) $\log (\sqrt{(x - a)(x - b)(x - c)})$

7. Solve $2^{x-4} = 10$.

8. Solve $\log (x + 2) + \log (x) = 2$.

9. Compute $\sqrt{\dfrac{(310.6)^2(16)(10^2)}{(7.29)(10^3)}}$.

10. Compute: (a) ln 4.36; (b) ln 0.0075.

11. Find N if: (a) ln $N = 2.1126$; (b) ln $N = 8.0064$.

12. Use the change of bases rule to compute:
    (a) $\log_7 5$ (b) $\log_2 37$ (c) ln 10

13. If the population of N_0 people grows at a constant rate of r per cent per year, then there will be $N = N_0(I + r)^t$ people in the population in t years. If the population grows at a rate of 3% a year, then how long will it take for the population to triple?

14. Suppose that $2000 is compounded continuously at a rate of 5 per cent. How much money will we have at the end of six years?

CHAPTER 5

CIRCULAR FUNCTIONS

Trigonometry deals with the measurements of triangles and problems involving triangles. The trigonometric functions are usually expressed using the various ratios of the sides of right triangles. However, the ideas developed in trigonometry can be viewed from a functional approach—in particular, "circular functions." We begin by constructing the "winding function."

5.1 THE WINDING FUNCTION

Consider the unit circle with center at the origin. An equation of this circle is $x^2 + y^2 = 1$. Recall that any circle centered at the origin is symmetric with respect to the x-axis, the y-axis, and the origin. Thus, symmetry enables us to find points on the circle.

EXAMPLE 1

Consider the point $\left(\frac{\sqrt{2}}{2}, \frac{\sqrt{2}}{2}\right)$ on the unit circle $x^2 + y^2 = 1$ (see Fig. 117).

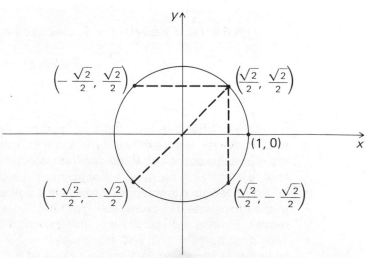

The reflection about the x-axis is $\left(\dfrac{\sqrt{2}}{2}, \dfrac{-\sqrt{2}}{2}\right)$.

The reflection about the y-axis is $\left(\dfrac{-\sqrt{2}}{2}, \dfrac{\sqrt{2}}{2}\right)$.

The reflection about the origin is $\left(\dfrac{-\sqrt{2}}{2}, \dfrac{-\sqrt{2}}{2}\right)$.

Until now, we have restricted outselves to discussing distances measured along a straight line. However, it seems possible that between any two points A and B on a circle there should exist a real number that corresponds to the length of the arc $\overset{\frown}{AB}$. Consider any real number $\theta > 0$, and let $P(\theta)$ be that point on the unit circle which is obtained when an arc of length θ is wrapped around the unit circle from the point $(1, 0)$ in the *counterclockwise* direction (see Fig. 118).

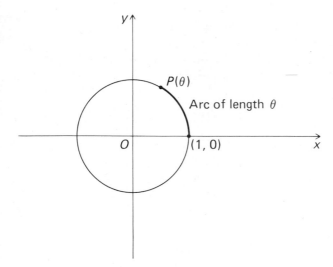

Figure 118

Note that if we take $\theta = 0$, then we have

$$P(0) = (1, 0)$$

This relation between numbers and points on the unit circle defines a function. However, we want this function P to have all the real numbers as its domain and the collection of all ordered pairs on the unit circle as its range. Therefore, we let $P(-\theta)$ represent that point on the unit circle which is obtained when an arc of length θ is wrapped around the unit circle starting from the point $(1, 0)$ in the clockwise direction (see Fig. 119). Thus, if the point P starts at $(1, 0)$ and moves $|\theta|$ units around the circumference of the unit circle in a counterclockwise direction if $\theta > 0$, and clockwise if $\theta < 0$, then we can determine the exact position of the point $P(\theta)$ for any value of θ.

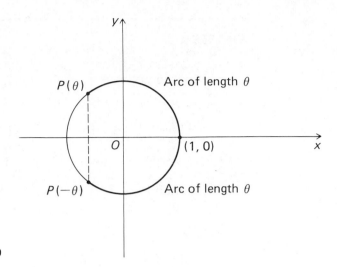

Figure 119

Definition: The function P, defined by $P(\theta) = (x, y)$, where θ is any real number and (x, y) is a point on the unit circle, is called a **winding function.** That is, $P : \theta \rightarrow P(\theta)$ where $P(\theta) = (x, y)$. The domain of P is the set of all real numbers and the range of P is $\{(x, y) | x^2 + y^2 = 1\}$.

Recall that the circumference of a circle of radius r is $2\pi r$. Thus, the unit circle has circumference 2π. If we start at the point $(1, 0)$ and travel counterclockwise around the circle, then we will travel a distance of 2π. Similarly, halfway around the circle will be a distance of π; one fourth of the way is a distance of $\frac{2\pi}{4} = \frac{\pi}{2}$ (see Fig. 120A, B, and C).

Note: $P(2\pi) = P(0) = (1, 0)$

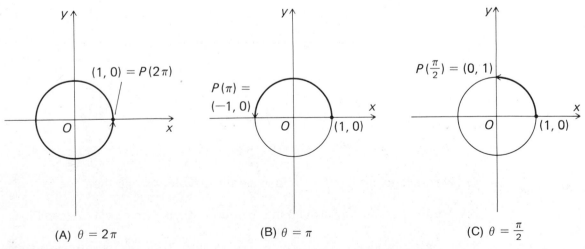

Figure 120

PROBLEM 1

Plot the points $P\left(-\dfrac{\pi}{2}\right)$, $P(-\pi)$, and $P\left(-\dfrac{3\pi}{2}\right)$.

Solution: $P\left(-\dfrac{\pi}{2}\right)$ is located one fourth of the way around the circle in a *clockwise* direction. Thus, $P\left(-\dfrac{\pi}{2}\right) = (0, -1)$. Similarly, $P(-\pi) = (-1, 0)$, and $P\left(-\dfrac{3\pi}{2}\right) = (1, 0)$ (see Fig. 121A, B, and C).

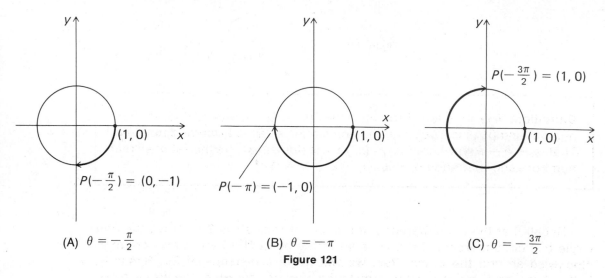

(A) $\theta = -\dfrac{\pi}{2}$ (B) $\theta = -\pi$ (C) $\theta = -\dfrac{3\pi}{2}$

Figure 121

Note: $P(-\pi) = P(\pi) = (-1, 0)$ and $-\pi = \pi - 2\pi$

$$P\left(-\dfrac{3\pi}{2}\right) = P\left(\dfrac{\pi}{2}\right) = (0, 1) \text{ and } -\dfrac{3\pi}{2} = \dfrac{\pi}{2} - 2\pi$$

What occurs when $|\theta| > 2\pi$?

PROBLEM 2

Plot $P\left(\dfrac{5\pi}{2}\right)$ and $P\left(-\dfrac{7\pi}{2}\right)$.

Solution: Since $\dfrac{5\pi}{2} = 2\pi + \dfrac{\pi}{2}$, then in order to travel a length of $\theta = \dfrac{5\pi}{2}$, we must go around the circle one entire revolution and then one fourth $\left(\text{or } \dfrac{\pi}{2} \text{ units}\right)$ of the way around in the counterclockwise direction (see Fig. 122A). Also, since $-\dfrac{7\pi}{2} = -2\pi - \dfrac{3\pi}{2}$, to find $P\left(-\dfrac{7\pi}{2}\right)$ we travel in a

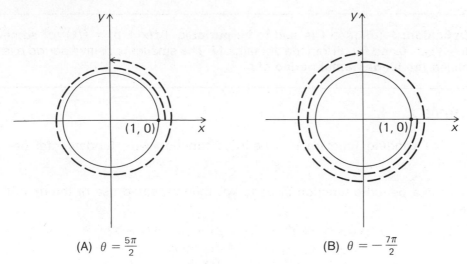

(A) $\theta = \dfrac{5\pi}{2}$ (B) $\theta = -\dfrac{7\pi}{2}$

Figure 122

clockwise direction around the circle once and then continue three fourths of the way around (see Fig. 122*B*).

Note: $P\left(\dfrac{5\pi}{2}\right) = P\left(-\dfrac{7\pi}{2}\right) = P\left(\dfrac{\pi}{2}\right) = (0, 1),$

$\dfrac{5\pi}{2} = 2\pi + \dfrac{\pi}{2},$ and $-\dfrac{7\pi}{2} = -2\pi - \dfrac{3\pi}{2}$

Problems 1 and 2 show us that $P\left(\dfrac{\pi}{2}\right) = P\left(-\dfrac{3\pi}{2}\right) = P\left(\dfrac{5\pi}{2}\right) =$ $P\left(-\dfrac{7\pi}{2}\right) = (0, 1)$. That is, $P\left(\dfrac{\pi}{2}\right) = P\left(\dfrac{\pi}{2} + 2n\pi\right)$, where $n = 0, \pm 1, \pm 2, \ldots$. By letting $n = 0, -1, 1,$ and -2, respectively, we get $P\left(\dfrac{\pi}{2}\right), P\left(-\dfrac{3\pi}{2}\right),$ $P\left(\dfrac{5\pi}{2}\right),$ and $P\left(-\dfrac{7\pi}{2}\right)$. Since the circumference of the unit circle is 2π, we have $P(\theta) = P(\theta + 2n\pi)$, where $n = 0, \pm1, \pm2, \ldots$. Thus, the winding function P associates more than one real number with the same point on the unit circle. For example, $P(0) = P(2\pi) = P(4\pi) = (1, 0)$. Also, $P(0) = P(-2\pi) = P(-4\pi) = (1, 0)$. Thus, the function values of P repeat every time we travel 2π units on the circle. We say that P is a **periodic function** with a period of 2π. That is,

$P(\theta) = P(\theta + 2n\pi), n = 0, \pm1, \pm2, \ldots$ and for all θ.

The winding function is just an example of a larger group of functions which have the property that they repeat in value at fixed intervals.

Definition: A function f is said to be **periodic** if $f(x + p) = f(x)$ for some fixed $p > 0$ and for all x in the domain of f. The smallest positive period p is called the **fundamental period** of f.

EXAMPLE 2

The winding function is a periodic function with fundamental period 2π.

If f is a periodic function of period p, then repeated use of the definition gives

$$f(x) = f(x + p) = f[(x + p) + p] = f(x + 2p)$$
$$= f[(x + 2p) + p] = f(x + 3p)$$
$$= f[(x + 3p) + p] = f(x + 4p)$$

Continuing this, we have the following theorem which is proved using mathematical induction as described in Appendix II.

Theorem 1: If f is a periodic function with fundamental period p, then $f(x + np) = f(x)$ for all integers n.

The fact that Theorem 1 holds for negative integers can be seen from

$$f(x - p) = f[(x - p) + p] = f(x)$$
$$f(x - 2p) = f[(x - 2p) + p] = f(x - p) = f(x)$$

PROBLEM 3

Suppose f is a periodic function with fundamental period 5. Use Theorem 1 to determine the values of: (a) $f(6)$ (b) $f(-2)$ (c) $f(13.5)$ (d) $f(-7)$

Solution: (a) $f(6) = f(1 + 5) = f(1)$
(b) $f(-2) = f(-2 + 5) = f(3)$
(c) $f(13.5) = f(3.5 + 2 \cdot 5) = f(3.5)$
(d) $f(-7) = f(-2 - 5) = f(-2) = f(3)$

PROBLEM 4

Show that if f is a periodic function, then f is not one-to-one.

Solution: Let f be a periodic function with fundamental period p. Because $p \neq 0$, $x + p \neq x$. However, $f(x + p) = f(x)$. Thus, $f(x + p) = f(x)$

does *not* imply $x + p = x$, and f is not one-to-one. Also, since f is not one-to-one, it does not have an inverse. Geometrically, the fact that a periodic function is not one-to-one can be seen by using the horizontal line test (see Fig. 123). The horizontal line l intersects the graph of f in more than one point. Thus, f is not one-to-one.

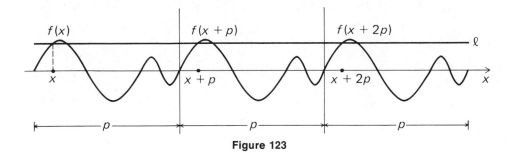

Figure 123

We now use the periodicity of the winding function and the symmetry of the unit circle to evaluate special values on the circumference of the unit circle.

PROBLEM 5

Find the coordinates of $P\left(\dfrac{\pi}{4}\right)$.

Solution: Let $P\left(\dfrac{\pi}{4}\right) = (x, y)$. Since $P\left(\dfrac{\pi}{4}\right) = P\left(\dfrac{1}{2} \cdot \dfrac{\pi}{2}\right)$, the point $P\left(\dfrac{\pi}{4}\right)$ is the midpoint of the arc $\overarc{AB}$ which joins $(1, 0)$ and $(0, 1)$ (see Fig. 124).

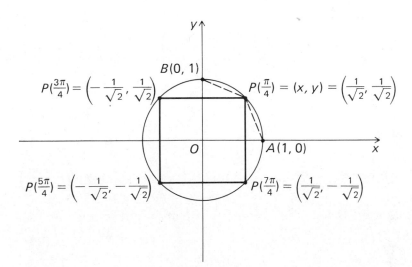

Figure 124

Since $P\left(\dfrac{\pi}{4}\right)$ is the midpoint of $\overset{\frown}{AB}$, $\overset{\frown}{AP} = \overset{\frown}{PB}$. We know from geometry that equal chords subtend equal arcs—that is, $AP = PB$. Using the distance formula, we have:

$$\text{chord } AP = \text{chord } PB$$
$$\sqrt{(x-1)^2 + y^2} = \sqrt{x^2 + (y-1)^2}$$

squaring both sides:

$$(x-1)^2 + y^2 = x^2 + (y-1)^2$$
$$x^2 - 2x + 1 + y^2 = x^2 + y^2 - 2y + 1$$
$$-2x = -2y$$
$$x = y$$

Since $x = y$ and all points on the circumference of the circle must satisfy $x^2 + y^2 = 1$, it follows that $x^2 + x^2 = 1$, or $2x^2 = 1$. Solving for x we obtain

$$x = \frac{1}{\sqrt{2}} \text{ or } x = \frac{-1}{\sqrt{2}}$$

But the point $P\left(\dfrac{\pi}{4}\right)$ is in the first quadrant and x must be positive. Therefore $x = y = \dfrac{1}{\sqrt{2}}$ and $P\left(\dfrac{\pi}{4}\right) = \left(\dfrac{1}{\sqrt{2}}, \dfrac{1}{\sqrt{2}}\right)$. Rationalizing the denominators we get $P\left(\dfrac{\pi}{4}\right) = \left(\dfrac{\sqrt{2}}{2}, \dfrac{\sqrt{2}}{2}\right)$.

We now use the symmetry of the unit circle to see that $P\left(\dfrac{3\pi}{4}\right) = \left(\dfrac{-\sqrt{2}}{2}, \dfrac{\sqrt{2}}{2}\right)$, $P\left(\dfrac{5\pi}{4}\right) = \left(\dfrac{-\sqrt{2}}{2}, \dfrac{-\sqrt{2}}{2}\right)$, and $P\left(\dfrac{7\pi}{4}\right) = \left(\dfrac{\sqrt{2}}{2}, \dfrac{-\sqrt{2}}{2}\right)$.

PROBLEM 6

Given that $P\left(\dfrac{\pi}{3}\right) = \left(\dfrac{1}{2}, \dfrac{\sqrt{3}}{2}\right)$ find $P\left(\dfrac{2\pi}{3}\right)$, $P\left(\dfrac{4\pi}{3}\right)$, and $P\left(\dfrac{5\pi}{3}\right)$.

Solution: Using the symmetry of the unit circle we see that $P\left(\dfrac{2\pi}{3}\right) = \left(-\dfrac{1}{2}, \dfrac{\sqrt{3}}{2}\right)$, $P\left(\dfrac{4\pi}{3}\right) = \left(-\dfrac{1}{2}, \dfrac{-\sqrt{3}}{2}\right)$, and $P\left(\dfrac{5\pi}{3}\right) = \left(\dfrac{1}{2}, \dfrac{-\sqrt{3}}{2}\right)$.

5.1 EXERCISES

1. Find the coordinates of the following points on the unit circle:
   (a) $P(3\pi)$
   (b) $P\left(\dfrac{-5\pi}{2}\right)$
   (c) $P(14\pi)$

2. Recall that $\pi = 3.14159.\ldots$ Use the unit circle to determine the quadrant in which the coordinates of $P(\theta)$ are found for each of the following:
   (a) $P(1.7)$ (b) $P(4.2)$
   (c) $P(-7)$ (d) $P(15)$

3. Suppose f and g are periodic functions of fundamental period 2π, and the graphs of each are given in Figure 125 for the interval $[0, 2\pi)$. Draw the graph of f and g for the intervals $[-2\pi, 0)$ and $[2\pi, 4\pi)$.

4. Let f be a periodic function with period 3. Express each of the following values in terms of values in the interval $[0, 3)$:
   (a) $f(4)$ (b) $f(10)$
   (c) $f(6.2)$ (d) $f(-2)$
   (e) $f(-3)$ (f) $f(-7)$

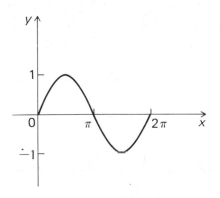

Graph of f

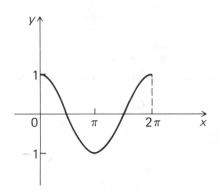

Graph of g

Figure 125

5. Let f and g be periodic functions with fundamental period p. Determine which of the following functions are periodic:
   (a) $f + g$ (b) fg
   (c) $\dfrac{f}{g}$ (d) $f \circ g$

6. Let P be the winding function. Show that if $P(\theta) = (x, y)$ then
   (a) $P(\theta + \pi) = (-x, -y)$
   (b) $P(\theta - \pi) = (-x, -y)$
   (c) $P\left(\theta + \dfrac{\pi}{2}\right) = (-y, x)$
   (d) $P\left(\dfrac{\pi}{2} - \theta\right) = (y, x)$

5.2 SINE AND COSINE

The winding function P introduced in Section 5.1 assigns to each real number θ an ordered pair of real numbers (x, y) on the unit circle $x^2 + y^2 = 1$. That is,

$$P : \theta \to (x, y) \text{ and } P(\theta) = (x, y)$$

However, the fact that the range of the winding function P consists of ordered pairs and not single real numbers limits the uses of the winding function.

Since it is more convenient to work with functions that map a set of real numbers to a set of real numbers, we will use the winding function P to construct two other functions. One of the functions maps θ into x, and the other takes θ into the y of the ordered pair (x, y).

Definition: Let P be the winding function on the unit circle defined by $P(\theta) = (x, y)$. Let $P(\theta) = (x, y)$ lie on the unit circle. Then

(a) the function converting θ into x is called the **cosine** function, and is written

$$x = \cos \theta$$

(b) the function converting θ into y is called the **sine** function, and is written

$$y = \sin \theta$$

Thus, $P(\theta) = (\cos \theta, \sin \theta)$.

EXAMPLE 1

Since $P(0) = (1, 0)$, we have $\cos 0 = 1$ and $\sin 0 = 0$. Note $(\cos 0)^2 + (\sin 0)^2 = (1)^2 + (0)^2 = 1$.

PROBLEM 1

Evaluate $\cos \dfrac{\pi}{4}$ and $\sin \dfrac{\pi}{4}$.

Solution: Since $P\left(\dfrac{\pi}{4}\right) = \left(\dfrac{\sqrt{2}}{2}, \dfrac{\sqrt{2}}{2}\right)$, we have $\cos \dfrac{\pi}{4} = \dfrac{\sqrt{2}}{2}$ and $\sin \dfrac{\pi}{4} = \dfrac{\sqrt{2}}{2}$. Again, note that $\left(\sin \dfrac{\pi}{4}\right)^2 + \left(\cos \dfrac{\pi}{4}\right)^2 = \left(\dfrac{\sqrt{2}}{2}\right)^2 + \left(\dfrac{\sqrt{2}}{2}\right)^2 = 1$.

Since the cosine and sine functions were defined using the winding function P, the domains of sine and cosine are all the real numbers θ. For any real value θ the cosine and sine are the first and second coordinates of P, respectively. Every ordered pair (x, y) of P must lie on the unit circle, $x^2 + y^2 = 1$, and it follows that $-1 \le x \le 1$ and $-1 \le y \le 1$. Thus, $-1 \le \cos \theta \le 1$ and $-1 \le \sin \theta \le 1$, and the range of cosine is $[-1, 1]$ and the range of sine is $[-1, 1]$. Figure 126 illustrates the cosine and sine functions.

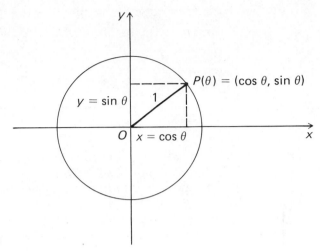

Figure 126

Recall the signs of the coordinates of ordered pairs (x, y) in the different quadrants. Using this information we can determine the signs of the sine and cosine functions in the quadrants.

TABLE 5

QUADRANT FOR $P(\theta)$	VALUES FOR θ	COS θ	SIN θ
I	$0 \leq \theta \leq \frac{\pi}{2}$	+	+
II	$\frac{\pi}{2} \leq \theta \leq \pi$	−	+
III	$\pi \leq \theta \leq \frac{3\pi}{2}$	−	−
IV	$\frac{3\pi}{2} \leq \theta \leq 2\pi$	+	−

The following theorem is a direct result of the definition of cosine and sine.

Theorem 2: For any real number θ, $\sin^2 \theta + \cos^2 \theta = 1$.

PROBLEM 2

If $\cos \theta = -\dfrac{3}{5}$ and $\pi < \theta < \dfrac{3\pi}{2}$, find $\sin \theta$.

Solution: Since $\pi < \theta < \dfrac{3\pi}{2}$, $P(\theta)$ lies in Quadrant III. From Theorem 2 we have $\cos^2 \theta + \sin^2 \theta = 1$, or $\sin \theta = \pm\sqrt{1 - \cos^2 \theta}$. However, because $P(\theta)$ is in Quadrant III, we use $\sin \theta = -\sqrt{1 - \cos^2 \theta} = -\sqrt{1 - \left(-\dfrac{3}{5}\right)^2} = -\sqrt{1 - \dfrac{9}{25}} = -\dfrac{4}{5}$.

Remark: The usual notation for the powers of sine and cosine is

$$(\sin \theta)^2 = \sin^2 \theta$$

$$(\cos \theta)^2 = \cos^2 \theta$$

The winding function P is a periodic function with fundamental period 2π; that is, $P(\theta) = P(\theta + 2n\pi)$, $n = 0, \pm 1, \pm 2, \ldots$. Since $\cos \theta$ and $\sin \theta$ are the x and y components of $P(\theta)$, then $\cos \theta = \cos(\theta + 2n\pi)$ and $\sin \theta = \sin(\theta + 2n\pi)$. Thus, sine and cosine are periodic functions with fundamental period 2π, and we can use this to determine some important values of $\cos \theta$ and $\sin \theta$. These values are listed in Table 6.

TABLE 6

θ	$P(\theta)$	$\cos \theta$	$\sin \theta$	θ	$P(\theta)$	$\cos \theta$	$\sin \theta$
0	$(1, 0)$	1	0	π	$(-1, 0)$	-1	0
$\dfrac{\pi}{6}$	$\left(\dfrac{\sqrt{3}}{2}, \dfrac{1}{2}\right)$	$\dfrac{\sqrt{3}}{2}$	$\dfrac{1}{2}$	$\dfrac{7\pi}{6}$	$\left(\dfrac{-\sqrt{3}}{2}, -\dfrac{1}{2}\right)$	$\dfrac{-\sqrt{3}}{2}$	$-\dfrac{1}{2}$
$\dfrac{\pi}{4}$	$\left(\dfrac{\sqrt{2}}{2}, \dfrac{\sqrt{2}}{2}\right)$	$\dfrac{\sqrt{2}}{2}$	$\dfrac{\sqrt{2}}{2}$	$\dfrac{5\pi}{4}$	$\left(\dfrac{-\sqrt{2}}{2}, \dfrac{-\sqrt{2}}{2}\right)$	$\dfrac{-\sqrt{2}}{2}$	$\dfrac{-\sqrt{2}}{2}$
$\dfrac{\pi}{3}$	$\left(\dfrac{1}{2}, \dfrac{\sqrt{3}}{2}\right)$	$\dfrac{1}{2}$	$\dfrac{\sqrt{3}}{2}$	$\dfrac{4\pi}{3}$	$\left(-\dfrac{1}{2}, \dfrac{-\sqrt{3}}{2}\right)$	$-\dfrac{1}{2}$	$\dfrac{-\sqrt{3}}{2}$
$\dfrac{\pi}{2}$	$(0, 1)$	0	1	$\dfrac{3\pi}{2}$	$(0, -1)$	0	-1
$\dfrac{2\pi}{3}$	$\left(-\dfrac{1}{2}, \dfrac{\sqrt{3}}{2}\right)$	$-\dfrac{1}{2}$	$\dfrac{\sqrt{3}}{2}$	$\dfrac{5\pi}{3}$	$\left(\dfrac{1}{2}, \dfrac{-\sqrt{3}}{2}\right)$	$\dfrac{1}{2}$	$\dfrac{-\sqrt{3}}{2}$
$\dfrac{3\pi}{4}$	$\left(\dfrac{-\sqrt{2}}{2}, \dfrac{\sqrt{2}}{2}\right)$	$\dfrac{-\sqrt{2}}{2}$	$\dfrac{\sqrt{2}}{2}$	$\dfrac{7\pi}{4}$	$\left(\dfrac{\sqrt{2}}{2}, \dfrac{-\sqrt{2}}{2}\right)$	$\dfrac{\sqrt{2}}{2}$	$\dfrac{-\sqrt{2}}{2}$
$\dfrac{5\pi}{6}$	$\left(\dfrac{-\sqrt{3}}{2}, \dfrac{1}{2}\right)$	$\dfrac{-\sqrt{3}}{2}$	$\dfrac{1}{2}$	$\dfrac{11\pi}{6}$	$\left(\dfrac{\sqrt{3}}{2}, -\dfrac{1}{2}\right)$	$\dfrac{\sqrt{3}}{2}$	$-\dfrac{1}{2}$

Using Table 6 and the fact that $\cos \theta$ and $\sin \theta$ are the coordinates of $P(\theta)$, we can prove some important properties of the cosine and sine function. First, we recall that a function f is said to be **even** if $f(-x) = f(x)$, and is said to be **odd** if $f(-x) = -f(x)$ for all values of x for which f is defined.

Theorem 3: The cosine function is an **even** function. That is, $\cos(-\theta) = \cos \theta$. The sine function is an **odd** function. That is, $\sin(-\theta) = -\sin \theta$.

Proof: Consider the unit circle given in Figure 127. Let $P(\theta) = (x, y)$. The symmetry of the unit circle gives $P(-\theta) = (x, -y)$. Thus, $(x, y) = (\cos \theta, \sin \theta)$ and $(x, -y) = (\cos(-\theta), \sin(-\theta))$. The following relationships result:

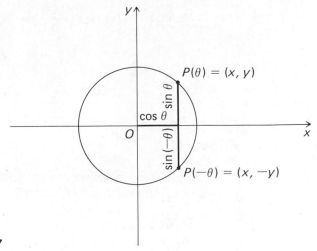

Figure 127

$$\cos(-\theta) = x = \cos \theta$$

and

$$\sin(-\theta) = -y = -\sin \theta$$

Hence, $\cos(-\theta) = \cos \theta$ and $\sin(-\theta) = -\sin \theta$.

PROBLEM 3

Find $\cos \theta$ and $\sin \theta$ for $\theta = \dfrac{-5\pi}{3}$ and $\theta = \dfrac{-3\pi}{4}$. (Make use of Table 6.)

Solution: (a) $\cos\left(\dfrac{-5\pi}{3}\right) = \cos\left(\dfrac{5\pi}{3}\right) = \dfrac{1}{2}$

$\sin\left(\dfrac{-5\pi}{3}\right) = -\sin\left(\dfrac{5\pi}{3}\right) = -\left(\dfrac{-\sqrt{3}}{2}\right) = \dfrac{\sqrt{3}}{2}$

(b) $\cos\left(\dfrac{-3\pi}{4}\right) = \cos\left(\dfrac{3\pi}{4}\right) = \dfrac{-\sqrt{2}}{2}$

$\sin\left(\dfrac{-3\pi}{4}\right) = -\sin\left(\dfrac{3\pi}{4}\right) = -\left(\dfrac{\sqrt{2}}{2}\right) = \dfrac{-\sqrt{2}}{2}$

PROBLEM 4

Evaluate $\cos\left(\dfrac{15\pi}{6}\right)$ and $\sin\left(\dfrac{15\pi}{6}\right)$.

Solution: (a) $\cos\left(\dfrac{15\pi}{6}\right) = \cos\left(\dfrac{3\pi}{6} + \dfrac{12\pi}{6}\right) = \cos\left(\dfrac{\pi}{2} + 2\pi\right)$

$$= \cos\frac{\pi}{2} = 0$$

(b) $\sin\left(\dfrac{15\pi}{6}\right) = \sin\dfrac{\pi}{2} = 1$

PROBLEM 5

Determine whether the functions f, g, and h are even, odd, or neither even nor odd, where $f(\theta) = \cos 2\theta$, $g(\theta) = \cos\theta + \sin\theta$, and $h(\theta) = \sin^3\theta$.

Solution: (a) $f(-\theta) = \cos(-2\theta) = \cos 2\theta$; therefore f is even.
(b) $g(-\theta) = \cos(-\theta) + \sin(-\theta) = \cos\theta - \sin\theta$. Thus, g is neither even nor odd because $g(-\theta) \neq g(\theta)$ and $g(-\theta) \neq -g(\theta) = -\cos\theta - \sin\theta$.
(c) $h(-\theta) = \sin^3(-\theta) = [\sin(-\theta)]^3 = [-\sin\theta]^3 = -\sin^3\theta = -h(\theta)$. Hence, h is odd.

PROBLEM 6

If the cosine function is a periodic function with fundamental period 2π, determine the period of the function f defined by $f(\theta) = \cos 2\theta$.

Solution: Since the cosine function has period 2π, we have

$$\begin{aligned} f(\theta) &= \cos 2\theta \\ &= \cos(2\theta + 2\pi) \\ &= \cos[2(\theta + \pi)] \\ &= f(\theta + \pi) \end{aligned}$$

Thus, $f(\theta) = f(\theta + \pi)$ and $f(\theta) = \cos 2\theta$ defines a periodic function with fundamental period π.

5.2 EXERCISES

1. Determine the value(s) of θ if $0 \leq \theta < 2\pi$, using Table 6.

(a) $\cos\theta = \dfrac{1}{2}$ (b) $\cos\theta = 0$

(c) $\cos\theta = \dfrac{\sqrt{3}}{2}$ (d) $\cos\theta = \dfrac{-\sqrt{2}}{2}$

(e) $\sin\theta = \dfrac{1}{2}$ (f) $\sin\theta = 1$

(g) $\sin\theta = \dfrac{-\sqrt{3}}{2}$ (h) $\sin\theta = \dfrac{\sqrt{2}}{2}$

2. Determine the quadrant(s) in which the following conditions occur:
(a) $\cos\theta < 0$ (b) $\cos\theta > 0$
(c) $\sin\theta < 0$ (d) $\sin\theta > 0$
(e) $\sin\theta > 0$ and $\cos\theta < 0$
(f) $\cos\theta > 0$ and $\sin\theta < 0$

3. (a) If $\sin \theta = \dfrac{3}{5}$ and $\cos \theta > 0$,
   find $\cos \theta$.

   (b) If $\cos \theta = \dfrac{-1}{2}$ and $\sin \theta < 0$,
   find $\sin \theta$.

4. Evaluate:

   (a) $\sin \left(\dfrac{-\pi}{4}\right)$ (b) $\cos \left(\dfrac{13\pi}{4}\right)$

   (c) $\sin \left(\dfrac{19\pi}{6}\right)$ (d) $\sin (-9\pi)$

   (e) $\cos (-4\pi)$ (f) $\sin \left(\dfrac{9\pi}{2}\right)$

5. Evaluate and simplify the following:

   (a) $\sin \left(\dfrac{\pi}{6}\right) \cdot \cos \left(\dfrac{\pi}{4}\right)$

   (b) $\cos \left(\dfrac{\pi}{6}\right) \cdot \sin \left(\dfrac{\pi}{4}\right)$

   (c) $\dfrac{1 - \cos \left(\dfrac{\pi}{3}\right)}{2}$

   (d) $\dfrac{1 + \cos \left(\dfrac{\pi}{3}\right)}{2}$

   (e) $\dfrac{1 - \cos \left(\dfrac{4\pi}{3}\right)}{1 + \cos \left(\dfrac{4\pi}{3}\right)}$

   (f) $\sin^2 \left(\dfrac{7\pi}{5}\right) + \cos^2 \left(\dfrac{7\pi}{5}\right)$

6. Use the fact that the cosine and sine functions are periodic with fundamental period 2π to determine the period of the following:
   (a) $\cos 4\theta$ (b) $\sin 3\theta$

   (c) $\sin \left(\dfrac{\theta}{2}\right)$ (d) $\cos (\theta + 1)$

7. Let h be the function defined by $h(\theta) = \dfrac{\sin \theta}{\cos \theta}$, where $\cos \theta \neq 0$. Determine whether h is even, odd, or neither.

8. Let f be the function defined by $f(\theta) = \cos \theta$ and g the function defined by $g(\theta) = 3\theta$. Does $f \circ g - g \circ f$?

9. Discuss whether $f(\theta) = \cos \theta$ and $g(\theta) = \sin \theta$ have inverses.

10. Evaluate:

    (a) $\cos \left(\dfrac{\pi}{3} - \dfrac{\pi}{6}\right)$. Does $\cos \left(\dfrac{\pi}{3} - \dfrac{\pi}{6}\right) = \cos \dfrac{\pi}{3} - \cos \dfrac{\pi}{6}$?

    (b) $\sin \left(\dfrac{5\pi}{6} + \dfrac{\pi}{6}\right)$. Does $\sin \left(\dfrac{5\pi}{6} + \dfrac{\pi}{6}\right) = \sin \dfrac{5\pi}{6} + \sin \dfrac{\pi}{6}$?

    (c) What can generally be said about $\cos(\theta_1 - \theta_2)$ and $\cos \theta_1$ and $\cos \theta_2$? Similarly, what can generally be said about $\sin(\theta_1 + \theta_2)$ and $\sin \theta_1$ and $\sin \theta_2$?

5.3 THE GRAPHS OF SINE AND COSINE

In this section we shall examine the graphs of $f(\theta) = \sin \theta$ and $g(\theta) = \cos \theta$. Since both the sine and cosine functions are periodic functions with fundamental period 2π, we can restrict the values of θ to the interval $[0, 2\pi)$. The entire graph of both functions can be obtained by repeating the graphs we obtain on the interval $[0, 2\pi)$.

Using the values of $\sin \theta$ and $\cos \theta$ given in Table 7, we can graph $\sin \theta$ and $\cos \theta$ on the interval $[0, 2\pi)$ (see Figs. 128 and 129, respectively).

TABLE 7

θ	SIN θ	COS θ
0	0	1
$\dfrac{\pi}{4}$	$\dfrac{\sqrt{2}}{2}$	$\dfrac{\sqrt{2}}{2}$
$\dfrac{\pi}{2}$	1	0
$\dfrac{3\pi}{4}$	$\dfrac{\sqrt{2}}{2}$	$\dfrac{-\sqrt{2}}{2}$
π	0	-1
$\dfrac{5\pi}{4}$	$\dfrac{-\sqrt{2}}{2}$	$\dfrac{-\sqrt{2}}{2}$
$\dfrac{3\pi}{2}$	-1	0
$\dfrac{7\pi}{4}$	$\dfrac{-\sqrt{2}}{2}$	$\dfrac{\sqrt{2}}{2}$

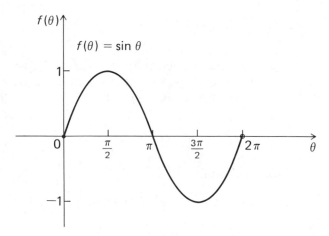

Figure 128

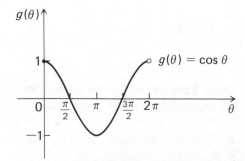

Figure 129

We now use the periodicity of sine and cosine to obtain the graphs of sin θ and cos θ for any real number θ (see Figs. 130 and 131).

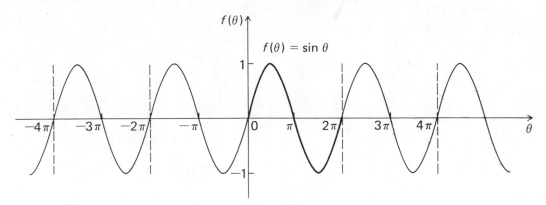

Figure 130

Similarly, the graph of $g(\theta) = \cos \theta$ is

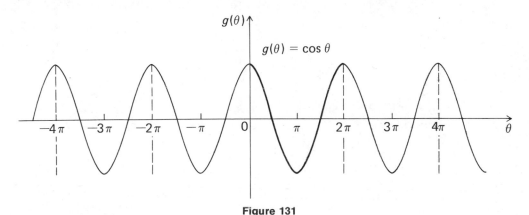

Figure 131

We now consider some variations on the graphs of sine and cosine. Let $f(\theta) = A + B \sin(k\theta + C)$ and $g(\theta) = A + B \cos(k\theta + C)$, where A, B, C, and k are constants. We shall see that these constants represent translations, stretchings, and contractions of the graphs of $f(\theta) = \sin \theta$ and $g(\theta) = \cos \theta$. The reader should take this opportunity to review the graphing techniques discussed in Section 2.3 (Chapter 2).

Before giving examples of these graphs, we define an important term used in the graphing of certain periodic functions. The **amplitude** of a periodic function is one half of the absolute value of the difference between the maximum and minimum values. Thus, both the sine and cosine functions have an amplitude equal to $\frac{1}{2} |1 - (-1)| = 1$.

EXAMPLE 1

The graph of $g(\theta) = 2 + \cos \theta$ is a vertical translation of $g(\theta) = \cos \theta$ upward 2 units (see Fig. 132).

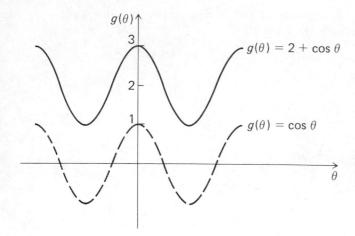

Figure 132

The graph of $f(\theta) = A + \sin \theta$ or $g(\theta) = A + \cos \theta$ is the graph of $\sin \theta$ or $\cos \theta$, respectively, translated vertically A units—upward if $A > 0$ and downward if $A < 0$. Note that A has no effect on the period or amplitude.

EXAMPLE 2

The graph of $f(\theta) = 3 \sin \theta$ is a vertical stretching of the graph of $f(\theta) = \sin \theta$. The amplitude of $f(\theta) = 3 \sin \theta$ is 3 (see Fig. 133).

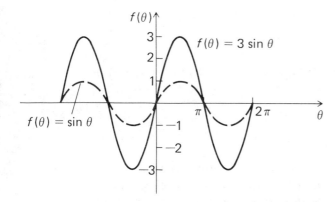

Figure 133

EXAMPLE 3

The graph of $f(\theta) = \left(-\dfrac{1}{3}\right) \sin \theta$ is a vertical contraction and a reflection of the graph of $f(\theta) = \sin \theta$. The amplitude is $\dfrac{1}{3}$ (see Fig. 134).

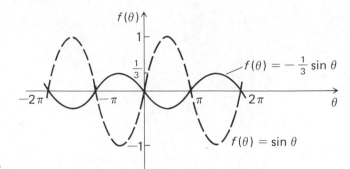

Figure 134

The graph of $f(\theta) = B \sin \theta$ or $g(\theta) = B \cos \theta$ is the graph of $\sin \theta$ or $\cos \theta$, respectively, stretched vertically if $|B| > 1$ and contracted vertically if $|B| < 1$. If B is negative, there is also a reflection about the x-axis. The amplitude is $|B|$. Note that B has no effect on the period.

EXAMPLE 4

The graph of $g(\theta) = \cos 2\theta$ is a horizontal contraction of the graph of $g(\theta) = \cos \theta$. The period is π (see Fig. 135).

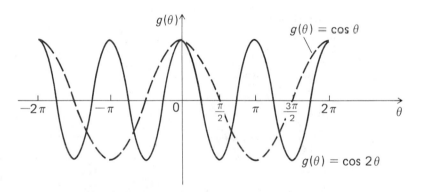

Figure 135

The graph of $f(\theta) = \sin k\theta$ or $g(\theta) = \cos k\theta$, $k \neq 0$, is the graph of $\sin \theta$ and $\cos \theta$, respectively, stretched horizontally if $|k| < 1$ and contracted horizontally if $|k| > 1$. If k is negative, there is also a reflection about the y-axis. Note that k has no effect on the amplitude. However, k does affect the period—that is, the new period is $\dfrac{2\pi}{|k|}$.

EXAMPLE 5

The graph of $g(\theta) = \cos \left(\theta - \dfrac{\pi}{4}\right)$ is a horizontal translation, to the right $\dfrac{\pi}{4}$ units, of the graph of $g(\theta) = \cos \theta$ (see Fig. 136).

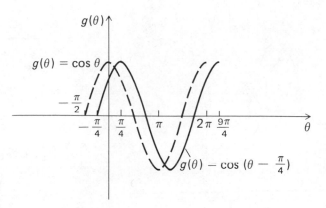

Figure 136

> The graph of $f(\theta) = \sin(\theta + C)$ or $g(\theta) = \cos(\theta + C)$ is the graph of $f(\theta) = \sin \theta$ or $g(\theta) = \cos \theta$, respectively, translated horizontally C units to the left if $C > 0$, and $|C|$ units to the right if $C < 0$. The constant C is called the **phase shift.**

Special attention must be taken if the functions are defined by $f(\theta) = \sin(k\theta + C)$ or $g(\theta) = \cos(k\theta + C)$. In such a case, the phase shift is $\dfrac{C}{k}$.

EXAMPLE 6

The graph of $f(\theta) = \sin(3\theta + \pi)$ is the graph of $\sin \theta$ shifted to the left $\dfrac{\pi}{3}$ units (see Fig. 137).

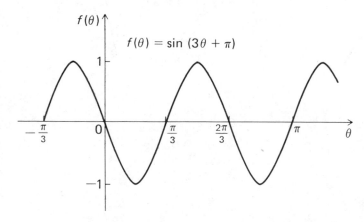

Figure 137

PROBLEM 1

Graph $f(\theta) = 3 \sin\left(2\theta + \dfrac{\pi}{3}\right)$.

Solution: We have $A = 0$, $B = 3$, $C = \dfrac{\pi}{3}$, and $k = 2$. Thus, the amplitude is 3. The phase shift $= \dfrac{C}{k} = \dfrac{\pi}{6}$. Finally, the new period is $\dfrac{2\pi}{k} = \pi$.

The graph of $f(\theta) = 3 \sin \left(2\theta + \dfrac{\pi}{3}\right)$ is given in Figure 138.

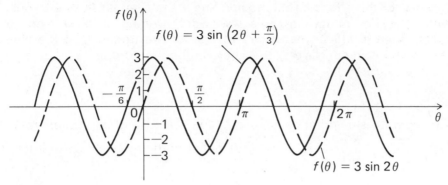

Figure 138

5.3 EXERCISES

1. Find the amplitude, period, and phase shift, and sketch the graph for each of the following functions:
   (a) $f(\theta) = \sin 2\theta$
   (b) $f(\theta) = \sin(2\theta + \pi)$
   (c) $f(\theta) = \dfrac{3}{4} \sin(2\theta + \pi)$
   (d) $f(\theta) = 1 + \left(\dfrac{3}{4}\right) \sin(2\theta + \pi)$

2. Follow the instructions in Exercise 1 for these functions:
   (a) $g(\theta) = \cos 4\theta$
   (b) $g(\theta) = \cos(4\theta - \pi)$
   (c) $g(\theta) = \left(\dfrac{1}{2}\right) \cos(4\theta - \pi)$
   (d) $g(\theta) = -\dfrac{1}{2} + \left(\dfrac{1}{2}\right) \cos(4\theta - \pi)$

3. Graph each of the following functions:
   (a) $f(\theta) = \cos \left(\dfrac{\theta}{2} + \pi\right)$
   (b) $f(\theta) = -3 \sin \left(2\theta - \dfrac{\pi}{2}\right)$
   (c) $f(\theta) = \left(\dfrac{1}{3}\right) \sin \left(3\theta + \dfrac{\pi}{2}\right)$
   (d) $f(\theta) = 2 \cos(2\theta - \pi)$

4. The **frequency** of a periodic function is the number of cycles the graph of the function completes in one time unit. It can be shown that the frequency is the reciprocal of the period. Determine the frequency for each of the following functions:
   (a) $f(\theta) = 4 \sin \left(2\theta + \dfrac{\pi}{3}\right)$
   (b) $f(\theta) = -2 \sin \theta\pi$
   (c) $f(\theta) = -2 \cos \left(4\theta - \dfrac{\pi}{4}\right)$
   (d) $f(\theta) = \cos \left(\theta - \dfrac{\pi}{2}\right)$

5. In an alternating current generator, the current generated is given by $I = 10 \sin 60\pi t$, where t is time measured in seconds. Determine the amplitude, period, phase shift, and frequency.

5.4 INVERSE SINE AND INVERSE COSINE

In Chapter 2 we introduced the concept of the inverse function. We showed that a function f has an inverse, f^{-1}, if and only if f is one-to-one. Geometrically, f^{-1} exists if the graph of $f(x)$ intersects any horizontal line in at most one point. The graphs of the sine and the cosine show that neither is one-to-one, and therefore, they do not have inverses. If we want inverses for the sine and cosine functions, then it will be necessary to restrict the domains of the sine and cosine functions in such a way that they will be one-to-one.

Inverse Sine

Consider the sine function defined by $f(\theta) = \sin \theta$ and the domain restricted to the interval $\left[-\frac{\pi}{2}, \frac{\pi}{2} \right]$ (Fig. 139). Note that even though we have restricted the

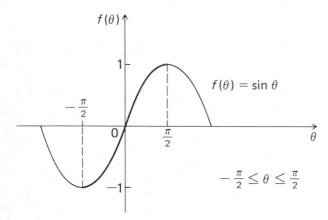

$$-\frac{\pi}{2} \leq \theta \leq \frac{\pi}{2}$$

Figure 139

domain of the sine function to $\left[-\frac{\pi}{2}, \frac{\pi}{2} \right]$, the range is still $[-1, 1]$. More impor-

tant, the sine function is strictly increasing on $\left[-\frac{\pi}{2}, \frac{\pi}{2} \right]$ and, consequently, is

one-to-one. Thus, sine has an inverse on the interval $\left[-\frac{\pi}{2}, \frac{\pi}{2} \right]$.

Definition: The **inverse sine** (also called **arcsine function**), denoted by $\sin^{-1}$ (or arcsin), is the inverse function of the sine function. The domain of $\sin^{-1}$ is $[-1, 1]$ and the range is $\left[-\frac{\pi}{2}, \frac{\pi}{2} \right]$. That is, $\theta = \sin^{-1} x$ if and only if $x = \sin \theta$, for $-\frac{\pi}{2} \leq \theta \leq \frac{\pi}{2}$.

It is very important to note that the notation $\sin^{-1}$ is *not* the same as $(\sin x)^{-1}$. That is,

$$\sin^{-1} x \neq (\sin x)^{-1}$$

$$\sin^{-1} x \neq \frac{1}{\sin x}$$

PROBLEM 1

Evaluate $\sin^{-1}\left(\frac{1}{2}\right)$, $\sin^{-1}\left(\frac{\sqrt{2}}{2}\right)$, $\sin^{-1}(0)$ and $\sin^{-1}(2)$.

Solution: (a) Let $\theta = \sin^{-1}\left(\frac{1}{2}\right)$. Equivalently, $\sin \theta = \frac{1}{2}$ and the value of θ which satisfies this relation is obtained by reading a table of values of $\sin \theta$ backwards, as in Table 6. We have $\sin\left(\frac{\pi}{6}\right) = \frac{1}{2}$ and $\sin\left(\frac{5\pi}{6}\right) = \frac{1}{2}$, but $\frac{5\pi}{6}$ does not belong to the range of $\sin^{-1}$; that is, $\frac{5\pi}{6}$ does not belong to $\left[-\frac{\pi}{2}, \frac{\pi}{2}\right]$. Thus, $\theta = \frac{\pi}{6}$.

(b) $\sin^{-1}\left(\frac{\sqrt{2}}{2}\right) = \frac{\pi}{4}$

(c) $\sin^{-1}(0) = 0$

(d) Let $\theta = \sin^{-1}(2)$. Then $\sin \theta = 2$, which is *not* possible.

We can use the values in Problem 1 and other values to obtain the graph of $\theta = \sin^{-1} x$.

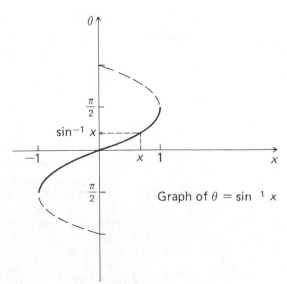

Graph of $\theta = \sin^{-1} x$

Figure 140

PROBLEM 2

Find $\sin^{-1}\left[\sin\left(-\frac{\pi}{4}\right)\right]$.

Solution: Let $\theta = \sin^{-1}\left[\sin\left(-\frac{\pi}{4}\right)\right]$. Then $\sin\theta = \sin\left(-\frac{\pi}{4}\right)$ and $\theta = -\frac{\pi}{4}$.

PROBLEM 3

For what values of θ, if any, does $\sin^{-1}[\sin\theta] = \theta$?

Solution: Note that $\sin^{-1}[\sin\theta] = \theta$ is equivalent to $(f \circ g)(\theta) = \theta$, where f is the $\sin^{-1}$ function and g is the sine function. Thus, we wish to find those values of θ for which the inverse sine is defined. Hence, $\theta \in \left[-\frac{\pi}{2}, \frac{\pi}{2}\right]$.

Inverse Cosine

The cosine function can be made a one-to-one function by restricting its domain to the interval $[0, \pi]$.

> **Definition:** The **inverse cosine** (or **arccosine function**), denoted by $\cos^{-1}$ (or arccos), is the inverse function of the cosine function. The domain of $\cos^{-1}$ is $[-1, 1]$ and the range is $[0, \pi]$. That is, $\theta = \cos^{-1} x$ if and only if $x = \cos\theta$, for $0 \le \theta \le \pi$.

PROBLEM 4

Evaluate $\cos^{-1}\left(-\frac{1}{2}\right)$, $\cos^{-1}(0)$, and $\cos^{-1}(1)$.

Solution: (a) Let $\theta = \cos^{-1}\left(-\frac{1}{2}\right)$; then $\cos\theta = -\frac{1}{2}$. Thus, to evaluate $\cos^{-1}\left(-\frac{1}{2}\right)$ we look for a value of θ whose cosine equals $-\frac{1}{2}$. Using Table 6 we have $\cos\left(\frac{2\pi}{3}\right) = -\frac{1}{2}$ and $\cos\left(\frac{4\pi}{3}\right) = -\frac{1}{2}$, but $\frac{4\pi}{3}$ does not belong to $[0, \pi]$. Therefore, $\theta = \frac{2\pi}{3}$.

(b) $\cos^{-1}(0) = \frac{\pi}{2}$

(c) $\cos^{-1}(1) = 0$

The graph of $\cos^{-1}$ is given in Figure 141.

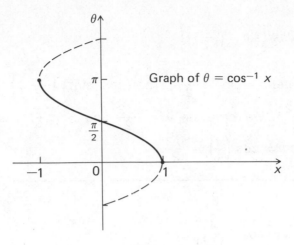

Graph of $\theta = \cos^{-1} x$

Figure 141

PROBLEM 5

Find $\cos^{-1}\left[\cos\left(\dfrac{3}{26}\right)\right]$.

Solution: Let $\theta = \cos^{-1}\left[\cos\left(\dfrac{3}{26}\right)\right]$. Then $\cos\theta = \cos\left(\dfrac{3}{26}\right)$ and $\theta = \dfrac{3}{26}$.

PROBLEM 6

For what values of θ, if any, does $\cos^{-1}[\cos\theta] = \theta$?

Solution: The equation $\cos^{-1}[\cos\theta] - \theta$ is equivalent to $(f \circ g)(\theta) = \theta$, where f is the $\cos^{-1}$ function and g is the cosine function, and the composition acts like the identity function. Thus, we wish to find those values of θ for which $\cos^{-1}$ is defined. Hence $\theta \in [0, \pi]$.

PROBLEM 7

Show that $\sin[\cos^{-1} x] = \sqrt{1 - x^2}$.

Solution: Let $\theta = \cos^{-1} x$; then $\cos\theta = x$, where $0 \leq \theta \leq \pi$. Since the sine function is positive in Quadrants I and II and $\sin^2\theta + \cos^2\theta = 1$, we have $\sin\theta = \sqrt{1 - \cos^2\theta}$. Substituting $x = \cos\theta$ we obtain $\sin\theta = \sqrt{1 - x^2}$. Finally, recall that we let $\theta = \cos^{-1} x$ and therefore we obtain $\sin[\cos^{-1} x] = \sqrt{1 - x^2}$.

PROBLEM 8

Evaluate $\cos\left[\sin^{-1}\left(\frac{\sqrt{2}}{2}\right)\right]$.

Solution: Let $\theta = \sin^{-1}\left(\frac{\sqrt{2}}{2}\right)$. Then $\cos\left[\sin^{-1}\left(\frac{\sqrt{2}}{2}\right)\right] = \cos\theta$. Since $\theta = \sin^{-1}\left(\frac{\sqrt{2}}{2}\right)$, we have $\sin\theta = \frac{\sqrt{2}}{2}$ and $\theta = \frac{\pi}{4}$. Thus, $\cos\left[\sin^{-1}\left(\frac{\sqrt{2}}{2}\right)\right] = \cos\theta = \cos\left(\frac{\pi}{4}\right) = \frac{\sqrt{2}}{2}$.

PROBLEM 9

Evaluate $\sin\left[\cos^{-1}\left(\frac{4}{5}\right)\right]$.

Solution: Let $\theta = \cos^{-1}\left(\frac{4}{5}\right)$. Then $\sin\left[\cos^{-1}\left(\frac{4}{5}\right)\right] = \sin\theta$. Since $\theta = \cos^{-1}\left(\frac{4}{5}\right)$, we have $\cos\theta = \frac{4}{5}$. We now use the identity $\sin^2\theta + \cos^2\theta = 1$ and the fact that $0 \le \theta \le \pi$ to find $\sin\theta$.

$$\sin^2\theta = 1 - \cos^2\theta$$

$$\sin\theta = \pm\sqrt{1 - \cos^2\theta} = \pm\sqrt{1 - \left(\frac{4}{5}\right)^2} = \pm\sqrt{1 - \frac{16}{25}} = \pm\sqrt{\frac{9}{25}}$$

$$= \pm\frac{3}{5}$$

Since $0 \le \theta \le \pi$, $P(\theta)$ must be in the first or second quadrant. Thus, $\sin\theta$ must be positive and $\sin\theta = \frac{3}{5}$. Finally, we have $\sin\left[\cos^{-1}\left(\frac{4}{5}\right)\right] = \sin\theta = \frac{3}{5}$.

5.4 EXERCISES

1. Evaluate the following:

   (a) $\sin^{-1}(0)$ (b) $\sin^{-1}\left(\frac{\sqrt{3}}{2}\right)$

   (c) $\sin^{-1}\left(\frac{1}{2}\right)$ (d) $\sin^{-1}(-1)$

   (e) $\cos^{-1}\left(\frac{-\sqrt{3}}{2}\right)$ (f) $\cos^{-1}\left(\frac{1}{2}\right)$

   (g) $\cos^{-1}(-1)$ (h) $\cos^{-1}\left(\frac{\sqrt{2}}{2}\right)$

2. Evaluate the following:

   (a) $\sin^{-1}\left[\sin\left(\frac{1}{17}\right)\right]$

   (b) $\cos\left[\cos^{-1}\left(\frac{5}{9}\right)\right]$

   (c) $\cos^{-1}\left[\sin\left(\frac{\pi}{6}\right)\right]$

   (d) $\sin^{-1}[\cos(\pi)]$

   (e) $\sin\left[\cos^{-1}\left(\frac{2}{5}\right)\right]$

   (f) $\cos\left[\sin^{-1}\left(\frac{3}{5}\right)\right]$

3. Prove that $\cos[\sin^{-1} x] = \sqrt{1 - x^2}$.

4. Determine the values of x, if any, for which:
   (a) $\sin[\sin^{-1}(\theta)] = x$
   (b) $\cos[\cos^{-1}(\theta)] = x$

5. Discuss the symmetry with respect to the x-axis, θ-axis, and origin of the graphs of:
   (a) $\theta = \sin^{-1} x$ (b) $\theta = \cos^{-1} x$

6. Determine whether the functions $\sin^{-1}$ and $\cos^{-1}$ are odd, even, or neither.

7. Determine whether the following statements are true or false:
   (a) $\sin^{-1}(1) + \sin^{-1}(-1) = 0$
   (b) $\cos^{-1}(1) + \cos^{-1}(-1) = 0$
   (c) $\sin^{-1}(1) + \cos^{-1}(1) = \dfrac{\pi}{2}$
   (d) $\cos^{-1}\left(\dfrac{1}{2}\right) + \sin^{-1}\left(\dfrac{1}{2}\right) = \cos^{-1}(0)$

8. Show that the following are true. (*Hint:* take sin or cos of both sides.)
   (a) $\sin^{-1}\left(\dfrac{12}{13}\right) = \cos^{-1}\left(\dfrac{5}{13}\right)$
   (b) $\cos^{-1}\left(\dfrac{2}{3}\right) = \sin^{-1}\left(\dfrac{\sqrt{5}}{3}\right)$

9. Graph:
   (a) $\theta = \dfrac{\pi}{2} + \sin^{-1}(x + 1)$
   (b) $\theta = \dfrac{\pi}{2} - \cos^{-1} x$

10. Determine the domain for the functions defined by:
    (a) $\sin^{-1}(2x + 1)$
    (b) $\cos^{-1}\left(\dfrac{x}{3} - 1\right)$

5.5 OTHER CIRCULAR FUNCTIONS

We now use the sine and cosine functions to define four other circular functions. These four new functions are called tangent, cotangent, secant, and cosecant and are abbreviated tan, cot, sec, and csc, respectively.

Definitions:

1. **tangent:** $\tan \theta \equiv \dfrac{\sin \theta}{\cos \theta}$, $\cos \theta \neq 0$

2. **cotangent:** $\cot \theta \equiv \dfrac{\cos \theta}{\sin \theta}$, $\sin \theta \neq 0$

3. **secant:** $\sec \theta \equiv \dfrac{1}{\cos \theta}$, $\cos \theta \neq 0$

4. **cosecant:** $\csc \theta \equiv \dfrac{1}{\sin \theta}$, $\sin \theta \neq 0$

The Tangent Function

The tangent function is not defined for every real number. For example,

$\tan\left(\dfrac{\pi}{2}\right) = \dfrac{\sin\left(\dfrac{\pi}{2}\right)}{\cos\left(\dfrac{\pi}{2}\right)} = \dfrac{1}{0}$, which is undefined. It is easy to see that the tangent

function is undefined for any real number whose cosine is 0. That is, the tangent is undefined for $\left(\dfrac{\pi}{2}\right) + n\pi$, where n is any integer. We can now give a more accurate definition of the tangent function.

Definition: The **tangent function,** written tan, is defined by the equation $\tan \theta = \dfrac{\sin \theta}{\cos \theta}$. The domain of tan is the set of real numbers $\theta \neq \left(n + \dfrac{1}{2}\right) \pi$, $n = 0, \pm 1, \pm 2, \pm 3, \ldots .$

PROBLEM 1

Evaluate $\tan(0)$, $\tan \left(\dfrac{\pi}{4}\right)$, and $\tan \left(\dfrac{3\pi}{4}\right)$.

Solution: (a) $\tan(0) = \dfrac{\sin(0)}{\cos(0)} = \dfrac{0}{1} = 0$

(b) $\tan \left(\dfrac{\pi}{4}\right) = \dfrac{\sin \left(\dfrac{\pi}{4}\right)}{\cos \left(\dfrac{\pi}{4}\right)} = \dfrac{\dfrac{\sqrt{2}}{2}}{\dfrac{\sqrt{2}}{2}} = 1$

(c) $\tan \dfrac{3\pi}{4} = \dfrac{\sin \left(\dfrac{3\pi}{4}\right)}{\cos \left(\dfrac{3\pi}{4}\right)} = \dfrac{\dfrac{\sqrt{2}}{2}}{\dfrac{-\sqrt{2}}{2}} = -1$

PROPERTIES OF THE TANGENT FUNCTION

1. The tangent function is periodic with period π. That is, $\tan(\theta + \pi) = \tan \theta$.

Proof: $\tan(\theta + \pi) = \dfrac{\sin(\theta + \pi)}{\cos(\theta + \pi)}$. We now use two formulas which will be developed more thoroughly in the next chapter. They are addition formulas.

$$\sin(\theta + \pi) = \sin \theta \cos \pi + \cos \theta \sin \pi = -\sin \theta$$

and $\qquad \cos(\theta + \pi) = \cos \theta \cos \pi - \sin \theta \sin \pi = -\cos \theta$

Thus, we have $\tan(\theta + \pi) = \dfrac{\sin(\theta + \pi)}{\cos(\theta + \pi)} = \dfrac{-\sin \theta}{-\cos \theta} = \tan \theta.$

2. The tangent function is an odd function. That is, $\tan(-\theta) = -\tan \theta$.

Proof: $\tan(-\theta) = \dfrac{\sin(-\theta)}{\cos(-\theta)} = \dfrac{-\sin \theta}{\cos \theta} = -\tan \theta.$

Before we graph the tangent function, we note that as we approach the value $\frac{\pi}{2}$, the cosine becomes very small and the tangent becomes very large. The vertical line $\theta = \frac{\pi}{2}$ is an asymptote to the tangent curve. The graph of the tangent function is given in Figure 142. The lines $\theta = \frac{\pi}{2} + n\pi$, $n = 0$, ± 1,

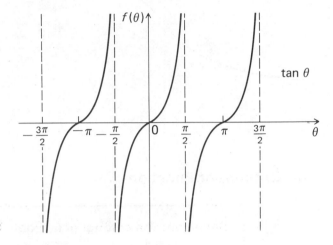

Figure 142

± 2, . . . , are asymptotes. The zeros of the tangent function are 0, $\pm\pi$, $\pm 2\pi$, $\pm 3\pi$, ...

We conclude our discussion of the tangent function by defining the inverse tangent function. Note that for $\frac{-\pi}{2} < \theta < \frac{\pi}{2}$ the tangent function is increasing. Therefore, the inverse can be defined over this interval.

Definition: The **inverse tangent** (or **arctangent**), denoted by $\tan^{-1}$ (or arctan), is the inverse function of the tangent function. The domain of $\tan^{-1}$ is the set of real numbers, and its range is $\left(\frac{-\pi}{2}, \frac{\pi}{2}\right)$. That is, $\theta = \tan^{-1} x$ if and only if $x = \tan\theta$, for $-\frac{\pi}{2} < \theta < \frac{\pi}{2}$.

PROBLEM 2

Evaluate $\tan^{-1}(-\sqrt{3})$, $\tan^{-1}(0)$, and $\tan^{-1}\left(\frac{\sqrt{3}}{3}\right)$.

Solution: (a) Let $\theta = \tan^{-1}(-\sqrt{3})$. *Then* $\tan\theta = -\sqrt{3} \approx -1.732$. Using Table D in Appendix III, we find $\theta = \frac{-\pi}{3}$.

(b) $\tan^{-1}(0) = 0$

(c) $\tan^{-1}\left(\frac{\sqrt{3}}{3}\right) = \frac{\pi}{6}$

The graph of inverse tangent is given in Figure 143.

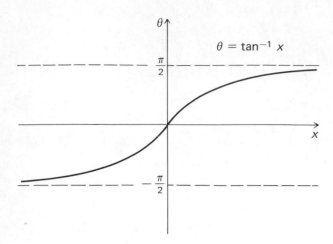

$$\theta = \tan^{-1} x$$

Figure 143

The Cotangent Function

> **Definition:** The **cotangent function,** written cot, is defined by the equation $\cot \theta = \dfrac{\cos \theta}{\sin \theta}$. The domain of cot is the set of real numbers $\theta \neq n\pi, n = 0, \pm 1, \pm 2, \pm 3, \ldots$. The range is the set of all real numbers.

PROBLEM 3

Evaluate $\cot \left(\dfrac{\pi}{2}\right)$, $\cot(0)$, and $\cot \left(\dfrac{\pi}{4}\right)$.

Solution: (a) $\cot \left(\dfrac{\pi}{2}\right) = \dfrac{\cos \left(\dfrac{\pi}{2}\right)}{\sin \left(\dfrac{\pi}{2}\right)} = \dfrac{0}{1} = 0$

(b) $\cot(0) = \dfrac{\cos(0)}{\sin(0)} = \dfrac{1}{0}$, which is undefined

(c) $\cot \left(\dfrac{\pi}{4}\right) = \dfrac{\cos \left(\dfrac{\pi}{4}\right)}{\sin \left(\dfrac{\pi}{4}\right)} = \dfrac{\dfrac{\sqrt{2}}{2}}{\dfrac{\sqrt{2}}{2}} = 1$

PROPERTIES OF THE COTANGENT FUNCTION

1. The cotangent function is periodic with period π. That is, $\cot(\theta + \pi) = \cot \theta$.

Proof: $\cot(\theta + \pi) = \dfrac{\cos(\theta + \pi)}{\sin(\theta + \pi)} = \dfrac{-\cos\theta}{-\sin\theta} = \cot\theta.$

2. The cotangent function is an odd function. That is, $\cot(-\theta) = -\cot\theta$.

Proof: $\cot(-\theta) = \dfrac{\cos(-\theta)}{\sin(-\theta)} = \dfrac{\cos\theta}{-\sin\theta} = -\cot\theta.$

The graph of $\cot\theta$ is given in Figure 144. The lines $\theta = n\pi$ are asymptotes. The zeros of the cotangent function are $\pm\dfrac{\pi}{2}, \pm\dfrac{3\pi}{2}, \pm\dfrac{5\pi}{2}, \ldots$.

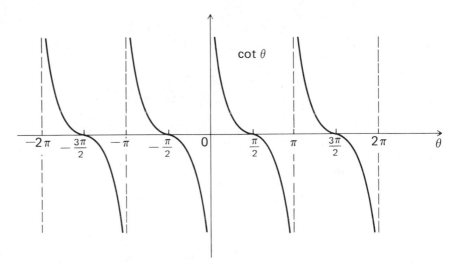

Figure 144

The cotangent is a decreasing function on the interval $(0, \pi)$. Therefore, we can define the inverse cotangent function in terms of $\tan^{-1}$.

Definition: The **inverse cotangent** (or **arccotangent**), denoted by $\cot^{-1}$, is defined by the equation

$$\cot^{-1} x = \frac{\pi}{2} - \tan^{-1} x$$

The domain of $\cot^{-1}$ is the set of real numbers, and its range is $(0, \pi)$.

PROBLEM 4

Evaluate $\cot^{-1}\left(\dfrac{-\sqrt{3}}{3}\right)$, $\cot^{-1}(0)$, and $\cot^{-1}(\sqrt{3})$.

Solution: (a) $\cot^{-1}\left(\dfrac{-\sqrt{3}}{3}\right) = \dfrac{\pi}{2} - \tan^{-1}\left(\dfrac{-\sqrt{3}}{3}\right) = \dfrac{\pi}{2} - \left(\dfrac{-\pi}{6}\right) = \dfrac{2\pi}{3}$

(b) $\cot^{-1}(0) = \dfrac{\pi}{2} - \tan^{-1}(0) = \dfrac{\pi}{2} - 0 = \dfrac{\pi}{2}$

(c) $\cot^{-1}(\sqrt{3}) = \dfrac{\pi}{2} - \tan^{-1}(\sqrt{3}) = \dfrac{\pi}{2} - \dfrac{\pi}{3} = \dfrac{\pi}{6}$

The graph of inverse cotangent is given in Figure 145.

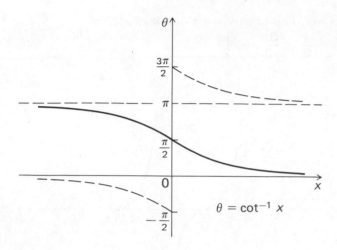

$\theta = \cot^{-1} x$

Figure 145

The Secant Function

> **Definition:** The **secant function,** written sec, is defined by the equation $\sec \theta = \dfrac{1}{\cos \theta}$. The domain of sec is the set of real numbers $\theta \neq \dfrac{\pi}{2} + n\pi$, $n = 0, \pm 1, \pm 2, \pm 3, \ldots$. The range consists of all real numbers greater than or equal to 1, and less than or equal to -1.

PROBLEM 5

Evaluate $\sec\left(\dfrac{\pi}{6}\right)$, $\sec\left(\dfrac{\pi}{2}\right)$, and $\sec(\pi)$.

Solution: (a) $\sec\left(\dfrac{\pi}{6}\right) = \dfrac{1}{\cos\left(\dfrac{\pi}{6}\right)} = \dfrac{1}{\dfrac{\sqrt{3}}{2}} = \dfrac{2}{\sqrt{3}}$

(b) $\sec\left(\dfrac{\pi}{2}\right) = \dfrac{1}{\cos\left(\dfrac{\pi}{2}\right)} = \dfrac{1}{0}$, which is undefined

(c) $\sec(\pi) = \dfrac{1}{\cos(\pi)} = \dfrac{1}{-1} = -1$

PROPERTIES OF THE SECANT FUNCTION

> 1. The secant function is periodic with a period of 2π. That is, $\sec(\theta + 2\pi) = \sec\theta$.

Proof: $\sec(\theta + 2\pi) = \dfrac{1}{\cos(\theta + 2\pi)} = \dfrac{1}{\cos\theta} = \sec\theta$.

> 2. The secant function is an even function. That is, $\sec(-\theta) = \sec\theta$.

Proof: $\sec(-\theta) = \dfrac{1}{\cos(-\theta)} = \dfrac{1}{\cos\theta} = \sec\theta$.

The graph of $\sec\theta$ is given in Figure 146. The lines $\theta = \dfrac{n\pi}{2}$ are asymptotes. Note there are no zeros for the secant function.

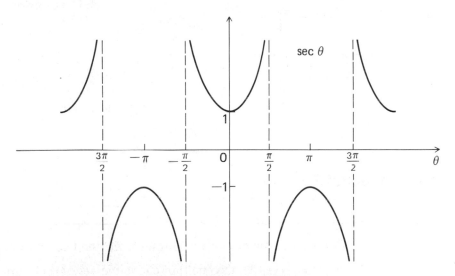

Figure 146

Just as the secant was defined in terms of the cosine, the inverse secant function is defined in terms of the inverse cosine function.

> **Definition:** The **inverse secant** (or **arcsecant**), denoted by $\sec^{-1}$, is defined by the equation
>
> $$\sec^{-1} x = \cos^{-1}\left(\frac{1}{x}\right)$$
>
> The domain of $\sec^{-1}$ is the set of all real numbers *except* $-1 < x < 1$. The range is $0 \le \sec^{-1} x < \dfrac{\pi}{2}$ for $x \ge 1$, and $\dfrac{\pi}{2} < \sec^{-1} x \le \pi$ for $x \le -1$.

PROBLEM 6

Find $\sec^{-1}(-2)$ and $\sec^{-1}(1)$.

Solution: (a) $\sec^{-1}(-2) = \cos^{-1}\left(\dfrac{1}{-2}\right) = \dfrac{2\pi}{3}$

(b) $\sec^{-1}(1) = \cos^{-1}\left(\dfrac{1}{1}\right) = \cos^{-1}(1) = 0$

The graph of the inverse secant consists of two branches and is given in Figure 147.

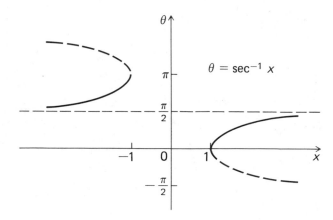

Figure 147

The Cosecant Function

Definition: The **cosecant function,** written csc, is defined by the equation $\csc \theta = \dfrac{1}{\sin \theta}$. The domain of csc is the set of real numbers $\theta \neq n\pi, n = 0,$ $\pm 1, \pm 2, \pm 3, \ldots$. The range of csc is the same as that for the secant function.

PROBLEM 7

Find $\csc(0)$, $\csc\left(\dfrac{\pi}{6}\right)$, and $\csc\left(\dfrac{\pi}{2}\right)$.

Solution: (a) $\csc(0) = \dfrac{1}{\sin(0)} = \dfrac{1}{0}$, which is undefined

(b) $\csc\left(\dfrac{\pi}{6}\right) = \dfrac{1}{\sin\left(\dfrac{\pi}{6}\right)} = \dfrac{1}{\dfrac{1}{2}} = 2$

(c) $\csc\left(\dfrac{\pi}{2}\right) = \dfrac{1}{\sin\left(\dfrac{\pi}{2}\right)} = \dfrac{1}{1} = 1$

PROPERTIES OF THE COSECANT FUNCTION

1. The cosecant function is periodic with period 2π. That is, $\csc(\theta + 2\pi) = \csc\theta$.

Proof: $\csc(\theta + 2\pi) = \dfrac{1}{\sin(\theta + 2\pi)} = \dfrac{1}{\sin\theta} = \csc\theta.$

2. The cosecant function is an odd function. That is, $\csc(-\theta) = -\csc\theta$.

Proof: $\csc(-\theta) = \dfrac{1}{\sin(-\theta)} = \dfrac{1}{-\sin\theta} = -\csc\theta.$

The graph of $\csc\theta$ is given in Figure 148. The lines $\theta = n\pi$ are asymptotes. There are no zeros for the cosecant function.

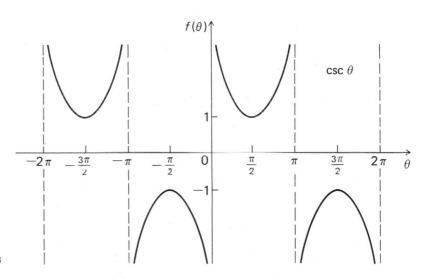

Figure 148

Definition: The **Inverse cosecant** (or **arccosecant**), denoted by $\csc^{-1}$, is defined by the equation

$$\csc^{-1} x = \sin^{-1}\left(\frac{1}{x}\right)$$

The domain of csc^{-1} is the set of all real numbers *except* for $-1 < x < 1$. The range is given by $0 < csc^{-1} x \le \frac{\pi}{2}$ for $x \ge 1$, and $-\frac{\pi}{2} \le csc^{-1} x < 0$ for $x \le -1$.

PROBLEM 8

Find, if possible, $csc^{-1}\left(\frac{1}{2}\right)$, $csc^{-1}(1)$, and $csc^{-1}(-1)$.

Solution: (a) $csc^{-1}\left(\frac{1}{2}\right)$ does not exist because $-1 < x < 1$ is not included in the domain of csc^{-1}.

(b) $csc^{-1}(1) = sin^{-1}\left(\frac{1}{1}\right) = sin^{-1}(1) = \frac{\pi}{2}$

(c) $csc^{-1}(-1) = sin^{-1}\left(\frac{1}{-1}\right) = sin^{-1}(-1) = -\frac{\pi}{2}$

The graph of the inverse cosecant function is given in Figure 149.

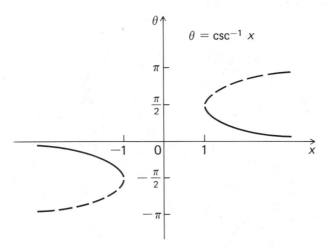

Figure 149

5.5 EXERCISES

1. Complete the following table:

θ	0	$\frac{\pi}{6}$	$\frac{\pi}{4}$	$\frac{\pi}{3}$	$\frac{\pi}{2}$	$-\frac{\pi}{6}$	$-\frac{\pi}{4}$	$-\frac{\pi}{3}$
$\tan \theta$								

2. Using the results of Exercise 1, complete the following table:

x	$-\sqrt{3}$	-1	$-\dfrac{\sqrt{3}}{3}$	0	$\dfrac{\sqrt{3}}{3}$	1	$\sqrt{3}$
$\tan^{-1} x$							

3. Graph:
   (a) $y = \tan 2\theta$
   (b) $y = 2 \tan \left(\theta + \dfrac{\pi}{2} \right)$.

4. Evaluate each of the following:
   (a) $\cot \left(\dfrac{\pi}{4} \right)$ (b) $\cot \pi$
   (c) $\cot \left(\dfrac{\pi}{3} \right)$ (d) $\cot \left(-\dfrac{\pi}{4} \right)$
   (e) $\cot \left(-\dfrac{\pi}{3} \right)$ (f) $\cot \left(-\dfrac{11\pi}{6} \right)$

5. Graph:
   (a) $y = \cot 2\theta$
   (b) $y = 2 \cot \theta$.

6. Evaluate each of the following:
   (a) $\sec \left(\dfrac{\pi}{4} \right)$ (b) $\sec \left(\dfrac{2\pi}{3} \right)$
   (c) $\sec \left(\dfrac{9\pi}{4} \right)$ (d) $\sec \left(-\dfrac{\pi}{4} \right)$
   (e) $\sec \left(\dfrac{-2\pi}{3} \right)$ (f) $\sec \left(\dfrac{-7\pi}{4} \right)$
   (g) $\csc \left(\dfrac{\pi}{4} \right)$ (h) $\csc \left(\dfrac{2\pi}{3} \right)$
   (i) $\csc \left(\dfrac{\pi}{6} \right)$

7. Graph:
   (a) $y = \sec \left(\theta + \dfrac{\pi}{2} \right)$
   (b) $y = 1 + \csc \theta$.

8. Evaluate each of the following:
   (a) $\cot^{-1} (-1)$ (b) $\cot^{-1} (\sqrt{3})$
   (c) $\sec^{-1} (\sqrt{2})$ (d) $\sec^{-1} (-1)$
   (e) $\csc^{-1} (2)$ (f) $\csc^{-1} (\sqrt{2})$

9. Show that $\tan^{-1} (-x) = -\tan^{-1} (x)$.

10. Express each of the following expressions in terms of x:
    (a) $\cot^{-1} (\cot x)$ (b) $\tan(\tan^{-1} x)$
    (c) $\sin(\tan^{-1} x)$ (d) $\cos(\tan^{-1} x)$

SUMMARY

Periodic Functions

1. A function f is **periodic** if $f(x + p) = f(x)$ for all values of x in the domain of f and for some fixed value p. The smallest positive period p is called the **fundamental period** of f.

2. If f is periodic with fundamental period p, then $f(x + np) = f(x)$ for all integers n.

Sine and Cosine

1. Let P be the winding function on the unit circle defined by $P(\theta) = (x, y)$. Then $x = \cos \theta$ and $y = \sin \theta$. Thus, $P(\theta) = (\cos \theta, \sin \theta)$.
2. For any real number θ, $\sin^2 \theta + \cos^2 \theta = 1$.
3. $\sin(-\theta) = -\sin \theta$; sin is an **odd** function.
   $\cos(-\theta) = \cos \theta$; cos is an **even** function.

Graphs of Sine and Cosine

1. Graph of **sine:**

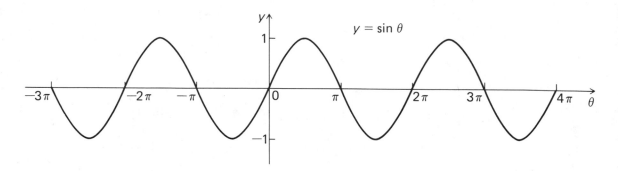

2. Graph of **cosine:**

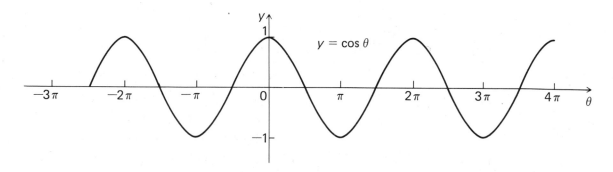

3. $\sin \theta = \sin(\theta + 2n\pi)$, n any integer.
   $\cos \theta = \cos(\theta + 2n\pi)$, n any integer.
   Thus, sine and cosine are periodic functions with the fundamental period 2π.
4. $y = A + B \sin(k\theta + C)$ or $y = A + B \cos(k\theta + C)$, $k > 0$. The graphs of these equations can be obtained from the graphs of $y = \sin \theta$ and $y = \cos \theta$, respectively, by some translation, stretching, or contraction.

$$\text{amplitude} = |B|, \text{ period } = \frac{2\pi}{|k|}$$

$$\text{phase shift} = \begin{cases} \dfrac{C}{k} \text{ units to right if } \dfrac{C}{k} < 0 \\ \dfrac{C}{k} \text{ units to left if } \dfrac{C}{k} > 0 \end{cases}$$

Inverse Sine and Inverse Cosine

1. $\theta = \sin^{-1} x$ if and only if $x = \sin \theta$ and $-\dfrac{\pi}{2} \le \theta \le \dfrac{\pi}{2}$; the domain of $\sin^{-1}$ is $[-1, 1]$ and the range is $\left[-\dfrac{\pi}{2}, \dfrac{\pi}{2}\right]$.

2. $\theta = \cos^{-1} x$ if and only if $x = \cos \theta$ and $0 \le \theta \le \pi$; the domain of $\cos^{-1}$ is $[-1, 1]$ and the range is $[0, \pi]$.

Other Circular Functions

1. $\tan \theta = \dfrac{\sin \theta}{\cos \theta}$, $\theta \ne \dfrac{\pi}{2} + n\pi$

   $\cot \theta = \dfrac{\cos \theta}{\sin \theta}$, $\theta \ne n\pi$

   $\sec \theta = \dfrac{1}{\cos \theta}$, $\theta \ne \dfrac{\pi}{2} + n\pi$

   $\csc \theta = \dfrac{1}{\sin \theta}$, $\theta \ne n\pi$

2. tan and cot have fundamental period $= \pi$.
   sec and csc have fundamental period $= 2\pi$.

3. Inverse:

   (a) $\theta = \tan^{-1} x$ if and only if $x = \tan \theta$ and $-\dfrac{\pi}{2} < \theta < \dfrac{\pi}{2}$; the domain of $\tan^{-1}$ is the set of real numbers and the range is $\left(-\dfrac{\pi}{2}, \dfrac{\pi}{2}\right)$.

   (b) $\cot^{-1} x = \dfrac{\pi}{2} - \tan^{-1} x$; the domain is all real numbers and the range is $(0, \pi)$.

   (c) $\sec^{-1} x = \cos^{-1} \left(\dfrac{1}{x}\right)$; the domain is all real numbers *except* $-1 < x < 1$ and the range consists of:

   $$\begin{cases} 0 \le \sec^{-1} x < \dfrac{\pi}{2}, \text{ for } x \ge 1 \\[2mm] \dfrac{\pi}{2} < \sec^{-1} x \le \pi, \text{ for } x \le -1 \end{cases}$$

   (d) $\csc^{-1} x = \sin^{-1} \left(\dfrac{1}{x}\right)$; the domain is all real numbers *except* $-1 < x < 1$ and the range consists of:

   $$\begin{cases} 0 < \csc^{-1} x \le \dfrac{\pi}{2}, \text{ for } x \ge 1 \\[2mm] -\dfrac{\pi}{2} \le \csc^{-1} x < 0, \text{ for } x \le -1 \end{cases}$$

CHAPTER 5 EXERCISES

1. Let P be the winding function on the unit circle where $P(\theta) = (\cos \theta, \sin \theta)$. Find $P\left(-\dfrac{3\pi}{4}\right)$.

2. Graph:

   (a) $y = \sin \left(\theta + \dfrac{\pi}{4}\right)$

   (b) $y = \pi + \cos \theta$

3. Prove the identities:
   (a) $\cos \theta = \sin \theta \cot \theta$
   (b) $\csc \theta = \cot \theta \sec \theta$

4. Evaluate each of the following:
   (a) $\sin \left[\sin^{-1} \left(\frac{\sqrt{2}}{2} \right) \right]$
   (b) $\cos \left[\sin^{-1} \left(-\frac{\sqrt{2}}{2} \right) \right]$
   (c) $\cos \left[\sin^{-1} \left(\sin \frac{\pi}{2} \right) \right]$
   (d) $\sin^{-1} \left[\cos(\sin^{-1} 1) \right]$

5. Prove:
   (a) $1 + \tan^2 \theta = \sec^2 \theta$, where $\cos \theta \neq 0$
   (b) $1 + \cot^2 \theta = \csc^2 \theta$, where $\sin \theta \neq 0$

6. Evaluate each of the following:
   (a) $\tan[\tan^{-1}(\theta + 2)]$
   (b) $\sin[\tan^{-1}(-1)]$
   (c) $\tan^{-1} \left[\sin \left(\frac{\pi}{2} \right) \right]$
   (d) $\tan \left[\sin^{-1} \left(\frac{\sqrt{3}}{2} \right) \right]$

7. Determine if each of the following functions is even, odd, or neither:
   (a) $f(\theta) = \sin \theta \cos \theta$
   (b) $f(\theta) = \cot 2\theta$

8. Show that $\sin^{-1}(-x) = -\sin^{-1}(x)$.

9. Find $\sin[\sin^{-1}(x)]$ in terms of x.

10. Determine the values of θ such that:
    (a) $\sin \theta = -\cos \theta$
    (b) $\sin \theta = \cos(-\theta)$

CHAPTER 6

TRIGONOMETRIC FUNCTIONS

In Chapter 5 we discussed circular functions and their properties. The domains of the circular functions consisted of real numbers θ. In this chapter we examine an application of the circular functions where the real number θ is the measure of an angle.

6.1 ANGLES, RADIANS, AND DEGREES

Recall from geometry that an angle is defined as a geometric configuration formed by the union of two half-lines (rays) with common origin in the plane. The angle is formed by one of the rays revolving about the point of intersection while the other ray remains fixed. The fixed ray is called the **initial side,** and the rotating ray is called the **terminal side** of the angle. The fixed point is called the **vertex** of the angle (see Fig. 150). The angle is said to be positive if the rotation

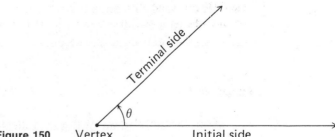

Figure 150   Vertex θ Initial side   Terminal side

is counterclockwise and negative if the rotation is clockwise. If we place the initial side so that it coincides with the positive x-axis and the vertex is at the origin, then the angle θ is said to be in **standard position** (see Fig. 151).

Along with direction, each angle has a magnitude. The magnitude of an angle is a numerical measure of the amount of rotation. We shall see that the measure of an angle is determined by measuring the length of the arc it intercepts on the unit circle.

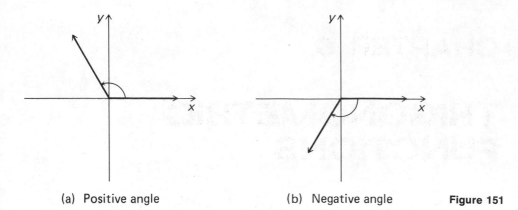

(a) Positive angle   (b) Negative angle   **Figure 151**

One method of measuring an angle is to divide the circumference of the unit circle into 2π equal arcs. The measure of each of these arcs is called 1 **radian.** To define a radian more formally, consider a unit circle with its center at the origin (see Fig. 152).

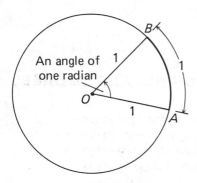

Figure 152

Definition: One **radian** is the measure of the central angle *AOB* which corresponds to a circular arc of length 1 and radius 1.

EXAMPLE 1

The circumference of the unit circle is 2π; therefore, the angle which subtends an arc of length 2π has a measure of 2π radians.

EXAMPLE 2

The angle whose initial and terminal sides form a straight line has a measure of π radians. This is because such an angle subtends an arc of length equal to one half the circumference of the circle. That is, $\frac{1}{2}(2\pi) = \pi$ (see Fig. 153).

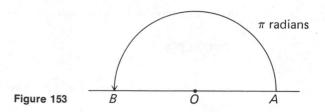

Figure 153 B O A

EXAMPLE 3

A right angle measures $\frac{\pi}{2}$ radians.

Another method of measuring an angle is to divide the circumference of the unit circle into 360 equal arcs. The measure of each of these arcs is defined to be 1 **degree,** denoted by 1°. Each degree is divided into 60 minutes, written 60′, and each minute is divided into 60 seconds, denoted by 60″. Thus, the measure of an angle is the number of degrees of the arc subtended by the angle (see Fig. 154).

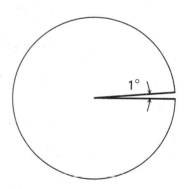

Figure 154

EXAMPLE 4

The circumference of a unit circle is 2π; therefore, the angle which subtends an arc of length 2π measures 360°.

EXAMPLE 5

The measure of the angle whose initial and terminal sides form a straight line is 180°. Also, the measure of a right angle is 90°.

The two methods of measuring an angle give us a relation between radians and degrees. Since 2π radians = 360 degrees, we have

$$1 \text{ radian} = \frac{180}{\pi} \text{ degrees} \approx 57.2958°$$

$$1° = \frac{\pi}{180} \text{ radian} \approx 0.0174533 \text{ radian}$$

It is conventional to omit the unit ''radian''; that is, we shall write $180° = \pi$.

PROBLEM 1

Convert $\theta = \dfrac{\pi}{4}$, $\theta = \dfrac{5\pi}{6}$, and $\theta = 2$ to degrees.

Solution: (a) $\dfrac{\pi}{4}$ radians $= \dfrac{\pi}{4}\left(\dfrac{180°}{\pi}\right) = 45°$

(b) $\dfrac{5\pi}{6}$ radians $= \dfrac{5\pi}{6}\left(\dfrac{180°}{\pi}\right) = 150°$

(c) 2 radians $= 2\left(\dfrac{180°}{\pi}\right) \approx 114°35'$

PROBLEM 2

Convert $\theta = 60°$, $\theta = -225°$, and $\theta = 315°$ to radians.

Solution: (a) $60° = 60\left(\dfrac{\pi}{180}\right) = \dfrac{\pi}{3}$

(b) $-225° = -225\left(\dfrac{\pi}{180}\right) = \dfrac{-5\pi}{4}$

(c) $315° = 315\left(\dfrac{\pi}{180}\right) = \dfrac{7\pi}{4}$

The following are the radian measures of some frequently used angles:

$$15° = \frac{\pi}{12} \qquad 30° = \frac{\pi}{6} \qquad 45° = \frac{\pi}{4} \qquad 60° = \frac{\pi}{3} \qquad 75° = \frac{5\pi}{12}$$

$$90° = \frac{\pi}{2} \qquad 105° = \frac{7\pi}{12} \qquad 120° = \frac{2\pi}{3} \qquad 135° = \frac{3\pi}{4} \qquad 150° = \frac{5\pi}{6}$$

$$180° = \pi \qquad 225° = \frac{5\pi}{4} \qquad 270° = \frac{3\pi}{2} \qquad 315° = \frac{7\pi}{4} \qquad 360° = 2\pi$$

Length of a Circular Arc

Suppose we wish to find the length of the arc subtended by an angle θ. Consider the unit circle $x^2 + y^2 = 1$ and the circle $x^2 + y^2 = r^2$, with radius $r > 1$ (see Fig. 155). Figure 155 shows that the arcs θ and s determine the same central

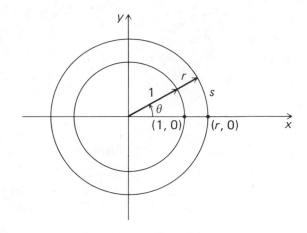

Figure 155

angle. Since the arcs determine the same angle, the respective arcs and radii are proportional. That is,

$$\frac{\theta}{s} = \frac{1}{r}$$

or, equivalently,

$$s = r\theta$$

Thus, we have the following theorem.

Theorem 1: The length s of the arc of a circle whose radius is r and whose central angle is equal to θ radians is given by

$$s = r\theta$$

PROBLEM 3

Find the length of the circular arc that is generated by a central angle of 120° and a circle of radius 4.

Solution: First, we convert 120° to radians: $120° = \frac{2\pi}{3}$. Thus,

$$s = 4 \cdot \left(\frac{2\pi}{3}\right) = \frac{8\pi}{3}$$

Area of a Circular Sector

Consider the circle $x^2 + y^2 = r^2$ (see Fig. 156). Suppose we wish to find the area A of the sector formed by the angle θ. We use the following proportion:

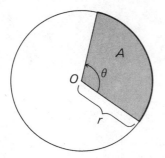

Figure 156

$$\frac{\text{area of sector}}{\text{area of circle}} = \frac{\theta}{2\pi}$$

Since the area of a circle with radius r is πr^2, we have

$$\frac{A}{\pi r^2} = \frac{\theta}{2\pi}$$

or, equivalently,

$$A = \frac{1}{2}r^2\theta$$

Therefore, we have the following theorem.

Theorem 2: The area A of a sector of a circle with radius r and central angle θ, measured in radians, is given by

$$A = \frac{1}{2}r^2\theta$$

PROBLEM 4

Find the area of the sector of a circle with radius 4 inches and subtended by an angle of 105°.

Solution: Since $105° = \frac{7\pi}{12}$, $A = \frac{1}{2}(4)^2\left(\frac{7\pi}{12}\right) = \frac{14\pi}{3}$ square inches.

Angular Speed

The term speed is defined as distance traveled per unit of time. Also, we can define the term **angular speed,** which is the amount of rotation per unit of time. Angular speed is usually measured in revolutions per minute, written rpm. If the terminal side of an angle moves through an angle θ at a constant speed in time t, then its **angular speed** is

$$w = \frac{\theta}{t}$$

Using the formula $s = r\theta$, which relates arc length to the subtended angle, we can establish a relationship between linear speed and angular speed.

Consider a point on the terminal side of an angle. The point moves a linear distance s during the same time t; therefore, its **linear speed** is $v = \frac{s}{t}$. We now divide both sides of $s = r\theta$ by t to get

$$\frac{s}{t} = \frac{r\theta}{t}$$

But $v = \frac{s}{t}$, the linear speed, and $w = \frac{\theta}{t}$, the angular speed. Thus,

$$v = rw$$

and we have the following theorem.

Theorem 3: The linear speed v of a point a distance r from the center of rotation is

$$v = rw$$

where w is the angular speed measured in radians per unit time.

Note: The units of time must be the same for v and w, and the units of distance for v and r must be the same.

PROBLEM 5

A car's wheels have a 12-inch radius, and are revolving at the rate of 350 rpm. How fast is the car traveling?

Solution: We begin by converting $w = 350$ rpm to radians per minute. Recall that one revolution is equal to 2π radians. Thus, 350 rpm = $350(2\pi) = 700\pi$ radians per minute. Then $v = rw = 12(700\pi) = 8400\pi$ inches per minute. Expressing this in terms of miles per hour we multiply by 60 minutes per hour and divide by (5280)(12) inches per mile to obtain

$$v = \frac{(8400\pi)(60)}{(5280)(12)} = (7.9\pi)\text{mph} \approx 25 \text{ mph}$$

6.1 EXERCISES

1. Convert to degrees:

   (a) $\dfrac{\pi}{3}$ (b) $\dfrac{5\pi}{6}$

   (c) $\dfrac{6\pi}{5}$ (d) $\dfrac{19\pi}{6}$

   (e) $\dfrac{\pi}{10}$ (f) 0

   (g) $\dfrac{2\pi}{9}$ (h) $\dfrac{5\pi}{3}$

2. Convert to radians:
   (a) $120°$ (b) $6°$
   (c) $-80°$ (d) $9°$
   (e) $15°$ (f) $67.5°$
   (g) $-315°$ (h) $357°$

3. Find the arc length of a circle of radius r which is subtended by a central angle θ for each of the following.
   (a) $r = 2$ inches and $\theta = 105°$

   (b) $r = 4$ inches and $\theta = \dfrac{5\pi}{4}$

   (c) $r = 5$ inches and $\theta = \pi°$

   (d) $r = \dfrac{1}{\pi}$ inches and $\theta = 135°$

4. Find the area of the sector of a circle of radius r which is subtended by a central angle θ for each of the following:

   (a) $r = 3$ inches and $\theta = \dfrac{\pi}{27}$

   (b) $r = 2$ inches and $\theta = 150°$

   (c) $r = 4$ inches and $\theta = 180°$

   (d) $r = 10$ inches and $\theta = \dfrac{5\pi}{4}$

5. The diameter of the wheels of a bike is 20 inches and the wheels are moving at 5 rpm. How fast is the bike moving?

6.2 TRIGONOMETRIC FUNCTIONS

In Chapter 5 we defined trigonometric functions for which the domains were the set of real numbers. In this section we shall show that these functions also are defined so that their domains are sets of angles. If the angle measurement is given in radians, the domains of the functions are the same. For example, we define the sine of $\dfrac{\pi}{4}$ radians to be the same as $\sin\dfrac{\pi}{4}$. If the angle measurement is given in degrees, we convert degrees to radians to determine the value of the trigonometric functions. In particular, for the sine and cosine functions we have:

> The sine of an angle is the same number as the sine of the radian measure of the angle, the cosine of an angle is the same number as the cosine of the radian measure of the angle, and similarly for the other functions—tangent, cotangent, secant, and cosecant.

EXAMPLE 1

Since $60° = \dfrac{\pi}{3}$ radians, $\sin 60° = \sin \dfrac{\pi}{3} = \dfrac{\sqrt{3}}{2}$. Similarly, $\cos 60° = \cos\dfrac{\pi}{3} = \dfrac{1}{2}$, $\tan 60° = \tan\dfrac{\pi}{3} = \sqrt{3}$, $\cot 60° = \cot\dfrac{\pi}{3} = \dfrac{\sqrt{3}}{3}$, $\sec 60° = \sec\dfrac{\pi}{3} = 2$, and $\csc 60° = \csc\dfrac{\pi}{3} = \dfrac{2\sqrt{3}}{3}$.

Consider an angle θ in standard position (see Fig. 157). Let $P_1(x_1, y_1)$ and

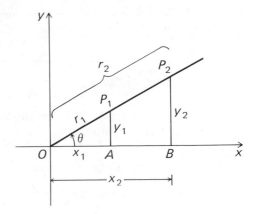

Figure 157

$P_2(x_2, y_2)$ be two distinct points on the terminal side of θ other than $(0, 0)$. The triangles OAP_1 and OBP_2 are similar triangles, and thus have proportional sides. That is,

$$\frac{y_1}{r_1} = \frac{y_2}{r_2}, \frac{x_1}{r_1} = \frac{x_2}{r_2}, \text{ and } \frac{y_1}{x_1} = \frac{y_2}{x_2}$$

where $r_1 = \sqrt{x_1^2 + y_1^2}$ and $r_2 = \sqrt{x_2^2 + y_2^2}$. Also, the similar triangles give

$$\frac{x_1}{y_1} = \frac{x_2}{y_2}, \frac{r_1}{x_1} = \frac{r_2}{x_2}, \text{ and } \frac{r_1}{y_1} = \frac{r_2}{y_2}$$

Thus, the ratios $\frac{x}{r}, \frac{y}{r}$, and $\frac{x}{y}$ are independent of the choice of two points taken on the terminal side of the angle θ. Their ratios are used to define the six trigonometric functions.

Definition: Let θ be an angle in standard position and let (x, y) be any point other than $(0, 0)$ on the terminal side of θ. Let $r = \sqrt{x^2 + y^2}$ be the distance between (x, y) and $(0, 0)$. Then the six *trigonometric functions* are defined as follows:

$$\text{sine} = \left\{ (\theta, \sin \theta) | \sin \theta = \frac{y}{r} \right\}$$

$$\text{cosine} = \left\{ (\theta, \cos \theta) | \cos \theta = \frac{x}{r} \right\}$$

$$\text{tangent} = \left\{ (\theta, \tan \theta) | \tan \theta = \frac{y}{x}, x \neq 0 \right\}$$

$$\text{cotangent} = \left\{ (\theta, \cot \theta) | \cot \theta = \frac{x}{y}, y \neq 0 \right\}$$

$$\text{secant} = \left\{ (\theta, \sec \theta) | \sec \theta = \frac{r}{x}, x \neq 0 \right\}$$

$$\text{cosecant} = \left\{(\theta,\ \csc\theta)|\csc\theta = \frac{r}{y},\ y \neq 0\right\}$$

PROBLEM 1

Determine the trigonometric functions of the angle θ if $(-2, 3)$ lies on its terminal side.

Solution: Since $(x,\ y) = (-2,\ 3)$, we have $r = \sqrt{(-2)^2 + (3)^2} = \sqrt{13}$ and $\sin\theta = \dfrac{y}{r} = \dfrac{3}{\sqrt{13}}$, $\cos\theta = \dfrac{x}{r} = \dfrac{-2}{\sqrt{13}}$, $\tan\theta = \dfrac{y}{x} = \dfrac{-3}{2}$, $\cot\theta = \dfrac{x}{y} = \dfrac{-2}{3}$, $\sec\theta = \dfrac{r}{x} = \dfrac{-\sqrt{13}}{2}$, $\csc\theta = \dfrac{r}{y} = \dfrac{\sqrt{13}}{3}$.

PROBLEM 2

Evaluate the trigonometric functions of θ when $\theta = 45°$.

Solution: Let θ be in standard position with the vertex situated at the center of the unit circle (see Fig. 158). We form the right triangle OAB by

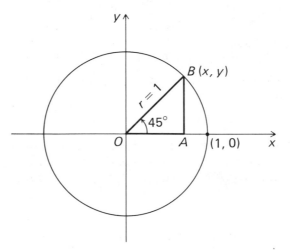

Figure 158

dropping a perpendicular line segment to the x-axis from the point B. The triangle OAB is an isosceles right triangle where $\overline{OA} = \overline{AB}$. Since the point B is on the circumference of the unit circle, $\overline{OB} = r = 1$. The Pythagorean Theorem gives

$$(\overline{OA})^2 + (\overline{AB})^2 = (\overline{OB})^2$$

$$2(\overline{OA})^2 = (1)^2,\ \text{because } \overline{OA} = \overline{AB},\ \text{and } \overline{OB} = 1$$

$$(\overline{OA})^2 = \frac{1}{2}, \text{ dividing by 2}$$

$$\overline{OA} = \frac{1}{\sqrt{2}}, \text{ taking square root and distance is positive}$$

$$\overline{OA} = \frac{\sqrt{2}}{2}, \text{ rationalizing the denominator}$$

Since $\overline{OA} = \overline{AB}$, we have $x = \overline{OA} = \frac{\sqrt{2}}{2}$ and $y = \overline{AB} = \frac{\sqrt{2}}{2}$. Thus, sin $45° = \frac{y}{r} = \frac{\sqrt{2}}{2}$, cos $45° = \frac{x}{r} = \frac{\sqrt{2}}{2}$, tan $45° = \frac{y}{x} = 1$, cot $45° = \frac{y}{x} = 1$, sec $45° = \frac{r}{x} = \frac{1}{\frac{\sqrt{2}}{2}} = \frac{2}{\sqrt{2}} = \sqrt{2}$, csc $45° = \frac{r}{y} = \frac{1}{\frac{\sqrt{2}}{2}} = \frac{2}{\sqrt{2}} = \sqrt{2}$.

PROBLEM 3

Determine in which quadrant the terminal side of θ lies if cos $\theta = -\frac{9}{10}$. Evaluate sin θ.

Solution: If cos $\theta = \frac{x}{r} = -\frac{9}{10}$, then $x = -9$ and $r = 10$ because r is always positive. Therefore, θ is in either Quadrant II or III. Since $r = \sqrt{x^2 + y^2}$, we have $10 = \sqrt{(-9)^2 + y^2} \Rightarrow 10 = \sqrt{81 + y^2} \Rightarrow 10^2 = 81 + y^2 \Rightarrow y^2 = 19 \Rightarrow y = \pm\sqrt{19}$. Thus, sin $\theta = \frac{y}{r} = \frac{\sqrt{19}}{10}$ if θ is in Quadrant II, and sin $\theta = \frac{y}{r} = \frac{-\sqrt{19}}{10}$ if θ is in Quadrant III. Actually there are infinitely many possible values of θ because of the periodicity of the trigonometric functions. In any case, such angles must have one of the two terminal sides given in Figure 159.

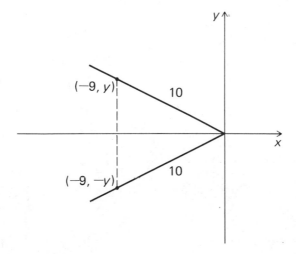

Figure 159

6.2 EXERCISES

1. Determine the values of the sine, cosine, and tangent of an angle θ for each of the given points on the terminal side of θ.
   (a) $(0, 6)$ (b) $(-1, -2)$
   (c) $(3, -\sqrt{5})$ (d) $(0, 1)$
   (e) $(-5, 12)$ (f) $(-2, -3)$

2. Using Figure 160, find the value of the six trigonometric functions of θ where $\theta = 30°$. (Hint: Use the relationships between the sides of a 30°, 60°, 90° triangle.)

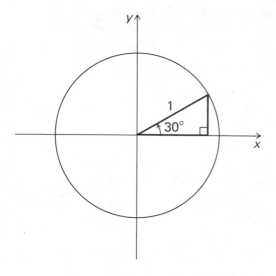

Figure 160

3. Use the information from Exercise 2 to find the values of the trigonometric functions for an angle of 60°.

4. Determine in which quadrant the terminal side of θ lies if $\sin \theta = -\dfrac{7}{9}$. Find $\cos \theta$ and $\tan \theta$.

5. Let $P(x, y)$ be any point in the xy-plane. Show that its coordinates can be written as $(r \cos \theta, r \sin \theta)$ where r is the distance from the origin to P and θ is in standard position. Locating points in this manner is called using the polar coordinate system (see Fig. 161).

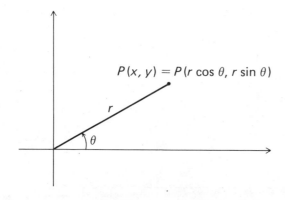

$$P(x, y) = P(r \cos \theta, r \sin \theta)$$

Figure 161

6.3 TRIGONOMETRIC IDENTITIES

Basic Identities

An identity is an equation that is true for all values of the variables for which both sides of the equation are defined. Since the six trigonometric functions are very closely related to each other, there are many identities relating the trigonometric functions. We begin with the basic identities. The first was discussed in Chapter 5.

$$\sin^2 \theta + \cos^2 \theta = 1 \tag{1}$$

If we divide both sides of (1) by $\cos^2 \theta$, we obtain $\dfrac{\sin^2 \theta}{\cos^2 \theta} + 1 = \dfrac{1}{\cos^2 \theta}$. Since $\tan \theta = \dfrac{\sin \theta}{\cos \theta}$ and $\sec \theta = \dfrac{1}{\cos \theta}$ we get a second identity.

$$\tan^2 \theta + 1 = \sec^2 \theta \tag{2}$$

To derive a third identity, we divide both sides of (1) by $\sin^2 \theta$ to get $1 + \dfrac{\cos^2 \theta}{\sin^2 \theta} = \dfrac{1}{\sin^2 \theta}$. Using $\cot \theta = \dfrac{\cos \theta}{\sin \theta}$ and $\csc \theta = \dfrac{1}{\sin \theta}$, we obtain

$$1 + \cot^2 \theta = \csc^2 \theta \tag{3}$$

In order to obtain the proof of a particular identity, many substitutions and manipulations may be necessary. One technique for proving identities is to express everything in terms of sines and cosines before performing algebraic manipulations.

PROBLEM 1

Prove that $\csc \theta = \sin \theta + \cos \theta \cot \theta$.

Solution: Using $\cot \theta = \dfrac{\cos \theta}{\sin \theta}$, and writing the right side in terms of sine and cosine, the right side becomes

$$= \sin \theta + \cos \theta \, \frac{\cos \theta}{\sin \theta}$$

$$= \sin \theta + \frac{\cos^2 \theta}{\sin \theta}$$

$$= \frac{\sin^2 \theta + \cos^2 \theta}{\sin \theta} \quad \text{(common denominator)}$$

$$= \frac{1}{\sin \theta} \quad \quad (\sin^2 \theta + \cos^2 \theta = 1)$$

$$= \csc \theta$$

PROBLEM 2

Prove that $\dfrac{\sin \theta}{1 - \cos \theta} - \cot \theta = \dfrac{1}{\sin \theta}$ is an identity.

Solution: Writing the left side in terms of sines and cosines, we obtain

$$\frac{\sin \theta}{1 - \cos \theta} - \frac{\cos \theta}{\sin \theta}$$

$$\|$$

$$\frac{\sin^2 \theta - (1 - \cos \theta) \cos \theta}{(1 - \cos \theta) \sin \theta}$$

$$\|$$

$$\frac{\sin^2 \theta - \cos \theta + \cos^2 \theta}{(1 - \cos \theta) \sin \theta}$$

$$\|$$

$$\frac{\sin^2 \theta + \cos^2 \theta - \cos \theta}{(1 - \cos \theta) \sin \theta}$$

$$\|$$

$$\frac{(1 - \cos \theta)}{(1 - \cos \theta) \sin \theta}$$

$$\|$$

$$\frac{1}{\sin \theta}$$

Converting to sines and cosines is not the only technique used to prove identities.

PROBLEM 3

Prove that $\dfrac{1 - \tan \theta}{1 + \tan \theta} = \dfrac{\cot \theta - 1}{\cot \theta + 1}.$

Solution: Since $\cot \theta = \dfrac{1}{\tan \theta}$, we have

$$\frac{\cot \theta - 1}{\cot \theta + 1} = \frac{\left(\dfrac{1}{\tan \theta}\right) - 1}{\left(\dfrac{1}{\tan \theta}\right) + 1} = \frac{\dfrac{(1 - \tan \theta)}{\tan \theta}}{\dfrac{(1 + \tan \theta)}{\tan \theta}} = \frac{1 - \tan \theta}{1 + \tan \theta}$$

6.3(A) EXERCISES

1. Simplify:
   (a) $\sin^4 \theta - \cos^4 \theta$ (c) $(\sin \theta + \cos \theta)^2 - 2 \sin \theta \cos \theta$

   (b) $\dfrac{1 - \sin^2 \theta}{\cos \theta}$ (d) $\dfrac{1 + \cot^2 \theta}{\cot^2 \theta}$

In Exercises 2 to 15 prove the identity:

2. $\tan \theta \cot \theta \sec \theta \cos \theta = 1$ 3. $\dfrac{1 + \cos \theta}{\sin \theta} = \dfrac{\sin \theta}{1 - \cos \theta}$

4. $\sec^2 \theta - \csc^2 \theta = \tan^2 \theta - \cot^2 \theta.$ 5. $\sec^2 \theta + \csc^2 \theta = \sec^2 \theta \csc^2 \theta$

6. $\sec^4 \theta - \tan^4 \theta = 1 + 2 \tan^2 \theta$ 7. $\tan \theta + \cot \theta = \sec \theta \csc \theta$

8. $(\cos \theta - \sin \theta)^2 + (\cos \theta + \sin \theta)^2 = 2$ 9. $(a \cos \theta + b \sin \theta)^2 + (-a \sin \theta + b \cos \theta)^2 = a^2 + b^2$

10. $\dfrac{1}{1 - \sin \theta} - \dfrac{1}{1 + \sin \theta} = 2 \tan \theta \sec \theta$ 11. $\dfrac{1 - \cos \theta}{\sin \theta} = \dfrac{\sin \theta}{1 + \cos \theta}$

12. $\sec \theta + \csc \theta = (\tan \theta + \cot \theta)(\cos \theta + \sin \theta)$ 13. $(\csc \theta - \cot \theta)^2 = \dfrac{1 - \cos \theta}{1 + \cos \theta}$

14. $\dfrac{\tan \theta}{1 - \cot \theta} + \dfrac{\cot \theta}{1 - \tan \theta} = 1 + \tan \theta + \cot \theta$ 15. $\dfrac{\tan^3 \theta - \cot^3 \theta}{\tan \theta - \cot \theta} = \tan^2 \theta + \csc^2 \theta$

The Addition Laws

The addition laws for the sine and cosine are formulas that state $\sin (\alpha + \beta)$ and $\cos (\alpha + \beta)$ in terms of $\sin \alpha$, $\sin \beta$, $\cos \alpha$, and $\cos \beta$.

$$\sin (\alpha + \beta) = \sin \alpha \cos \beta + \cos \alpha \sin \beta$$
$$\cos (\alpha + \beta) = \cos \alpha \cos \beta - \sin \alpha \sin \beta$$
$$\sin (\alpha - \beta) = \sin \alpha \cos \beta - \cos \alpha \sin \beta$$
$$\cos (\alpha - \beta) = \cos \alpha \cos \beta + \sin \alpha \sin \beta$$

Proof of $\cos (\alpha - \beta)$: Consider the points $P(\alpha)$, $P(\beta)$, $P(\alpha - \beta)$, and $P(0)$ on the unit circle (see Fig. 162). By definition, $P(\alpha) = (\cos \alpha, \sin \alpha)$, $P(\beta) = (\cos \beta,$

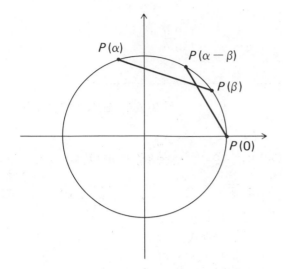

Figure 162

sin β), $P(\alpha - \beta) = [\cos (\alpha - \beta), \sin (\alpha - \beta)]$, and $P(0) = (1, 0)$. If $\alpha > \beta$, the length of the arc from $P(\beta)$ to $P(\alpha)$ is $\alpha - \beta$, and is the same as the length of the arc from $P(0)$ to $P(\alpha - \beta)$. Therefore, the length of the chord from $P(\beta)$ to $P(\alpha)$ must equal the length of the chord from $P(0)$ to $P(\alpha - \beta)$. Using the distance formula, we have

$$\text{dist } [P(\alpha - \beta), P(0)] = \text{dist } [P(\alpha), P(\beta)]$$

$$\sqrt{[\cos (\alpha - \beta) - 1]^2 + [\sin (\alpha - \beta)]^2} = \sqrt{[\cos \alpha - \cos \beta]^2 + [\sin \alpha - \sin \beta]^2}$$

$$[\cos (\alpha - \beta) - 1]^2 + [\sin (\alpha - \beta)]^2 = [\cos \alpha - \cos \beta]^2 + [\sin \alpha - \sin \beta]^2$$

$$\cos^2 (\alpha - \beta) - 2 \cos (\alpha - \beta) + 1 + \sin^2 (\alpha - \beta)$$
$$= \cos^2 \alpha - 2 \cos \alpha \cos \beta + \cos^2 \beta + \sin^2 \alpha - 2 \sin \alpha \sin \beta + \sin^2 \beta$$

Using $\sin^2 x + \cos^2 x = 1$, we simplify the above equation to get $2 - 2 \cos$ $(\alpha - \beta) = 2 - 2 \cos \alpha \cos \beta - 2 \sin \alpha \sin \beta$. Multiply both sides by $\left(-\dfrac{1}{2}\right)$ and add 1 to get

$$\cos (\alpha - \beta) = \cos \alpha \cos \beta + \sin \alpha \sin \beta$$

Proof of cos $(\alpha + \beta)$: If we replace β by $-\beta$, we obtain cos $[\alpha - (-\beta)] = \cos \alpha \cos (-\beta) + \sin \alpha \sin (-\beta)$. Then recall that $\cos (-\beta) = \cos (\beta)$ and $\sin (-\beta) = -\sin \beta$, and we see that

$$\cos (\alpha + \beta) = \cos \alpha \cos \beta - \sin \alpha \sin \beta$$

PROBLEM 4

Use the cos $(\alpha + \beta)$ formula to evaluate $\cos \left(\theta + \dfrac{\pi}{2}\right)$.

Solution: $\cos \left(\theta + \dfrac{\pi}{2}\right) = \cos \theta \cos \dfrac{\pi}{2} - \sin \theta \sin \dfrac{\pi}{2}$. Since $\cos \dfrac{\pi}{2} = 0$

and sin $\frac{\pi}{2} = 1$ we have cos $\left(\theta + \frac{\pi}{2}\right) = -\sin \theta$. Equivalently, sin $\theta =$ $-\cos \left(\theta + \frac{\pi}{2}\right)$.

We can use the results of Problem 4 to prove the addition formula for the sine function.

Proof of sin $(\alpha + \beta)$: Letting $\theta = \alpha + \beta$ in Problem 4 we obtain

$$\sin (\alpha + \beta) = -\cos \left(\alpha + \beta + \frac{\pi}{2}\right) = -\cos \left[\alpha + \left(\beta + \frac{\pi}{2}\right)\right]$$

$$= -\left[\cos \alpha \cos \left(\beta + \frac{\pi}{2}\right) - \sin \alpha \sin \left(\beta + \frac{\pi}{2}\right)\right]$$

Since cos $\left(\beta + \frac{\pi}{2}\right) = -\sin \beta$ and sin $\left(\beta + \frac{\pi}{2}\right) = \cos \beta$, we have sin $(\alpha + \beta) =$ sin α cos β + cos α sin β.

Proof of sin $(\alpha - \beta)$: If we replace β with $-\beta$ and use the facts that $\cos(-\beta) = \cos \beta$ and sin $(-\beta) = -\sin \beta$, we obtain

$$\sin (\alpha - \beta) = \sin \alpha \cos \beta - \cos \alpha \sin \beta$$

PROBLEM 5

Derive an addition formula for the tangent function.

Solution:

$$\tan (\alpha + \beta) = \frac{\sin (\alpha + \beta)}{\cos (\alpha + \beta)} - \frac{\sin \alpha \cos \beta + \cos \alpha \sin \beta}{\cos \alpha \cos \beta - \sin \alpha \sin \beta}$$

In order to obtain our formula in terms of the tangent function, we divide both numerator and denominator by cos α cos β, giving

$$\tan (\alpha + \beta) = \frac{\left(\dfrac{\sin \alpha \cos \beta}{\cos \alpha \cos \beta}\right) + \left(\dfrac{\cos \alpha \sin \beta}{\cos \alpha \cos \beta}\right)}{\left(\dfrac{\cos \alpha \cos \beta}{\cos \alpha \cos \beta}\right) - \left(\dfrac{\sin \alpha \sin \beta}{\cos \alpha \cos \beta}\right)}$$

$$= \frac{\tan \alpha + \tan \beta}{1 - \tan \alpha \tan \beta}$$

Thus,

$$\tan (\alpha + \beta) = \frac{\tan \alpha + \tan \beta}{1 - \tan \alpha \tan \beta}$$

PROBLEM 6

Evaluate tan 195° without using tables.

Solution:

$$\tan 195° = \tan(135° + 60°) = \frac{\tan 135° + \tan 60°}{1 - \tan 135° \tan 60°}$$

$$= \frac{-1 + \sqrt{3}}{1 - (-1)(\sqrt{3})} = \frac{-1 + \sqrt{3}}{1 + \sqrt{3}}$$

Letting $\beta = -\beta$ in Problem 5, we have

$$\tan(\alpha - \beta) = \frac{\tan \alpha - \tan \beta}{1 + \tan \alpha \tan \beta}$$

Double-Angle Laws

Many times we are interested in expressing the trigonometric functions of multiples of θ in terms of the trigonometric functions of θ. Recall that $\sin(\alpha + \beta) = \sin \alpha \cos \beta + \cos \alpha \sin \beta$. If we let $\alpha = \beta = \theta$, then we get

$$\sin 2\theta = \sin(\theta + \theta) = \sin \theta \cos \theta + \cos \theta \sin \theta$$

or

$$\sin 2\theta = 2 \sin \theta \cos \theta$$

Similarly, letting $\alpha = \beta = \theta$ in $\cos(\alpha + \beta)$ we get

$$\cos 2\theta = \cos^2 \theta - \sin^2 \theta$$

The double-angle formula for cosine has two alternate forms. Replacing $\sin^2 \theta$ by $1 - \cos^2 \theta$, then $\cos^2 \theta$ by $1 - \sin^2 \theta$, we obtain

(1) $\cos 2\theta = \cos^2 \theta - \sin^2 \theta = \cos^2 \theta - (1 - \cos^2 \theta) = 2 \cos^2 \theta - 1$
(2) $\cos 2\theta = \cos^2 \theta - \sin^2 \theta = (1 - \sin^2 \theta) - \sin^2 \theta = 1 - 2 \sin^2 \theta$

Thus,

$$\cos 2\theta = \cos^2 \theta - \sin^2 \theta = 2 \cos^2 \theta - 1 = 1 - 2 \sin^2 \theta$$

PROBLEM 7

Derive the double-angle law for the tangent function. That is, find tan 2θ.

Solution: Letting $\alpha = \beta = \theta$ in $\tan(\alpha + \beta)$ gives

$$\tan 2\theta = \frac{2 \tan \theta}{1 - \tan^2 \theta}$$

PROBLEM 8

Express $\cos 3\theta$ in terms of $\sin \theta$ and $\cos \theta$.

Solution: Write $3\theta = \theta + 2\theta$ and use the addition laws and double-angle formulas:

$$\begin{aligned}
\cos 3\theta = \cos(\theta + 2\theta) &= \cos \theta \cos 2\theta - \sin \theta \sin 2\theta \\
&= \cos \theta(\cos^2 \theta - \sin^2 \theta) - \sin \theta(2 \sin \theta \cos \theta) \\
&= \cos^3 \theta - \cos \theta \sin^2 \theta - 2 \sin^2 \theta \cos \theta \\
&= \cos^3 \theta - 3 \sin^2 \theta \cos \theta
\end{aligned}$$

Half-Angle Laws

We can now use the double-angle formulas to derive the half-angle formulas for the sine and cosine. We substitute $\frac{\theta}{2}$ for θ in $\cos 2\theta = 2 \cos^2 \theta - 1$ and $\cos 2\theta = 1 - 2 \sin^2 \theta$ and then solve for $\cos \frac{\theta}{2}$ and $\sin \frac{\theta}{2}$.

Substituting $\frac{\theta}{2}$ for θ in $\cos 2\theta = 2 \cos^2 \theta - 1$, we obtain

$$\cos \left[2 \cdot \frac{\theta}{2} \right] = 2 \cos^2 \left(\frac{\theta}{2} \right) - 1$$

or

$$\cos \theta = 2 \cos^2 \left(\frac{\theta}{2} \right) - 1$$

Solving for $\cos^2 \left(\frac{\theta}{2} \right)$, we get $\cos^2 \left(\frac{\theta}{2} \right) = \frac{1 + \cos \theta}{2}$, or

$$\cos \left(\frac{\theta}{2} \right) = \pm \sqrt{\frac{1 + \cos \theta}{2}}$$

where the sign of the radical depends on the quadrant in which $\frac{\theta}{2}$ is located.

By substituting $\frac{\theta}{2}$ for θ into $\cos 2\theta = 1 - 2\sin^2 \theta$ we obtain the half-angle formula for sine.

$$\sin\left(\frac{\theta}{2}\right) = \pm \sqrt{\frac{1 - \cos \theta}{2}}$$

We can obtain a formula for $\tan\left(\frac{\theta}{2}\right)$ by using $\sin\left(\frac{\theta}{2}\right)$ and $\cos\left(\frac{\theta}{2}\right)$.

$$\tan\left(\frac{\theta}{2}\right) = \frac{\sin\left(\frac{\theta}{2}\right)}{\cos\left(\frac{\theta}{2}\right)} = \pm \frac{\sqrt{1 - \cos \theta}}{\sqrt{1 + \cos \theta}}$$

Thus,

$$\tan\frac{\theta}{2} = \pm \sqrt{\frac{1 - \cos \theta}{1 + \cos \theta}}$$

Figure 163 gives the geometric interpretation of $\tan\left(\frac{\theta}{2}\right)$.

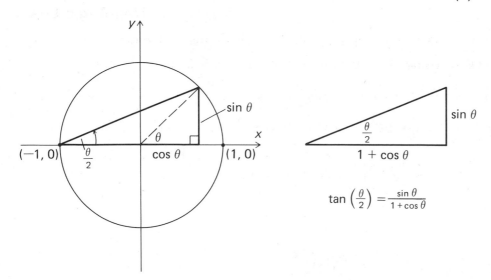

Figure 163

PROBLEM 9

Evaluate $\cos 15°$.

Solution: From the half-angle formula, we have $\cos^2 15° = \cos^2$

$$\left(\frac{30}{2}\right)^{\circ} = \frac{1}{2}(1 + \cos 30°) = \frac{1}{2}\left(1 + \frac{\sqrt{3}}{2}\right). \quad \text{Thus,} \quad \cos^2 15° = \frac{2 + \sqrt{3}}{4}, \text{ or}$$

$$\cos 15° = \frac{\sqrt{2 + \sqrt{3}}}{2}.$$

6.3(B) EXERCISES

Use the addition laws to compute:

1. $\sin(\theta + 2\pi)$ 2. $\cos(\theta + 2\pi)$ 3. $\sin(\theta + \pi)$

4. $\cos(\theta + \pi)$ 5. $\sin\left(\theta + \frac{\pi}{2}\right)$ 6. $\cos\left(\theta + \frac{\pi}{2}\right)$

Compute without tables:

7. $\cos 75°$ 8. $\sin 75°$ 9. $\tan 75°$

10. $\sin 15°$ 11. $\tan 15°$ 12. $\sin 195°$

13. Express $\cot(\alpha + \beta)$ in terms of $\cot \alpha$ and $\cot \beta$.

14. Prove the following identities:
 (a) $\sin(\alpha + \beta) + \sin(\alpha - \beta) = 2 \sin \alpha \cos \beta$
 (b) $\sin(\alpha + \beta) - \sin(\alpha - \beta) = 2 \cos \alpha \sin \beta$
 (c) $\cos(\alpha + \beta) + \cos(\alpha - \beta) = 2 \cos \alpha \cos \beta$
 (d) $\cos(\alpha - \beta) - \cos(\alpha + \beta) = 2 \sin \alpha \sin \beta$

15. (a) Express $\sin 3\theta$ In terms of $\sin \theta$.
 (b) Express $\tan 3\theta$ in terms of $\tan \theta$.
 (c) Express $\cot 3\theta$ in terms of $\cot \theta$.

16. Use double-angle laws to prove:
 (a) $\sin 4\theta = 8 \cos^3 \theta \sin \theta - 4 \cos \theta \sin \theta$
 (b) $\cos 4\theta = 8 \cos^4 \theta - 8 \cos^2 \theta + 1$
 (c) $\cot \theta = \cot 2\theta + \csc 2\theta$
 (d) $\tan \theta = \csc 2\theta - \cot 2\theta$

17. Use half-angle laws to compute:
 (a) $\sin 22.5°$ (b) $\tan 22.5°$
 (c) $\tan 67.5°$ (d) $\sin \frac{\pi}{2}$
 (e) $\tan \frac{\pi}{12}$ (f) $\cos \frac{5\pi}{8}$

18. Prove:
 (a) $\tan\left(\frac{\theta}{2}\right) = \frac{1 - \cos \theta}{\sin \theta}$
 (b) $\cot\left(\frac{\theta}{2}\right) = \frac{1 + \cos \theta}{\sin \theta}$

19. Prove that $\left[\sin\left(\frac{\theta}{2}\right) + \cos\left(\frac{\theta}{2}\right)\right]^2 - \sin \theta = 1$

20. Prove the following identities:
 (a) $\cos(\alpha + \beta) \cos(\alpha - \beta) = \cos^2 \alpha - \sin^2 \beta$
 (b) $\cot^2 \theta = \frac{1 + \cos 2\theta}{1 - \cos 2\theta}$
 (c) $\frac{\sin(\beta - \alpha)}{\sin \alpha \sin \beta} = \cot \alpha - \cot \beta$

21. Prove the following identities:

(a) $\sin \alpha + \sin \beta = 2 \sin \left[\frac{1}{2} (\alpha + \beta) \right]$

$\cos \left[\frac{1}{2} (\alpha - \beta) \right]$

(b) $\sin \alpha - \sin \beta = 2 \cos \left[\frac{1}{2} (\alpha + \beta) \right]$

$\sin \left[\frac{1}{2} (\alpha - \beta) \right]$

(c) $\cos \alpha + \cos \beta = 2 \cos \left[\frac{1}{2} (\alpha + \beta) \right]$

$\cos \left[\frac{1}{2} (\alpha - \beta) \right]$

(d) $\cos \alpha - \cos \beta = -2 \sin \left[\frac{1}{2} (\alpha + \beta) \right]$

$\sin \left[\frac{1}{2} (\alpha - \beta) \right]$

(Hint: Since $\alpha = \frac{1}{2} (\alpha + \beta) + \frac{1}{2} (\alpha - \beta) = u + v$ and $\beta = \frac{1}{2} (\alpha + \beta) - \frac{1}{2} (\alpha - \beta) = u - v$, we can substitute $\alpha = u + v$ and $\beta = u - v$.)

22. Use the results of Exercise 21 to prove

$$\frac{\sin \alpha - \sin \beta}{\cos \alpha + \cos \beta} = \tan \left[\frac{1}{2} (\alpha - \beta) \right]$$

6.4 TRIGONOMETRIC EQUATIONS

Until now, we have discussed equations involving trigonometric functions that were true for all values of the variable. Such equations we called identities. In this section we discuss equations involving trigonometric functions that are not true for all values assigned to the variables. However, since the trigonometric functions are periodic functions, these equations have many solutions. They are called **conditional equations.**

PROBLEM 1

Solve $\sin \theta = 1$.

Solution: The solutions are those values of θ which have a sine of 1. Thus, $\sin \theta = 1$ when $\theta = \frac{\pi}{2}$. Since sine is periodic with period 2π, the solutions are $\frac{\pi}{2}$ and $\frac{\pi}{2}$ plus multiples of 2π. That is, the solutions are $\frac{\pi}{2} + 2k\pi$, $k = 0, \pm 1, \pm 2, \ldots$.

In solving trigonometric equations, it is usually sufficient to find solutions on the interval $[0, 2\pi]$. Then any multiple of 2π can be added to obtain all the solutions.

PROBLEM 2

Solve $2 \cos \theta - \sqrt{3} = 0$.

Solution:
$$2 \cos \theta - \sqrt{3} = 0$$
$$2 \cos \theta = \sqrt{3}$$
$$\cos \theta = \frac{\sqrt{3}}{2}$$

$\theta = \frac{\pi}{6}$ is one solution, another is $\theta = \frac{-\pi}{6}$.

Solutions are $\theta = \pm \frac{\pi}{6} + 2k\pi$, $k = 0, \pm 1, \pm 2, \ldots$.

PROBLEM 3

Solve $\sin 2\theta = 1$.

Solution: Let $u = 2\theta$. Then we have $\sin u = 1$. Using the results of Problem 1, we have $u = \frac{\pi}{2} + 2k\pi$, where k is any integer. Now, since $u = 2\theta$, we have $2\theta = \frac{\pi}{2} + 2k\pi$, or the solution set is given by $\theta = \frac{\pi}{4} + k\pi$, where k is any integer.

Note: We added $2k\pi$ first, and then divided by 2.

In solving trigonometric equations, we may find it necessary to apply some algebraic operations before solving the equation.

PROBLEM 4

Solve $2 \sin^2 \theta + 3 \sin \theta - 2 = 0$.

Solution: Since this equation is quadratic in $\sin \theta$, we can factor the equation to obtain:

$$(2 \sin \theta - 1)(\sin \theta + 2) = 0$$

Thus, $2 \sin \theta - 1 = 0$ or $\sin \theta + 2 = 0$, or $\sin \theta = \frac{1}{2}$ or $\sin \theta = -2$. The solution set for $\sin \theta = \frac{1}{2}$ is $\theta = \frac{\pi}{6} + 2k\pi$ or $\frac{5\pi}{6} + 2k\pi$, where k is any integer. The equation $\sin \theta = -2$ has no solution since $|\sin \theta| \le 1$ for all θ. Thus, the solution set for the original equation is the union of these sets, $\left\{\frac{\pi}{6} + 2k\pi\right\} \cup \left\{\frac{5\pi}{6} + 2k\pi\right\}$, where k is any integer.

PROBLEM 5

Solve $\sin^2 \theta + \sin \theta - 3 = 0$.

Solution: Since the equation is quadratic in $\sin \theta$ and not easily factorable, we use the Quadratic Formula. Note that the unknown is $\sin \theta$.

$$\sin \theta = \frac{-1 \pm \sqrt{1 - 4(1)(-3)}}{2} = \frac{-1 \pm \sqrt{13}}{2} \approx \frac{-1 \pm 3.6}{2}$$

$\sin \theta \approx 1.3$ or $\sin \theta \approx -2.3$. Both of these have empty solution sets since $|\sin \theta| \leq 1$.

When a trigonometric equation involves more than one trigonometric function, we may find it useful to use identities to put it into a form involving one trigonometric function.

PROBLEM 6

Solve $\cos \theta + 1 = \sin \theta$.

Solution: $\cos \theta + 1 = \sin \theta$
$\cos^2 \theta + 2 \cos \theta + 1 = \sin^2 \theta$ (squaring both sides)
$\cos^2 \theta + 2 \cos \theta + 1 = 1 - \cos^2 \theta$ ($\sin^2 \theta = 1 - \cos^2 \theta$)
$2 \cos^2 \theta + 2 \cos \theta = 0$
$2 \cos \theta(\cos \theta + 1) = 0$

$$2 \cos \theta = 0 \quad \text{or} \quad \cos \theta + 1 = 0$$
$$\cos \theta = 0 \qquad \qquad \cos \theta = -1$$
$$\theta = \frac{\pi}{2}, \frac{3\pi}{2} \qquad \qquad \theta = \pi$$

Note that we squared both sides and therefore must check for extraneous roots. Checking by substituting into $\cos \theta + 1 = \sin \theta$, we find that only $\frac{\pi}{2}$ and π satisfy the equation. Thus, the solution set is $\left\{\frac{\pi}{2} + 2k\pi\right\} \cup \{\pi + 2k\pi\}$, where k is any integer.

PROBLEM 7

Solve $\sin 2\theta = \sin \theta$.

Solution: Since the equation involves trigonometric functions of two different angles, θ and 2θ, we use the double-angle law to derive an equation involving a single angle:

$$\sin 2\theta = \sin \theta$$
$$2 \sin \theta \cos \theta = \sin \theta \quad \text{(double-angle law)}$$
$$2 \sin \theta \cos \theta - \sin \theta = 0$$
$$\sin \theta (2 \cos \theta - 1) = 0$$
$$\sin \theta = 0 \quad \text{or} \quad 2 \cos \theta - 1 = 0$$
$$\theta = 0, \pi \qquad \cos \theta = \frac{1}{2}$$
$$\theta = \frac{\pi}{3}, \frac{5\pi}{3}$$

Thus, the solution set is $\{k\pi\} \cup \left\{ \pm \dfrac{\pi}{3} + 2k\pi \right\} \cup \left\{ \dfrac{5\pi}{3} + 2k\pi \right\} = \{k\pi\} \cup$ $\left\{ \pm \dfrac{\pi}{3} + 2k\pi \right\}$, where k is any integer.

PROBLEM 8

Solve $\cos(\sin \theta) = 1$.

Solution: Let $v = \sin \theta$. Then $\cos(\sin \theta) = \cos v = 1$, and $\cos v = 1$ when $v = 0$ and multiples of 2π. However, since $v = \sin \theta$ and $-1 \leq v \leq 1$, $v = 0$ is the only possible solution. That is, the other solutions, $2k\pi$, do not lie in the interval $-1 \leq v \leq 1$. Therefore, the solutions of the original equation are given by $v = 0$; that is, $\sin \theta = 0$. Thus, $\theta = 0$ and $\theta = \pi$ are solutions. The solution set is $\{2k\pi\} \cup \{\pi + 2k\pi\} = \{k\pi\}$, where k is any integer.

PROBLEM 9

Solve $\cos 2\theta \cos \theta + \sin 2\theta \sin \theta = 1$.

Solution: We use the addition law $\cos(\alpha - \beta) = \cos \alpha \cos \beta + \sin \alpha \sin \beta$ to obtain $\cos(2\theta - \theta) = 1$. Thus, we have $\cos \theta = 1$ and $\theta = 0 + 2k\pi$, where k is any integer.

PROBLEM 10

Solve $\cos(\pi - \theta) + \sin \left(\theta - \dfrac{\pi}{2} \right) = 1$.

Solution: Using the addition laws we obtain

$$\cos \pi \cos \theta + \sin \pi \sin \theta + \sin \theta \cos \left(\frac{\pi}{2} \right) - \cos \theta \sin \left(\frac{\pi}{2} \right) = 1$$

$$-\cos \theta + 0 + 0 - \cos \theta = 1$$
$$-2 \cos \theta = 1$$

$$\cos \theta = -\frac{1}{2}$$

Thus, $\theta = \dfrac{2\pi}{3}$ or $\dfrac{4\pi}{3}$ and the solution set is $\left\{\dfrac{2\pi}{3} + 2k\pi\right\} \cup \left\{\dfrac{4\pi}{3} + 2k\pi\right\}$, where k is any integer.

6.4   EXERCISES

Solve the following trigonometric equations:

1. $\cos \theta - 1 = 0$

2. $2 \sin \theta = 1$

3. $2 \sin \theta = \sqrt{2}$

4. $2 \cos \theta - 1 = 0$

5. $\tan \theta = \sqrt{3}$

6. $\sin \theta = \cos \theta$

7. $\sin 3\theta = 1$

8. $2 \sin 3\theta = 1$

9. $\tan 3\theta = 1$

10. $\cos \theta = 2$

Solve:

11. $\sin^2 \theta - 2 \sin \theta = 0$

12. $2 \cos^2 \theta + \cos \theta - 1 = 0$

13. $4 \cos^2 \theta + 4 \cos \theta = 3$

14. $2 \sin^2 \theta + 3 \sin \theta + 1 = 0$

15. $2 \sin^2 \theta + \sin \theta - 1 = 0$

Solve:

16. $\sin 2\theta - \cos \theta = 0$

17. $\sin \theta + \cos \theta = 1$

18. $\cos 2\theta + \sin \theta = 1$

19. $\cos 2\theta = \cos \theta$

20. $2 \sin 2\theta - \tan \theta = 0$

Solve:

21. $\sin(\cos \theta) = 1$

22. $\sin 5\theta \cos 2\theta - \cos 5\theta \sin 2\theta = \dfrac{\sqrt{2}}{2}$ for $0 \le \theta \le 2\pi$

23. $\sin 2\theta \sin \theta - \cos 2\theta \cos \theta = -\cos \theta$

24. $\sin \theta = \sin(2\theta - \pi)$

25. $\cot \theta = \tan(2\theta - 3\pi)$

6.5   TRIANGLES, TRIGONOMETRY, AND APPLICATIONS

As we stated in the beginning of Chapter 5, trigonometry deals with the study of triangles. In this section we study how trigonometry is used to solve for parts of triangles. We begin by discussing the relationship between the trigonometric functions and the ratios of the lengths of the sides of a right triangle. Then we derive standard formulas that apply to any plane triangle, including right triangles. We conclude this section with some applications of trigonometric functions.

Let us establish some notation for representing angles and sides of a triangle. We shall denote each side of a triangle by its length—a, b, or c. The measure of the angle opposite each side is denoted by the corresponding Greek letter α, β, or γ (see Fig. 164).

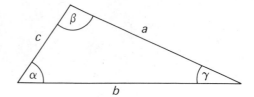

Figure 164

Right Triangles

Consider a right triangle in the plane with the angle α in standard position and the angle $\gamma = \dfrac{\pi}{2}$ (see Fig. 165).

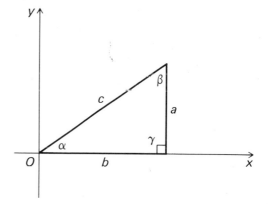

Figure 165

The six trigonometric functions can be given in terms of the ratios of the lengths of the sides of the right triangle. Note that the side c is the hypotenuse of the right triangle.

$$\sin \alpha = \frac{a}{c} = \frac{\text{length of side opposite } \alpha}{\text{length of hypotenuse}}$$

$$\cos \alpha = \frac{b}{c} = \frac{\text{length of adjacent side of } \alpha}{\text{length of hypotenuse}}$$

$$\tan \alpha = \frac{a}{b} = \frac{\text{length of side opposite } \alpha}{\text{length of adjacent side of } \alpha}$$

$$\cot \alpha = \frac{b}{a} = \frac{\text{length of adjacent side of } \alpha}{\text{length of side opposite } \alpha}$$

$$\sec \alpha = \frac{c}{b} = \frac{\text{length of hypotenuse}}{\text{length of adjacent side of } \alpha}$$

$$\csc \alpha = \frac{c}{a} = \frac{\text{length of hypotenuse}}{\text{length of side opposite } \alpha}$$

PROBLEM 1

Determine the missing sides and angles for the right triangle in Figure 166.

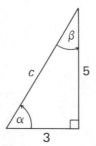

Figure 166

Solution: Using the Pythagorean Theorem we have $c^2 = (3)^2 + (5)^2 = 34$ and $c = \sqrt{34}$. Also, $\tan \alpha = \frac{5}{3} = 1.67$, so from Table D in Appendix III we see that α is approximately 59°. Since $\alpha + \beta = 90°$, $\beta = 31°$.

PROBLEM 2

Find a, b, and β in the right triangle shown in Figure 167.

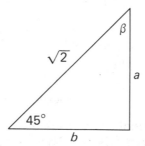

Figure 167

Solution: Since $45° + \beta = 90°$, $\beta = 45°$. Also, we know $\sin 45° = \frac{a}{\sqrt{2}}$ or $a = (\sqrt{2}) \sin 45°$. Thus, $a = (\sqrt{2}) \sin 45° = (\sqrt{2}) \left(\frac{1}{\sqrt{2}}\right) = 1$. Finally, using the Pythagorean Theorem, $(\sqrt{2})^2 = a^2 + b^2$, $(\sqrt{2})^2 = (1)^2 + b^2$, $b^2 = 1$, and $b = 1$. Thus $a = b = 1$ and $\beta = \alpha = 45°$. Since two sides are equal in length, the triangle is an **isosceles right triangle.**

PROBLEM 3

A 25-foot ladder leans against a wall with its bottom 15 feet from the wall. What angle does the ladder make with the ground? (See Fig. 168.)

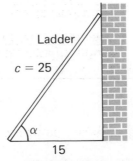

Figure 168

Solution: We have a right triangle with hypotenuse = 25 feet and b = 15 feet, and we are asked to find α. $\cos \alpha = \dfrac{\text{adjacent side}}{\text{hypotenuse}} = \dfrac{b}{c} = \dfrac{15}{25} = \dfrac{3}{5} = .60$. If $\cos \alpha = .60$, then the table of cosines gives $\theta \approx 53°$.

PROBLEM 4

Consider the flagpole in Figure 169. If the angle of elevation of the sun

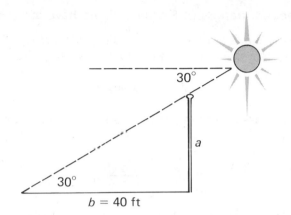

Figure 169

is 30° and the flagpole casts a 40-foot shadow, then find the height of the pole.

Solution: Again, we have a right triangle with α = 30° and b = 40. We are asked to find a. Since $\tan \alpha = \dfrac{a}{b}$, we have $a = (b) \tan \alpha$. Thus, $a = (40) \tan 30° = (40)(.5774) = 23.1$ feet.

Note: The angle α in Problem 4 is an **angle of elevation** since it is measured above the horizontal. An angle measured below the horizontal is called an **angle of depression.**

The Law of Sines and the Law of Cosines

We now derive a formula which relates the sides and angles of any triangle. Consider the triangles in Figure 170. We drop a perpendicular line from the

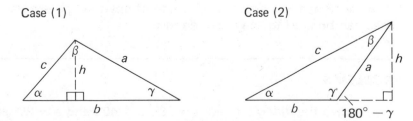

Figure 170

vertex opposite b to side b or its extension. If h is the length of this perpendicular, we can use right triangles to derive the Law of Sines. Both possibilities are given in Figure 170. In either triangle we have $\sin \alpha = \dfrac{h}{c}$ or $h = c \sin \alpha$. Now we consider the triangle in *Case 1:* We see that $\sin \gamma = \dfrac{h}{a}$ or $h = a \sin \gamma$. Therefore, we have $h = c \sin \alpha$ and $h = a \sin \gamma$. Thus, it follows that

$$c \sin \alpha = a \sin \gamma$$

In *Case 2,* the second triangle in Figure 170, we have $\sin (180° - \gamma) = \dfrac{h}{a}$ or $h = a \sin (180° - \gamma)$. Since $\sin (180° - \gamma) = \sin \gamma$, we obtain $h = a \sin \gamma$. Therefore, in *Case 2* we also have $h = c \sin \alpha$ and $h = a \sin \gamma$. Thus it follows that

$$c \sin \alpha = a \sin \gamma$$

Hence, in either case, we have $c \sin \alpha = a \sin \gamma$. Dividing both sides by $\sin \alpha$ $\sin \gamma$ gives

$$\frac{a}{\sin \alpha} = \frac{c}{\sin \gamma}$$

Note that there is no danger of dividing by 0 because triangles never have angles of 0° or 180°. If we were to drop a perpendicular from the vertex opposite a to side a or its extension in Figure 170, the same argument gives

$$\frac{c}{\sin \gamma} = \frac{b}{\sin \beta}$$

Combining these results gives the following theorem.

Theorem: The Law of Sines
In any triangle with sides a, b, and c, and the angles α, β, and γ opposite these respective sides, we have

$$\frac{\sin \alpha}{a} = \frac{\sin \beta}{b} = \frac{\sin \gamma}{c}$$

That is, the sides are proportional to the sines of the opposite angles.

Note: When two angles and a side of any triangle are known, the Law of Sines can be used to solve the triangle.

PROBLEM 5

Solve the triangle given $\alpha = 30°$, $\beta = 40°$, and $a = 20$ (see Fig. 171).

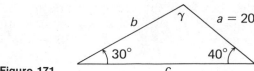

Figure 171

Solution:
(1) Since the sum of the angles of a triangle is 180°, we have
$\gamma + 30° + 40° = 180°$, or $\gamma = 110°$.
(2) To find b, we use

$$b = \frac{a \sin \beta}{\sin \alpha} = \frac{20 \sin 40°}{\sin 30°} = \frac{20(.6428)}{.5000} = 26$$

(3) To find c, we use

$$c = \frac{a \sin \gamma}{\sin \alpha} = \frac{20 \sin 110°}{\sin 30°} = \frac{20(.9397)}{.5000} = 38$$

The Ambiguous Case: SSA

If two sides and the angle opposite one of them are given, the triangle is not always uniquely determined. Given two sides and an angle opposite one of them, we may be able to construct one triangle, two triangles, or no triangle. Since there is a possibility of two triangles, this is usually called the ambiguous case.

PROBLEM 6: One Triangle

We are given $a = 100$, $b = 70$, and $\alpha = 80°$. Find β, γ, and c.

Solution: (1) To find β, we use

$$\sin \beta = \frac{b \sin \alpha}{a} = \frac{70 \sin 80°}{100} = .7(.9848) = .6894$$

$\beta = 43°35'$
(2) $\gamma = 180° - (\alpha + \beta) = 180° - 123°35' = 56°25'$

$$(3) \ c = \frac{a \sin \gamma}{\sin \alpha} = \frac{100 \sin (56°25')}{\sin 80°} = \frac{100(.8331)}{.9848} \approx 85$$

PROBLEM 7: Two Triangles

Given $a = 46.85$, $b = 76.62$, and $\alpha = 30°$, solve the triangle(s) for β, γ, and c.

Solution: To find β, we use

$$\sin \beta = \frac{b \sin \alpha}{a} = \frac{(76.62)(\sin 30°)}{46.85} = .8177$$

There are two angles less than 180° having a sine of .8177. They are 54°51′ and 180° − 54°51′ = 125°9′. This gives two possible triangles:

(1) If $\beta = 54°51′$, then $\gamma = 180° − (\alpha + \beta) = 180° − (84°51′) = 95°9′$.
Also,

$$c = \frac{a \sin \gamma}{\sin \alpha} = \frac{(46.85)(.9960)}{.5000} = 93.33$$

(2) If $\beta = 125°9′$, then $\gamma = 180° − (155°9′) = 24°51′$.
Thus,

$$c = \frac{a \sin \gamma}{\sin \alpha} = \frac{(46.85)(.4202)}{.5000} = 39.37$$

Both triangles are given in Figure 172.

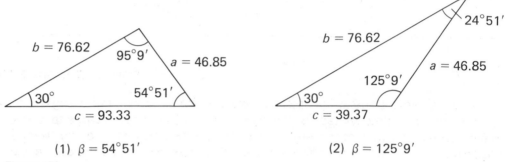

(1) $\beta = 54°51′$ (2) $\beta = 125°9′$

Figure 172

PROBLEM 8: No Triangle

Given $a = 3$, $b = 4$, and $\alpha = 60°$, find β, γ, and c.

Solution: The Law of Sines gives $\sin \beta = \frac{b \sin \alpha}{a} = \frac{4(.8660)}{3} = $ 1.155. However, $|\sin \beta| \leq 1$; therefore, there can be no such triangle.

We can summarize the conditions which determine each of the possibilities.

Let a and b be two sides of a triangle and α the angle opposite the side a. The following summarizes the possible cases:
I. If $\alpha \geq 90°$ and
  (1) $a > b$, then there is **one triangle.**
  (2) $a \leq b$, then there is **no triangle.**
II. If $\alpha < 90°$ and
  (1) $a \geq b$, then there is **one triangle.**
  (2) $a < b$ and $\frac{b \sin \alpha}{a} = 1$, then there is **one triangle.**

(3) $a < b$ and $\dfrac{b \sin \alpha}{a} < 1$, then there are **two triangles.**

(4) $a < b$ and $\dfrac{b \sin \alpha}{a} > 1$, then there is **no triangle.**

In order to use the Law of Sines we must know both the length of one side and the angle opposite it, in addition to either one other side or angle. However, if we are given two sides and the included angle, then the Law of Sines will not give the remaining parts of the uniquely determined triangle. We state a law which solves this problem. The formula is called the **Law of Cosines** and is essentially a generalization of the Pythagorean Theorem.

Theorem: The Law of Cosines
    In any triangle with sides a, b, and c, and the angles α, β, and γ opposite these respective sides, we have

$a^2 = b^2 + c^2 - 2bc \cos \alpha$
$b^2 = c^2 + a^2 - 2ca \cos \beta$
$c^2 = a^2 + b^2 - 2ab \cos \gamma$

That is, the square of any side of a triangle is equal to the sum of the squares of the other two sides minus twice the product of those sides and the cosine of the included angle.

Proof: Consider the triangle in Figure 173 with the vertex of angle α placed at the origin and side c along the positive x-axis.

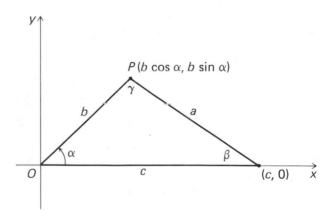

Figure 173

The coordinates of the point P are $(b \cos \alpha, b \sin \alpha)$ and the end point of side c has coordinates $(c, 0)$. Using the distance formula to determine a^2, we obtain

$$a^2 = (b \cos \alpha - c)^2 + (b \sin \alpha - 0)^2$$
$$= b^2 \cos^2 \alpha - 2bc \cos \alpha + c^2 + b^2 \sin^2 \alpha$$
$$= b^2(\cos^2 \alpha + \sin^2 \alpha) + c^2 - 2bc \cos \alpha$$
$$= b^2 + c^2 - 2bc \cos \alpha$$

If we place the origin at the other vertices, we obtain $b^2 = c^2 + a^2 - 2ac \cos \beta$ and $c^2 = a^2 + b^2 - 2ab \cos \gamma$.

PROBLEM 9

Given a triangle with $a = 50$, $c = 60$, and $\beta = 100°$, find b.

Solution:
$$
\begin{aligned}
b^2 &= c^2 + a^2 - 2ac \cos \beta \\
&= (60)^2 + (50)^2 - 2(50)(60) \cos 100° \\
&= 3600 + 2500 - 6000(-.1736) \\
&= 7141.6
\end{aligned}
$$

Thus, $b \approx 85$.

PROBLEM 10

Given $a = 8$, $b = 5$, and $c = 4$, find the smallest and largest angles.

Solution: The smallest and largest angles lie opposite the smallest and largest sides (see Fig. 174).

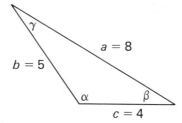

Figure 174

(1) To find α, we use:

$$
a^2 = b^2 + c^2 - 2bc \cos \alpha
$$
$$
(8)^2 = (5)^2 + (4)^2 - 2(5)(4) \cos \alpha
$$
$$
40 \cos \alpha = -23
$$

$$
\cos \alpha = -\frac{23}{40} = -.5750
$$

$$
\alpha = 125°16'
$$

(2) To find γ we use:

$$
c^2 = a^2 + b^2 - 2ab \cos \gamma
$$
$$
(4)^2 = (8)^2 + (5)^2 - 2(8)(5)\cos \gamma
$$
$$
80 \cos \gamma = 73
$$

$$
\cos \gamma = \frac{73}{80} = .9125
$$

$$
\gamma = 24°19'
$$

Applications

We now give several applications of the Law of Sines and Law of Cosines.

EXAMPLE 1: Area of Triangle Given Two Sides and an Angle

Consider the triangles in Figure 175. Suppose we wish to find the area

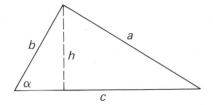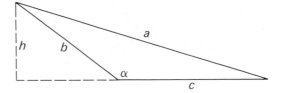

Figure 175

of a triangle given two sides and the included angle. Recall from plane geometry that the area K of a triangle is one half base times height. That is, $K = \frac{1}{2} hc = \frac{1}{2} c(b \sin \alpha) = \frac{1}{2} bc \sin \alpha$. Thus,

(1) Given b, c, and α, $K = \frac{1}{2} bc \sin \alpha$.

(2) Given c, a, and β, $K = \frac{1}{2} ac \sin \beta$.

(3) Given a, b, and γ, $K = \frac{1}{2} ab \sin \gamma$.

EXAMPLE 2: Area of Triangle Given Two Angles and a Side

Using the results from Example 1, we know that given sides b and c and angle α, the area K is given by $K = \frac{1}{2} bc \sin \alpha$. We use the Law of Sines to replace b:

$$\frac{\sin \beta}{b} = \frac{\sin \gamma}{c} \text{ gives } b = \frac{c \sin \beta}{\sin \gamma}$$

Substituting this expression for b into $K = \frac{1}{2} bc \sin \alpha$, we obtain

$K = \frac{c^2 \sin \alpha \sin \beta}{2 \sin \gamma}$. Similar results can be obtained when any two angles and side are given and may be written in any one of the following forms:

$$K = \frac{a^2 \sin \beta \sin \gamma}{2 \sin \alpha}, \ K = \frac{b^2 \sin \alpha \sin \gamma}{2 \sin \beta}, \ K = \frac{c^2 \sin \alpha \sin \beta}{2 \sin \gamma}$$

EXAMPLE 3: Heron's Formula

Suppose we are given the three sides a, b, and c of a triangle and we wish to derive a formula expressing the area K in terms of the sides. Example 1 and the Law of Sines give $K = \frac{1}{2} bc \sin \alpha$. Therefore,

$$K^2 = \frac{1}{4} b^2 c^2 \sin^2 \alpha = \frac{1}{4} b^2 c^2 (1 - \cos^2 \alpha) = \frac{bc}{2} (1 + \cos \alpha) \frac{bc}{2} (1 - \cos \alpha)$$

From the Law of Cosines we have $\cos \alpha = \dfrac{b^2 + c^2 - a^2}{2bc}$. Substituting for $\cos \alpha$, we have

$$\begin{aligned}
K^2 &= \frac{bc}{2} \left(1 + \frac{b^2 + c^2 - a^2}{2bc} \right) \frac{bc}{2} \left(1 - \frac{b^2 + c^2 - a^2}{2bc} \right) \\
&= \left(\frac{2bc + b^2 + c^2 - a^2}{4} \right) \left(\frac{2bc + b^2 - c^2 + a^2}{4} \right) \\
&= \left(\frac{(b + c)^2 - a^2}{4} \right) \left(\frac{(a^2 - (b - c)^2)}{4} \right) \\
&= \left(\frac{b + c + a}{2} \right) \left(\frac{b + c - a}{2} \right) \left(\frac{a - b + c}{2} \right) \left(\frac{a + b - c}{2} \right)
\end{aligned}$$

Let $s = \frac{1}{2} (a + b + c)$; then $\dfrac{b + c - a}{2} = s - a$, $\dfrac{a - b + c}{2} = s - b$, $\dfrac{a + b - c}{2} = s - c$. Thus, $K^2 = s(s - a)(s - b)(s - c)$ and

$$K = \sqrt{s(s - a)(s - b)(s - c)}$$

Heron's Formula

PROBLEM 11

Suppose a surveyor is at a point 3 miles from one end of a lake and 4 miles from the other end. Also, suppose the lake subtends an angle 150° (see Fig. 176). Find the length of the lake.

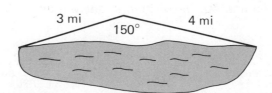

Figure 176

Solution: If we let *a* be the length of the lake, then the Law of Cosines gives

$$a^2 = (3)^2 + (4)^2 - 2(3)(4) \cos 150°$$

$$a^2 = 9 + 16 - 24 \left(-\frac{\sqrt{3}}{2} \right)$$

$$a^2 = 25 + 12\sqrt{3} = 45.78$$
$$a = 6.77$$

PROBLEM 12

Suppose two ships leave from the same point *O* and travel in directions N 30° E and N 55° E, respectively. If the first ship *S* is 35 miles from point *O* and 50 miles from the second ship *T*, how far is the second ship from *O* (see Fig. 177)?

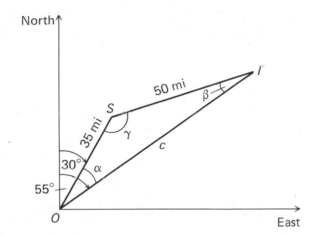

Figure 177

Solution: Let *c* be the distance from *O* to ship *T*. The angle $\alpha = 55° - 30° = 25°$ and from the Law of Sines

$$\frac{\sin \beta}{35} = \frac{\sin \alpha}{50}$$

or,

$$\sin \beta = \frac{35 \sin \alpha}{50} = \frac{35 \sin 25°}{50} = \frac{35(.4226)}{50}$$

$$= .2958$$

and $\beta \approx 17°10'$. Then $\gamma = 180° - (\alpha + \beta) \approx 137°50'$. Again, from the Law of Sines,

$$\frac{c}{\sin \gamma} = \frac{50}{\sin \alpha}$$

or,

$$c = \frac{50 \sin \gamma}{\sin \alpha} = \frac{50 \sin(137°50')}{\sin 25°} = \frac{50(.6713)}{(.4226)}$$

Thus, $c \approx 79.42$ miles.

6.5 EXERCISES

1. Find the missing parts of the right triangle in Figure 178, given:

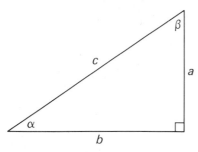

Figure 178

  (a) $\alpha = 60°$, $b = 10$
  (b) $\alpha = 31°$, $a = 73$
  (c) $\alpha = 41°$, $b = 15$
  (d) $a = 5$, $b = 12$
  (e) $b = 5$, $c = 7$
  (f) $a = 33$, $c = 40$
  (g) $\beta = 43°$, $b = 7.3$
  (h) $\beta = 49°$, $a = 12$

2. A tower casts a shadow 130 feet long when the angle of elevation of the sun is 38°. How tall is the tower?

3. Suppose a ladder which is leaning against a wall forms an angle of elevation of 55° with the ground. If the bottom of the ladder is 32 feet from the wall, how far up the wall is the top of the ladder?

4. From the top of a tower 175 feet high the angle of depression of a car is 27°10'. What is the distance from the car to the base of the tower?

5. If a triangle has sides a, b, c and angles α, β, γ opposite these respective sides, find the missing sides and angles using the Law of Sines, given:
  (a) $b = 6$, $\alpha = 80°$, $\gamma = 46°$
  (b) $\alpha = 50°$, $\beta = 60°$, $a = 4$
  (c) $a = 400$, $\alpha = 31°20'$, $\beta = 70°40'$
  (d) $a = 5$, $b = 7$, $c = 6$

6. Given the following information, determine whether one, two, or no triangles exist:
  (a) $a = 8$, $b = 8$, $\alpha = 120°$
  (b) $a = 5$, $b = 10$, $\beta = 60°$
  (c) $a = 3$, $b = 2$, $\beta = 30°$
  (d) $a = 3$, $b = 2$, $\beta = 45°$

7. Solve a triangle(s) with $a = 8$, $b = 10$, and $\alpha = 35°$.

8. Use the Law of Cosines to solve the following triangles:
  (a) $\alpha = 60°$, $b = 3$, $c = 4$
  (b) $\alpha = 100°$, $b = 8$, $c = 10$
  (c) $\alpha = 50°$, $b = 5$, $c = 8$
  (d) $a = 9$, $b = 6$, $c = 10$
  (e) $a = 5$, $b = 10$, $\gamma = 120°$
  (f) $a = 10$, $b = 20$, $c = 12$

9. Find the largest and smallest angles of a triangle whose sides are $a = 5$, $b = 10$, $c = 12$.

10. The sides of a parallelogram are 11 inches and 17 inches in length. If the diagonals form an angle of 60° with each other, then find the lengths of the diagonals.

11. Find the areas of the triangles with:
    (a) $a = 15$, $b = 20$, $\gamma = 37°$
    (b) $\alpha = 100°$, $b = 10$, $c = 30$
    (c) $a = 117$, $b = 85$, $c = 40$
    (d) $a = 14$, $b = 21$, $c = 25$

12. Use Heron's Formula to show that the area of an isosceles triangle with sides a, a, b is $K = \dfrac{b}{4} \sqrt{4a^2 - b^2}$.

13. Two surveyors are on opposite sides of a hill and are 600 feet apart. If the angles of elevation from the surveyors to the top of the hill are 19° and 21°, how high is the hill?

14. Two bikers leave the same point at the same time. One travels at a speed of 17 miles per hour in the direction N 20° E, and the second biker at a speed of 23 miles per hour in the direction N 78° E. How far apart are they in two hours?

15. The Earth is 93 million miles from the sun and the planet Mars is 141 million miles from the sun. How far is Mars from the Earth when the sun has just set in the west and Mars is 54° above the horizon in the east?

SUMMARY

Angles, Radians, and Degrees

1. 1 radian $= \dfrac{180}{\pi}$ degrees $\approx 57.2958°$

   $1° = \dfrac{\pi}{180}$ radian ≈ 0.0174533 radian

2. The length of the arc of a circle with a radius of r and a central angle equal to θ radians is given by $s = r\theta$.

3. The area A of a sector of a circle with radius r and central angle θ, measured in radians, is given by $A = \dfrac{1}{2} r^2\theta$.

Trigonometric Functions

1. Let θ be an angle in standard position and let (x, y) be any point other than $(0, 0)$ on the terminal side of θ. Let $r = \sqrt{x^2 + y^2}$. Then:

   $\sin \theta = \dfrac{y}{r}$ $\tan \theta = \dfrac{y}{x}$ $\sec \theta = \dfrac{r}{x}$

   $\cos \theta = \dfrac{x}{r}$ $\cot \theta = \dfrac{x}{y}$ $\csc \theta = \dfrac{r}{y}$

Trigonometric Formulas and Identities

1. REDUCTION FORMULAS

$$\sin(-\theta) = -\sin\theta \qquad\qquad \cos(-\theta) = \cos\theta$$
$$\sin(\pi - \theta) = \sin\theta \qquad\qquad \cos(\pi - \theta) = -\cos\theta$$
$$\sin(\pi + \theta) = -\sin\theta \qquad\qquad \cos(\pi + \theta) = -\cos\theta$$
$$\sin(2\pi - \theta) = -\sin\theta \qquad\qquad \cos(2\pi - \theta) = \cos\theta$$
$$\sin(2\pi + \theta) = \sin\theta \qquad\qquad \cos(2\pi + \theta) = \cos\theta$$

2. ADDITION FORMULAS

$$\sin(\alpha + \beta) = \sin\alpha\cos\beta + \cos\alpha\sin\beta$$
$$\cos(\alpha + \beta) = \cos\alpha\cos\beta - \sin\alpha\sin\beta$$
$$\tan(\alpha + \beta) = \frac{\tan\alpha + \tan\beta}{1 - \tan\alpha\tan\beta}$$

3. DIFFERENCE FORMULAS

$$\sin(\alpha - \beta) = \sin\alpha\cos\beta - \cos\alpha\sin\beta$$
$$\cos(\alpha - \beta) = \cos\alpha\cos\beta + \sin\alpha\sin\beta$$
$$\tan(\alpha - \beta) = \frac{\tan\alpha - \tan\beta}{1 + \tan\alpha\tan\beta}$$

4. COFUNCTION FORMULAS

$$\sin\left(\frac{\pi}{2} - \theta\right) = \cos\theta$$
$$\cos\left(\frac{\pi}{2} - \theta\right) = \sin\theta$$

5. PYTHAGOREAN IDENTITIES

$$\sin^2\theta + \cos^2\theta = 1$$
$$\tan^2\theta + 1 = \sec^2\theta$$
$$\cot^2\theta + 1 = \csc^2\theta$$

6. DOUBLE-ANGLE FORMULAS

$$\sin 2\theta = 2\sin\theta\cos\theta$$
$$\cos 2\theta = \cos^2\theta - \sin^2\theta = 2\cos^2\theta - 1 = 1 - 2\sin^2\theta$$

7. HALF-ANGLE FORMULAS

$$\sin\tfrac{1}{2}\theta = \pm\sqrt{\frac{1 - \cos\theta}{2}} \qquad \cos\tfrac{1}{2}\theta = \pm\sqrt{\frac{1 + \cos\theta}{2}}$$
$$\tan\tfrac{1}{2}\theta = \pm\sqrt{\frac{1 - \cos\theta}{1 + \cos\theta}} = \frac{\sin\theta}{1 + \cos\theta} = \frac{1 - \cos\theta}{\sin\theta}$$

8. SUM-PRODUCT FORMULAS

$$\sin \alpha \cos \beta = \frac{1}{2} \sin(\alpha + \beta) + \frac{1}{2} \sin(\alpha - \beta)$$

$$\cos \alpha \sin \beta = \frac{1}{2} \sin(\alpha + \beta) - \frac{1}{2} \sin(\alpha - \beta)$$

$$\cos \alpha \cos \beta = \frac{1}{2} \cos(\alpha + \beta) + \frac{1}{2} \cos(\alpha - \beta)$$

$$\sin \alpha \sin \beta = \frac{1}{2} \cos(\alpha + \beta) + \frac{1}{2} \cos(\alpha - \beta)$$

9. FACTORING FORMULAS

$$\sin \alpha + \sin \beta = 2 \sin \left[\frac{1}{2}(\alpha + \beta)\right] \cos \left[\frac{1}{2}(\alpha - \beta)\right]$$

$$\sin \alpha - \sin \beta = 2 \cos \left[\frac{1}{2}(\alpha + \beta)\right] \sin \left[\frac{1}{2}(\alpha - \beta)\right]$$

$$\cos \alpha + \cos \beta = 2 \cos \left[\frac{1}{2}(\alpha + \beta)\right] \cos \left[\frac{1}{2}(\alpha - \beta)\right]$$

$$\cos \alpha - \cos \beta = -2 \sin \left[\frac{1}{2}(\alpha + \beta)\right] \sin \left[\frac{1}{2}(\alpha - \beta)\right]$$

10. LAW OF SINES

$$\frac{a}{\sin \alpha} = \frac{b}{\sin \beta} = \frac{c}{\sin \gamma}$$

11. LAW OF COSINES

$$a^2 = b^2 + c^2 - 2bc \cos \alpha$$
$$b^2 = c^2 + a^2 - 2ac \cos \beta$$
$$c^2 = a^2 + b^2 - 2ab \cos \gamma$$

CHAPTER 6 EXERCISES

1. Convert the following degree measures to radian measures:
   (a) 105°   (b) 135°   (c) 330°
   (d) 120°   (e) 210°   (f) −315°
   (g) 300°   (h) 36°   (i) 108°

2. Convert the following radian measures to degree measures:
   (a) $\dfrac{2\pi}{3}$ (b) $-\dfrac{\pi}{4}$ (c) $\dfrac{5\pi}{2}$

   (d) $\dfrac{7\pi}{4}$ (e) -4π (f) $-\dfrac{\pi}{3}$

3. Find the length of the circular arc and the area of the circular sector with central angle θ in a circle with radius r if:

   (a) $\theta = \dfrac{\pi}{2}$, $r = 4$ inches

   (b) $r = 100$ inches, $\theta = 72°$
   (c) $\theta = 135°$, $r = 2$ inches
   (d) $\theta = \pi$, $r = 1$

4. A flywheel 6 feet in diameter turns 40 rpm. (a) Find its angular velocity in radians per second. (b) Find the speed of the belt that drives the flywheel.

5. Determine all six trigonometric functions of the angle θ if the terminal side of θ in its standard position contains the following points:
   (a) (3, 4)   (b) ($\sqrt{3}$, −1)
   (c) (1, 2)   (d) (−1, 0)
   (e) (−2, −3)   (f) (1, $2\sqrt{3}$)

6. Given each of the following trigonometric functions, find the remaining five trigonometric functions:
   (a) $\cos \theta = \dfrac{8}{17}$, θ in the first quadrant
   (b) $\tan \theta = \dfrac{1}{3}$, θ in the first quadrant
   (c) $\sin \theta = \dfrac{-1}{\sqrt{2}}$, θ in the third quadrant

7. Write the following as a trigonometric function of a single angle:
   (a) $\sin 30° \cos 45° + \cos 30° \sin 45°$
   (b) $\cos 35° \cos 70° − \sin 35° \sin 70°$
   (c) $\sin \dfrac{\pi}{8} \cos \dfrac{\pi}{4} − \cos \dfrac{\pi}{8} \sin \dfrac{\pi}{4}$
   (d) $\cos 7° \cos 7° + \sin 7° \sin 7°$

8. Prove that if α and β are complementary angles, then $\cos(6\alpha + \beta) = -\sin 5\alpha$.

9. Express $\cos 20\theta$ in terms of $\sin 5\theta$.

10. Prove the following identities:
   (a) $\dfrac{\cos 6\theta}{\sin 2\theta} + \dfrac{\sin 6\theta}{\cos 2\theta} = 2 \cot 4\theta$
   (b) $\cos^2 2\theta − \cos^2 \theta = \sin^2 \theta − \sin^2 2\theta$
   (c) $\tan \dfrac{\theta}{2} = \dfrac{1 − \cos \theta}{\sin \theta} = \dfrac{\sin \theta}{1 + \cos \theta}$

11. Solve for θ, where $0 \le \theta < 360°$:
   (a) $4 \cos^2 \theta = 3$
   (b) $4 \sin^2 \theta = 3 \sin \theta$
   (c) $\sin^2 \theta − \cos^2 \theta − \cos \theta = 1$
   (d) $\cos 2\theta = 3 \sin \theta + 2$

12. From the top of a building, the angle of elevation of the top of a flagpole across the street is 30° and the angle of depression of the bottom of the pole is 45°. If the horizontal distance from the observer to the pole is 60 feet, find the height of the pole.

13. Suppose a fire tower is situated on a hillside which has an inclination of 15°. If the tower casts a shadow of 250 feet down the slope of the hillside when the angle of elevation of the sun is 30°, how tall is the tower?

14. Show that if $\gamma = \dfrac{\pi}{2}$ in a triangle, then the Law of Cosines gives $c^2 = a^2 + b^2$. Also show that if $c^2 = a^2 + b^2$, then $\gamma = \dfrac{\pi}{2}$.

15. The resultant of two forces is the diagonal of a parallelogram in which the given forces are the sides. Suppose two forces of 75 pounds and 100 pounds, respectively, have a resultant of 140 pounds. Find the angle between the forces (see Fig. 179).

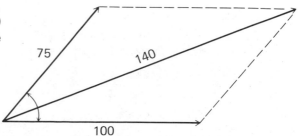

Figure 179

CHAPTER 7

INTRODUCTION TO ANALYTIC GEOMETRY

In this chapter, we introduce some of the fundamental concepts of analytic geometry. This area of mathematics deals with the relationships between the geometry of graphs and the algebraic representations of the graphs. In particular, we shall study a **conic (conic section).** A **conic** is a curve of intersection of a plane with a right-circular cone of two nappes; the three types of curves that result are the **parabola,** the **ellipse,** and the **hyperbola.**

Geometrically, a cone can be thought of as having two nappes, extending indefinitely in both directions. A **generator** of a cone is any line lying in the cone, and all the generators of a cone intersect in a common point, V, called the **vertex** of the cone. The line OV is called the **axis** (see Fig. 180).

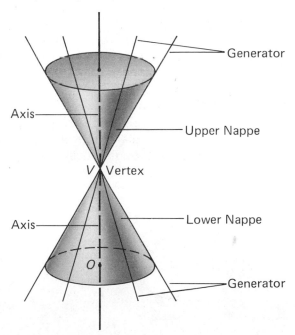

Figure 180

If the plane cutting a right-circular cone is parallel to one and only one generator of the cone, then the conic obtained is a **parabola.** If the cutting plane is parallel to two generators, it intersects both nappes of the cone, and a **hyper-**

bola is obtained. An **ellipse** results if the cutting plane is parallel to no genera-tor. A special case of the ellipse is a **circle,** which results if the cutting plane is also perpendicular to the axis of the cone. The graphs of these conics are shown in Figure 181.

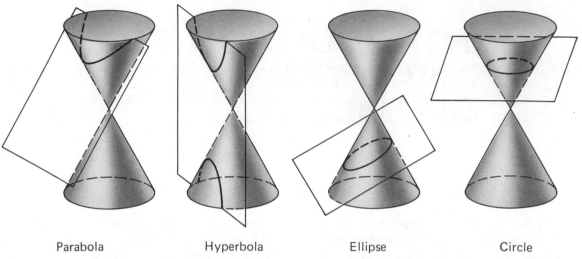

Parabola   Hyperbola   Ellipse   Circle

Figure 181

We shall see that all conic sections can be algebraically represented by the equation

$$Ax^2 + Cy^2 + Dx + Ey + F = 0 \tag{I}$$

where not both A and C are zero. Note that the second degree term Bxy does not appear in the equation of a conic section. In each section of this chapter we shall give the conditions on the constants in equation (I) which determine whether the graph of that equation is a parabola, hyperbola, ellipse, or circle.

7.1   CIRCLE

As a conic section, a circle is the curve obtained from the intersection of a right-circular cone by a plane perpendicular to the axis of the cone (see Fig. 182). Geometrically, the definition of a circle is given as follows.

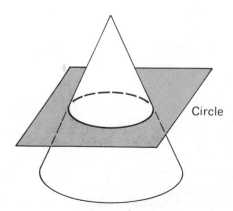

Circle

Figure 182

> **Definition:** A **circle** is the set of all points in a plane at a given distance from a fixed point. The fixed point is called the **center** of the circle and the fixed distance is called the **radius** of the circle.

If the center C is taken at the point (h, k) and the radius is r (Fig. 183), then

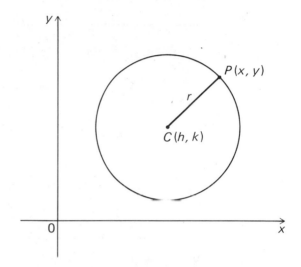

Figure 183

the equation of the circle can be obtained by using the distance formula. Let $P(x, y)$ by any point on the circle. Then the distance from the center C to the point P, $\overline{CP}$, must equal r. That is,

$$\sqrt{(x - h)^2 + (y - k)^2} = r$$

Squaring both sides gives

$$(x - h)^2 + (y - k)^2 = r^2$$

and we can state the following theorem.

> **Theorem 1:** The equation in standard form of the circle whose center is (h, k) and whose radius equals r is
>
> $$(x - h)^2 + (y - k)^2 = r^2$$

PROBLEM 1

Find an equation of the circle with center $(-2, 3)$ and radius equal to 5.

Solution: We have $h = -2$, $k = 3$, and $r = 5$. Substituting into the equation $(x - h)^2 + (y - k)^2 = r^2$ we have $(x - (-2))^2 + (y - 3)^2 = (5)^2$ or

$(x + 2)^2 + (y - 3)^2 = 25$. Squaring and then combining terms, we obtain

$$x^2 + 4x + 4 + y^2 - 6y + 9 = 25$$

or

$$x^2 + y^2 + 4x - 6y - 12 = 0$$

PROBLEM 2

Find an equation of the circle with center at the origin and radius equal to r.

Solution: If the center is at the origin, then we have $h = 0$ and $k = 0$. Thus, we have

$$x^2 + y^2 = r^2$$

PROBLEM 3

Given the standard form of a circle, $(x + 1)^2 + y^2 = 9$, find the center and radius.

Solution: The standard form of a circle is $(x - h)^2 + (y - k)^2 = r^2$. Therefore, for $(x + 1)^2 + y^2 = 9$ we have $x - h = x + 1$, or $h = -1$, and $y - k = y$, or $k = 0$. Thus, the center $(h, k) = (-1, 0)$ and the radius $r = 3$.

Following Problem 1, if we square and combine terms in the equation $(x - h)^2 + (y - k)^2 = r^2$, then we obtain

$$x^2 + y^2 - 2hx - 2ky + (h^2 + k^2 - r^2) = 0 \qquad \text{(II)}$$

Equation (II) is of the form

$$x^2 + y^2 + ax + by + c = 0 \qquad \text{(III)}$$

where $a = -2h$, $b = -2k$, and $c = h^2 + k^2 - r^2$.

Equation (III) is called the **general form** of an equation of a circle. Therefore, the two forms, general and standard, are equivalent. How do we put an equation in general form into standard form? The method used is that of completing the square and is illustrated in the following problems.

PROBLEM 4

Express $x^2 + y^2 - 4x + 8y - 5 = 0$ in standard form.

Solution: Grouping the equation as $(x^2 - 4x) + (y^2 + 8y) = 5$, we now complete the square for x terms and y terms on the left side. To complete the square for the x terms, we take half the x coefficient and square, getting $\left(\dfrac{-4}{2}\right)^2 = 4$. Similarly, for the y terms we have $\left(\dfrac{8}{2}\right)^2 = 16$. Adding these numbers to both sides of the equation gives

$$(x^2 - 4x + 4) + (y^2 + 8y + 16) = 5 + 4 + 16$$

or

$$(x - 2)^2 + (y + 4)^2 = 25$$

Comparing this with equation (I) we have center $(2, -4)$ and radius equal to 5.

PROBLEM 5

Express $4x^2 + 4y^2 - 8x + 24y + 4 = 0$ in standard form.

Solution: To complete the square for the x terms and y terms the coefficients of the x^2 and y^2 terms must be 1. Thus, dividing each side of the equation by 4, we obtain $x^2 + y^2 - 2x + 6y + 1 = 0$. Completing the square gives

$$(x^2 - 2x) + (y^2 + 6y) = -1$$
$$(x^2 - 2x + 1) + (y^2 + 6y + 9) = -1 + 1 + 9$$
$$(x - 1)^2 + (y + 3)^2 = 9$$

Thus, we have a circle with center $(1, -3)$ and radius 3.

PROBLEM 6

Express $x^2 + y^2 + 2x - 4y + 5 = 0$ in standard form.

Solution: Completing the square yields

$$(x^2 + 2x) + (y^2 - 4y) = -5$$
$$(x^2 + 2x + 1) + (y^2 - 4y + 4) = -5 + 1 + 4$$
$$(x + 1)^2 + (y - 2)^2 = 0$$

Therefore, we have a circle with center $(-1, 2)$ and radius 0. This represents a point, and such a circle is called a **degenerate circle** (or **point circle**).

PROBLEM 7

Express $x^2 + y^2 - 6x + 4y + 14 = 0$ in standard form.

Solution: Completing the square we obtain

$$(x - 3)^2 + (y + 2)^2 = -1$$

There are no real numbers (x, y) which satisfy this equation. Thus, the solution set is empty. Because of the form of the equation we say that this equation represents an **imaginary circle.**

Problems 4 to 7 lead us to the following theorem.

Theorem 2: The graph of the equation $x^2 + y^2 + ax + by + c = 0$ is
(i) a circle with center $\left(\dfrac{-a}{2}, \dfrac{-b}{2}\right)$ and radius $\dfrac{1}{2}\sqrt{a^2 + b^2 - 4c}$ if $\dfrac{a^2}{4} + \dfrac{b^2}{4} - c > 0$;

(ii) a point if $\dfrac{a^2}{4} + \dfrac{b^2}{4} - c = 0$; or

(iii) an imaginary circle if $\dfrac{a^2}{4} + \dfrac{b^2}{4} - c < 0$.

Proof: The equation $x^2 + y^2 + ax + by + c = 0$ can be put in standard form by completing the square. Therefore, we have

$$\left(x + \frac{a}{2}\right)^2 + \left(y + \frac{b}{2}\right)^2 = \frac{a^2}{4} + \frac{b^2}{4} - c$$

Thus, if $\dfrac{a^2}{4} + \dfrac{b^2}{4} - c > 0$, the graph is a circle with center $\left(\dfrac{-a}{2}, \dfrac{-b}{2}\right)$ and radius $\sqrt{\dfrac{a^2}{4} + \dfrac{b^2}{4} - c} = \dfrac{1}{2}\sqrt{a^2 + b^2 - 4c}$. When $\dfrac{a^2}{4} + \dfrac{b^2}{4} - c = 0$ we have a point. Finally, if $\dfrac{a^2}{4} + \dfrac{b^2}{4} - c < 0$, then we have an imaginary circle.

Note: The equation $x^2 + y^2 + ax + by + c = 0$ is a special form of $Ax^2 + Cy^2 + Dx + Ey + F$, where $A = C = 1$. Before we can use Theorem 2, the coefficients of x^2 and y^2 must both be 1. This can be achieved if $A = C$.

PROBLEM 8

Determine whether the following equations represent a real circle, a point, or an imaginary circle.

(i) $3x^2 + 3y^2 - 4x + 2y + 6 = 0$
(ii) $x^2 + y^2 - 6x - 4y - 3 = 0$
(iii) $x^2 + y^2 + 2x - 4y + 5 = 0$

Solution: (i) First, we make the coefficients of x^2 and y^2 equal to 1 by dividing both sides of the equation $3x^2 + 3y^2 - 4x + 2y + 6 = 0$ by 3. Thus, we have

$$x^2 + y^2 - \frac{4}{3}x + \frac{2}{3}y + 2 = 0$$

Now we use the results of Theorem 2 where $a = \frac{-4}{3}$, $b = \frac{2}{3}$, and $c = 2$. The expression $\frac{a^2}{4} + \frac{b^2}{4} - c = \frac{1}{4} \cdot \frac{16}{9} + \frac{1}{4} \cdot \frac{4}{9} - 2 = \frac{4}{9} + \frac{1}{9} - 2 = -\frac{13}{9} < 0$. Therefore, $3x^2 + 3y^2 - 4x + 2y + 6 = 0$ represents an imaginary circle.

(ii) The equation $x^2 + y^2 - 6x - 4y - 3 = 0$ gives $a = -6$, $b = -4$, and $c = -3$. Thus, $\frac{a^2}{4} + \frac{b^2}{4} - c = 9 + 4 + 3 = 16 > 0$ and we have a circle centered at (3, 2) and radius 4.

(iii) For the equation $x^2 + y^2 + 2x - 4y + 5 = 0$ we have $a = 2$, $b = -4$, and $c = 5$. Then $\frac{a^2}{4} + \frac{b^2}{4} - c = 1 + 4 - 5 = 0$ and the graph consists of the point $(-1, 2)$.

In all examples of circles, whether real, a point, or imaginary, we saw that the equation of a circle could be put in standard form by completing the square. However, to achieve the standard form, the coefficients of x^2 and y^2 must be equal to 1. Thus, we have the following theorem.

Theorem 3: The graph of the equation $Ax^2 + Cy^2 + Dx + Ey + F = 0$ is a circle if $A = C \neq 0$.

Proof. If $A = C \neq 0$, then we have $Ax^2 + Ay^2 + Dx + Ey + F = 0$, and dividing by A we obtain $x^2 + y^2 + \frac{D}{A}x + \frac{E}{A}y + \frac{F}{A} = 0$. This is the general form of a circle.

PROBLEM 9

Determine whether the graph of each of the following equations is a circle. If it is a circle, determine if it is real, a point, or imaginary.
(a) $2x^2 + 3y^2 + 12x - 8y + 31 = 0$
(b) $4x^2 + 4y^2 + 24x - 4y + 1 = 0$

Solution: (a) For $2x^2 + 3y^2 + 12x - 8y + 31 = 0$ we have $A = 2$ and $C = 3$. Therefore, by Theorem 3, the graph of the equation $2x^2 + 3y^2 + 12x - 8y + 31 = 0$ is *not* a circle.

(b) For $4x^2 + 4y^2 + 24x - 4y + 1 = 0$ we have $A = C = 4$, and the graph is a circle. To determine the nature of the circle, we use Theorem 2. Note that we must make the coefficients of x^2 and y^2 equal to 1. To achieve this we divide the equation by 4 to obtain $x^2 + y^2 + 6x - y + \frac{1}{4} = 0$. Thus, the values of a, b, and c in Theorem 2 are $a = 6$, $b = -1$, and $c = \frac{1}{4}$, and $\frac{a^2}{4} + \frac{b^2}{4} - c = \frac{(6)^2}{4} + \frac{(-1)^2}{4} - \frac{1}{4} = 9 + \frac{1}{4} - \frac{1}{4} = 9 > 0$. Hence, the circle is real with center $\left(\frac{-a}{2}, \frac{-b}{2} \right) = \left(-3, \frac{1}{2} \right)$ and radius 3.

Circles Determined by Three Conditions

Since the general equation of the circle $x^2 + y^2 + ax + by + c = 0$ contains three undetermined constants, three conditions are necessary to determine these coefficients. Suppose that (x_1, y_1), (x_2, y_2), and (x_3, y_3) are three points not in the same straight line. If the circle passes through the three given points, then the equation $x^2 + y^2 + ax + by + c = 0$ must be satisfied by the coordinates of each point. Substituting the coordinates of the three points into the general equation we get the equations

$$\begin{cases} x_1^2 + y_1^2 + ax_1 + by_1 + c = 0 \\ x_2^2 + y_2^2 + ax_2 + by_2 + c = 0 \\ x_3^2 + y_3^2 + ax_3 + by_3 + c = 0 \end{cases} \tag{IV}$$

We use this system of equations to determine a, b, and c.

PROBLEM 10

Find an equation of the circle passing through $(0, 1)$, $(0, 6)$, and $(3, 0)$.

Solution: The three equations expressing the condition that the circle $x^2 + y^2 + ax + by + c = 0$ pass through the points $(0, 1)$, $(0, 6)$, and $(3, 0)$ are

$$\begin{cases} 1 + b + c = 0 \\ 36 + 6b + c = 0 \\ 9 + 3a + c = 0 \end{cases}$$

Solving simultaneously, we find $a = -5$, $b = -7$, and $c = 6$. Thus, an equation of the circle passing through (0, 1), (0, 6), and (3, 0) is $x^2 + y^2 - 5x - 7y + 6 = 0$.

Similarly, an equation of the circle may be used in the form

$$(x - h)^2 + (y - k)^2 = r^2$$

in which case we have

$$\begin{cases} (x_1 - h)^2 + (y_1 - k)^2 = r^2 \\ (x_2 - h)^2 + (y_2 - k)^2 = r^2 \\ (x_3 - h)^2 + (y_3 - k)^2 = r^2 \end{cases} \tag{V}$$

The simultaneous solution of these equations determines the three unknown constants h, k, and r.

PROBLEM 11

Find an equation of the circle that contains the three points (9, −7), (−3, −1), and (6, 2).

Solution: Substituting the coordinates of the points (9, −7), (−3, −1), and (6, 2) into the system (*V*), we obtain

$$\begin{cases} (9 - h)^2 + (-7 - k)^2 = r^2 \\ (-3 - h)^2 + (-1 - k)^2 = r^2 \\ (6 - h)^2 + (2 - k)^2 = r^2 \end{cases}$$

Solving simultaneously, we find that $h = 3$, $k = -4$, and $r = \sqrt{45}$. Thus, an equation of the circle passing through (9, −7), (−3, −1), and (6, 2) is

$$(x - 3)^2 + (y + 4)^2 = 45$$

PROBLEM 12

Find the equation of the circle with center at (1, 2) and tangent to the line $y = x - 1$.

Solution: We are given $h = 1$ and $k = 2$. If we find r, we can obtain an equation of the circle by using the standard form. Let P be the point of tangency of the line $y = x - 1$ with the circle. Thus, $r = |\overline{PC}|$, and we must find the coordinates of P. This can be done by finding an equation of l_1 and then finding the point of intersection of l_1 with the line $y = x - 1$.

Since l_1 is along a diameter of the circle, and the line $y = x - 1$ is tangent to the circle, l_1 is perpendicular to $y = x - 1$. Since the slope of $y = x - 1$ is equal to 1, the slope of l_1 is equal to -1. Thus, using the point-slope form of an equation of a line, we obtain an equation of l_1:

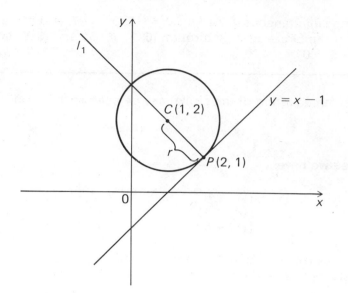

Figure 184

$$y - 2 = -1(x - 1) \text{ or } y = -x + 3$$

To obtain the point of intersection, P, we solve this equation simultaneously with the equation $y = x - 1$. Solving, we obtain P (2, 1). Thus,

$$r = |\overline{PC}| = \sqrt{(2 - 1)^2 + (1 - 2)^2} = \sqrt{2}$$

Therefore, an equation of the circle which has center (1, 2) and which is tangent to the line $y = x - 1$ is

$$(x - 1)^2 + (y - 2)^2 = (\sqrt{2})^2$$

or

$$(x - 1)^2 + (y - 2)^2 = 2$$

7.1 EXERCISES

1. Find an equation of the circle
   (a) with center at origin and radius 5.
   (b) with center (3, 2) and radius 4.
   (c) with center at $\left(-\dfrac{1}{2}, 0\right)$ and radius 1.
   (d) with center at (0, r) and radius r.
   (e) with ends of a diameter at (2, -3) and (6, 5).

2. Determine the center and radius of each of the following circles:
   (a) $(x - 1)^2 + (y + 6)^2 = 36$
   (b) $(x + 3)^2 + y^2 = 9$
   (c) $x^2 + \left(y + \dfrac{1}{3}\right)^2 = \dfrac{1}{4}$
   (d) $x^2 + y^2 = 5$
   (e) $(x + \pi)^2 + (y - 2)^2 = 16$
   (f) $(x - b)^2 + (y + a)^2 = c^2$

3. Write the following equations of circles in standard form:
   (a) $x^2 + y^2 - 49 = 0$
   (b) $x^2 + y^2 - 6x - 2y - 6 = 0$
   (c) $x^2 + y^2 - 10x - 11 = 0$
   (d) $x^2 + y^2 + 4x - 5y + 4 = 0$
   (e) $3x^2 + 3y^2 - 4y - 7 = 0$
   (f) $4x^2 + 4y^2 + 28x + 13 = 0$

4. Determine whether each of the following circles is a real circle, a point, or an imaginary circle.
   (a) $x^2 + y^2 + 5 = 0$
   (b) $x^2 + y^2 - 2x + 10y + 19 = 0$
   (c) $x^2 + y^2 + 2x - 4y + 5 = 0$
   (d) $4x^2 + 4y^2 + 24x - 4y + 1 = 0$
   (e) $(x + 1)^2 + (y - 2)^2 = 0$
   (f) $3x^2 + 3y^2 - 4x + 2y + 6 = 0$

5. Derive an equation of the circle passing through the three points:
   (a) $(8, -2)$, $(6, 2)$, and $(3, -7)$
   (b) $(0, 0)$, $(0, 8)$, and $(-6, 0)$
   (c) $(1, 5)$, $(4, 2)$, and $(-2, -1)$
   (d) $(9, 2)$, $(1, 1)$, and $(1, 3)$

6. Find an equation of the circle with center $(-1, -3)$ and tangent to the line $4y = -3x + 10$.

7. Find an equation of the circle whose y-intercepts are 6 and -2 and whose center is on the line $y = \dfrac{x}{2}$.

8. Graph the equations $y = \sqrt{25 - x^2}$, $y = -\sqrt{25 - x^2}$, and $x^2 + y^2 = 25$. Find the range and domain of each. Determine which represent functions. If any does represent a function, does it have an inverse?

9. Show that the determinant equation

$$\begin{vmatrix} x - h & y - k \\ -(y - k) & x - h \end{vmatrix} = r^2$$

   is an equation of a circle with center (h, k) and radius r.

10. Find the equations of the two circles tangent to the lines $4x + 3y - 7 = 0$ and $3x - 4y + 1 = 0$ and passing through the point $(2, 3)$.

7.2 TRANSLATION OF AXES

The shape of a curve is not affected by the position of the coordinate axes. However, the equation of the curve is affected. For example, a circle with center at $(-1, 3)$ and radius equal to 5 is represented by the equation

$$(x + 1)^2 + (y - 3)^2 = 25$$

or
$$x^2 + y^2 + 2x - 6y - 15 = 0$$

If we take the origin as the center, then the equation of the circle is simpler:

$$x^2 + y^2 = 25$$

When dealing with geometric problems in terms of coordinates, it may be helpful to choose the coordinate axes so as to make our equations as simple as possible. One method of changing the coordinate axes is to perform a **translation of axes.** A translation of axes is simply a new set of coordinate axes (x', y') chosen parallel to the x-axis and y-axis. Geometrically, this results in the movement of the origin O to the point O' (h, k) (see Fig. 185).

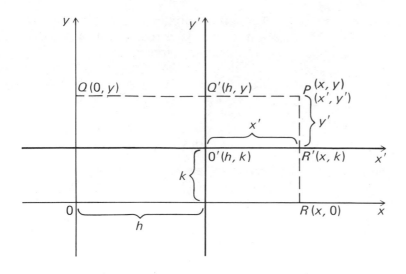

Figure 185

Suppose P is any point in the plane having coordinates (x, y) with respect to the given coordinate axes. Then P will have coordinates (x', y') with respect to the new axes. To obtain the relationships between these two sets of coordinates, we draw a line through P parallel to the x-axis and the x'-axis. Also, a line is drawn through P parallel to the y-axis and the y'-axis. We see that the first line through P intersects the x-axis at the point R and the x'-axis at the point R'. Similarly, the second line intersects the y-axis at the point Q and the y'-axis at the point Q'.

The coordinates of P with respect to the x and y axes are (x, y), the coordinates of R are (x, O), and the coordinates of R' are (x, k). To determine x and y in terms of x', y', h, and k we observe that

$$x = \overline{QP} = \overline{QQ'} + \overline{Q'P} = h + x'$$
$$y = \overline{RP} = \overline{RR'} + \overline{R'P} = k + y'$$

Thus

$$x = x' + h \qquad x' = x - h$$
$$\text{or}$$
$$y = y' + k \qquad y' = y - k$$

These equations are called the **equations of translation.** We can state these results as a theorem.

Theorem 4: If (x, y) represents a point P with respect to a given set of axes, and (x', y') is a representation of P after the axes are translated to a new origin having coordinates (h, k) with respect to the given axes, then

$$x = x' + h \text{ and } y = y' + k$$

or

$$x' = x - h \text{ and } y' = y - k$$

Therefore, any equation in x and y can be translated to an equation in x' and y' by replacing x by $(x' + h)$ and y by $(y' + k)$. The graph of the equation in x and y, with respect to the x- and y-axes, is exactly the same as the graph of the corresponding equation in x' and y' with respect to the x'- and y'-axes.

PROBLEM 1

Transform the equation $x^2 + y^2 - 6x + 4y - 12 = 0$ by translating the axes so that the new origin is at $(3, -2)$.

Solution: We have $h = 3$ and $k = -2$, and the equations of translation become

$$x = x' + 3 \text{ and } y = y' - 2$$

Substituting in the equation $x^2 + y^2 - 6x + 4y - 12 = 0$ we get

$$(x' + 3)^2 + (y' - 2)^2 - 6(x' + 3) + 4(y' - 2) - 12 = 0$$

or

$$(x')^2 + (y')^2 = 25$$

thus changing the form of the equation but not changing the graph (see Fig. 186).

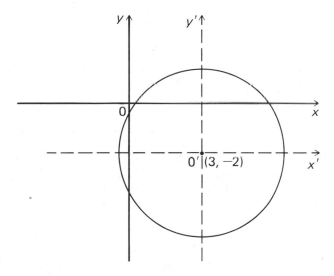

Figure 186

PROBLEM 2

Determine a translation of axes that will transform the equation $x^2 + y^2 - 4x + 6y - 12 = 0$ in such a way as to remove the x and y terms.

Solution: First Method
Let $x = x' + h$ and $y = y' + k$. We substitute these values of x and y in

the equation $x^2 + y^2 - 4x + 6y - 12 = 0$:

$$(x' + h)^2 + (y' + k)^2 - 4(x' + h) + 6(y' + k) - 12 = 0$$

Expanding and collecting terms, we obtain

$$(x')^2 + (y')^2 + (2h - 4)x' + (2k + 6)y' + h^2 + k^2 - 4h + 6k - 12 = 0$$

For the coefficients of x' and y' to be zero we must have $2h - 4 = 0$ and $2k + 6 = 0$. Thus, $h = 2$ and $k = -3$. Substituting these values, we obtain

$$(x')^2 + (y')^2 - 25 = 0$$

as the equation of $x^2 + y^2 - 4x + 6y - 12 = 0$ using new coordinate axes and having no x' or y' terms in the equation.

Solution: Second Method
The method given above always works. However, the method of completing the square can be easier, provided that the technique of completing the square applies to the given equation.

Completing the squares on x and y in the equation $x^2 + y^2 - 4x + 6y - 12 = 0$ gives

$$(x - 2)^2 + (y + 3)^2 = 25$$

If we replace $x - 2$ by x' and $y + 3$ by y', then we have

$$(x')^2 + (y')^2 - 25 = 0$$

As noted above, the method of completing the square does not always apply to every equation.

PROBLEM 3

Find a translation of axes that removes first degree terms in the equation $2xy - 3x = 1$.

Solution: If we let $x = x' + h$ and $y = y' + k$, then the equation $2xy - 3x = 1$ becomes

$$2(x' + h)(y' + k) - 3(x' + h) = 1$$
$$2x'y' + (2k - 3)x' + 2hy' + 2hk - 3h = 1$$

To remove first degree terms x' and y' we set $2k - 3 = 0$ and $2h = 0$. Thus, $k = \dfrac{3}{2}$ and $h = 0$. Substituting these values, we obtain

$$2x'y' = 1$$

The graph of this equation can be obtained using the translation equations $x = x' + h = x'$ and $y = y' + k = y + \frac{3}{2}$. Therefore, the new origin (h, k) is at the point $\left(0, \frac{3}{2}\right)$, and the graph of $2x'y' = 1$ is given in Figure 187.

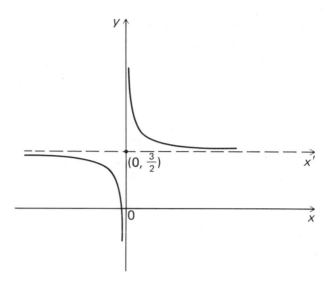

Figure 187

7.2 EXERCISES

1. Given $x' = x - 2$ and $y' = y - 4$, find the new coordinates of
   (a) (3, 1)   (b) (0, 0)
   (c) (0, 4)   (d) (−4, 4)
   (e) (−1, 6)   (f) (5, 0)
   (g) (−2, −3)   (h) (5, 7)

2. Find the equations of each of the following if the axes are translated to the new origin indicated:
   (a) $12x - 5y + 7 = 0$; $(-3, -3)$
   (b) $x^2 + y^2 = 9$; $(-4, 3)$
   (c) $x^2 - 6y = -9$; $\left(0, \frac{3}{2}\right)$
   (d) $y^2 - 2kx + k^2 = 0$; $\left(\frac{k}{2}, 0\right)$
   (e) $y^2 - x^3 + 6x^2 - 12x - 6y + 17 = 0$; $(2, 3)$

3. Simplify the following equations by determining a suitable translation of axes:
   (a) $y = (x - 1)^2$
   (b) $(y + 1)(x - 2) = 3$
   (c) $(y - 1)^2 = (x + 4)^3$
   (d) $(x - 2)^2 + (y + 1)^2 = 49$

4. Using a translation of axes remove the first degree terms from the following equations:
   (a) $x^2 + y^2 - 2x - 4y - 20 = 0$
   (b) $x^2 + y^2 - 6x - 10y + 18 = 0$
   (c) $x^2 + 2y^2 - 4x + 6y - 8 = 0$
   (d) $x^2 + 4y^2 + 4x - 24y + 24 = 0$
   (e) $4x^2 - 9y^2 - 16x - 18y - 29 = 0$

5. Using a translation of axes remove the first degree terms from the following equations:
   (a) $2xy - x - y + 4 = 0$
   (b) $xy - 2y - 4x = 0$
   (c) $xy + 3x - 4y = 5$

6. Determine the new origin which will remove the x term and the constant term from the equation $y = ax^2 + bx + c$.

7. Translate the curve $y = x^3 - 6x^2 + 7x + 5$ to the point $(2, 3)$. Show that the curve is symmetric with respect to this point.

7.3 PARABOLA

As a conic section, a parabola is the curve obtained by the intersection of a right-circular cone by a plane which is parallel to one and only one generator of the cone (see Fig. 188). Thus, a parabola intersects only one nappe of the cone.

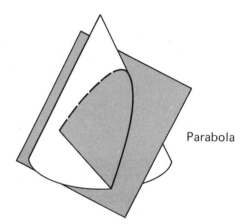

Parabola

Figure 188

Geometrically, the definition of a parabola is as follows.

> **Definition:** A **parabola** is the set of all points in a plane equidistant from a fixed point and a fixed line. The fixed point is called the **focus,** and the fixed line is called the **directrix.**

Using the definition we can derive an equation of a parabola. We begin by choosing the x-axis as perpendicular to the directrix and also we choose the focus so that it lies on the x-axis. The origin is taken at the point on the x-axis midway between the directrix and the focus. Then if the focal point F has coordinates $(p, 0)$, an equation of the directrix line is $x = -p$ (see Fig. 189). Let $P(x, y)$ be any point on the parabola. Then point P is equidistant from the point F and the directrix. We draw a line from P perpendicular to the directrix, and let $R(-p, y)$ be the point where the perpendicular line meets the directrix. Then,

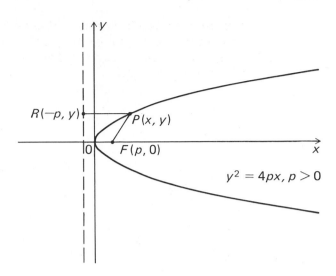

Figure 189

$$\text{distance } FP = \text{distance } RP$$

$$\sqrt{(x - p)^2 + y^2} = \sqrt{(x + p)^2 + (y - y)^2}$$

$$\sqrt{(x - p)^2 + y^2} = \sqrt{(x + p)^2}$$

Squaring both sides of the equation, we obtain

$$(x - p)^2 + y^2 = (x + p)^2$$
$$x^2 - 2px + p^2 + y^2 = x^2 + 2px + p^2$$

or

$$y^2 = 4px$$

Thus, we have the following theorem.

Theorem 5: An equation of the parabola having its focus at $(p, 0)$ and its directrix the line $x = -p$ is

$$y^2 = 4px$$

The line through the focal point drawn perpendicular to the directrix is called the **axis of symmetry** of the parabola. The point of intersection of this line and the curve is called the **vertex.** The parabola in Figure 189 has the x-axis as its axis of symmetry and the origin $(0, 0)$ as its vertex.

If $p < 0$, then the graph of the parabola with focus $(p, 0)$ and line $x = -p$ as directrix is given in Figure 190. If we interchange the x-axis and y-axis in the above analysis, then the focus is at the point $F(0, p)$, and the directrix is the line having the equation $y = -p$. The following theorem gives the form of the equation of such parabolas.

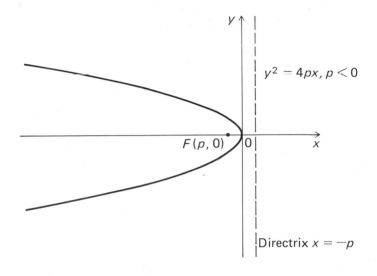

$$y^2 = 4px, p < 0$$

$F(p, 0)$

Directrix $x = -p$

Figure 190

Theorem 6: An equation of the parabola having its focus at $(0, p)$ and as its directrix the line $y = -p$ is

$$x^2 = 4py$$

If $p > 0$, the parabola opens upward. If $p < 0$, then the parabola opens downward [see Fig. 191(a) and (b)].

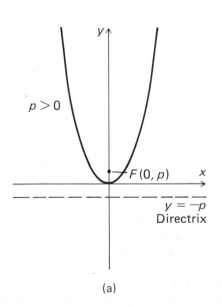

$p > 0$

$F(0, p)$

$y = -p$
Directrix

(a)

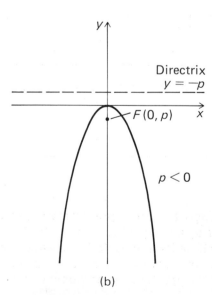

Directrix
$y = -p$

$F(0, p)$

$p < 0$

(b)

Figure 191

PROBLEM 1

Find an equation of the parabola whose focus is $(-2, 0)$ and whose directrix is the line $x = 2$.

Solution: Since the focus is on the x-axis and to the left of the directrix, the parabola opens to the left, and $p = -2$. Thus, an equation of the parabola is

$$y^2 = -8x$$

PROBLEM 2

Find the coordinates of the focus and an equation of the directrix of the parabolas. Graph: (a) $y^2 = -4x$; and (b) $x^2 = 2y$.

Solution: (a) The equation $y^2 = -4x$ is of the form $y^2 = 4px$. Therefore, $4p = -4$ or $p = -1$. Since $p < 0$, the parabola opens to the left. The focus is at the point $F(-1, 0)$. An equation of the directrix is $x = 1$. The graph of $y^2 = -4x$ is given in Figure 192.

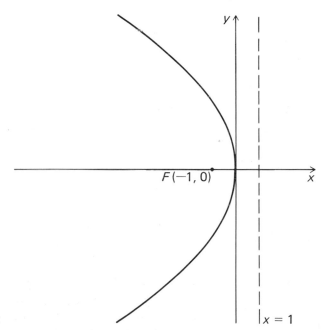

Figure 192

(b) The equation $x^2 = 2y$ is of the form $x^2 = 4py$. Thus, $4p = 2$ or $p = \frac{1}{2}$. Since $p > 0$, the parabola opens upward. The focus is $F\left(0, \frac{1}{2}\right)$ and the directrix $y = -\frac{1}{2}$ (see Fig. 193).

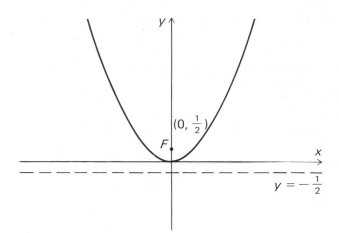

Figure 193

Theorems 5 and 6 give equations of a parabola with its vertex at the origin $(0, 0)$. If the vertex of the parabola is at the point (h, k), then we can use the translation equations to derive equations of a parabola with its vertex at (h, k). Consider the parabola in Figure 194. If we introduce new coordinate axes x'

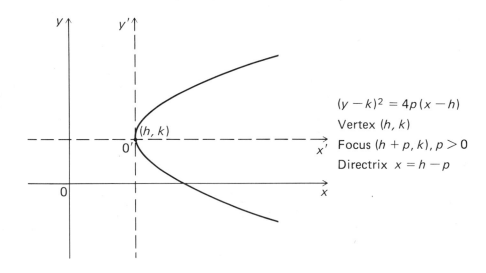

$(y - k)^2 = 4p(x - h)$

Vertex (h, k)

Focus $(h + p, k), p > 0$

Directrix $x = h - p$

Figure 194

and y' parallel to the original axes and having their origin at (h, k), then the vertex of the parabola is at the origin of the $x'y'$-system. An equation of the parabola is $(y')^2 = 4px'$, where p is the distance from the vertex to the focus.

The translation equations give us a relationship between the xy-system and the $x'y'$-system. That is, $x = x' + h, y = y' + k$ or $x' = x - h, y' = y - k$. Substituting for x' and y' in $(y')^2 = 4px'$, we obtain

$$(y - k)^2 = 4p(x - h), p > 0$$

The following theorems state the effect of a translation of axes on the standard form of the parabola.

Theorem 7: An equation of a parabola with focus $F(h + p, k)$ and directrix $x = -p + h$ is

$$(y - k)^2 = 4p(x - h)$$

(i) The vertex is $V(h, k)$.
(ii) If $p > 0$, then the parabola opens to the right.
(iii) If $p < 0$, then the parabola opens to the left.
(iv) The axis of symmetry is $y = k$.

The graphs of these parabolas are given in Figure 195.

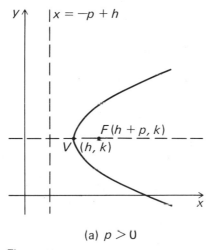 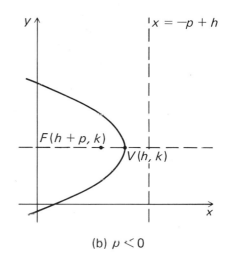

(a) $p > 0$ (b) $p < 0$

Figure 195

Theorem 8: An equation of a parabola with focus $F(h, k + p)$ and directrix $y = -p + k$ is

$$(x - h)^2 = 4p(y - k)$$

(i) The vertex is $V(h, k)$.
(ii) If $p > 0$, then the parabola opens upward.
(iii) If $p < 0$, then the parabola opens downward.
(iv) The axis of symmetry is $x = h$.

The graphs are shown in Figure 196.

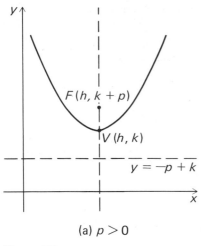

(a) $p > 0$

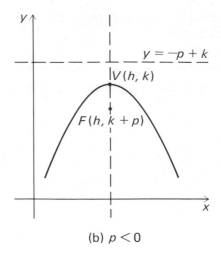

(b) $p < 0$

Figure 196

PROBLEM 3

Discuss the curve $(x - 1)^2 = 8(y + 3)$.

Solution: We rewrite the equation as

$$(x - 1)^2 = 8[y - (-3)]$$

Since the equation is of the form $(x - h)^2 = 4p(y - k)$, we have $h = 1$, $k = -3$, and $4p = 8$, or $p = 2$. Therefore, the vertex is $(h, k) = (1, -3)$, the focus $(h, k + p) = (1, -3 + 2) = (1, -1)$, and the directrix is the line $y = -p + k = -2 + (-3) = -5$. The graph is given in Figure 197.

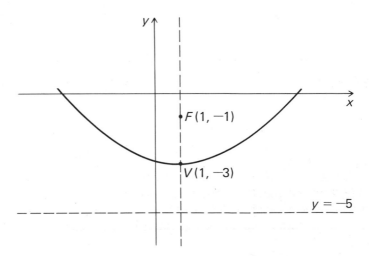

Figure 197

PROBLEM 4

Discuss the curve $(y + 1)^2 = -12(x - 4)$.

Solution: The equation $(y + 1)^2 = -12(x - 4)$ is equivalent to $[y - (-1)]^2 = -12(x - 4)$. Since the equation is of the form $(y - k)^2 = 4p(x - h)$, we have $h = 4$, $k = -1$, and $4p = -12$ or $p = -3$. The fact that $p < 0$ implies that the curve is a parabola which opens to the left. The vertex has coordinates $(4, -1)$, the focus is the point $(1, -1)$, and the directrix is the line $x = -p + h = -(-3) + 4 = 7$ (see Fig. 198).

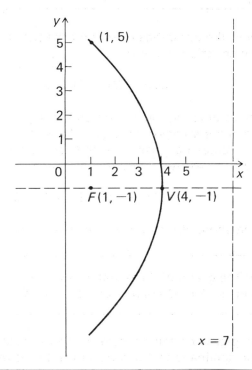

Figure 198

PROBLEM 5

Discuss the curve $y = x^2 + 2x - 2$.

Solution: We complete the square on the right side. First we take the constant to the left side to obtain

$$y + 2 = x^2 + 2x$$

Now, add $\left(\frac{1}{2} \text{ coefficient of } x\right)^2 = \left(\frac{1}{2} \cdot 2\right)^2 = (1)^2 = 1$ to both sides:

$$y + 3 = x^2 + 2x + 1 = (x + 1)^2$$

Thus, we have

$$(x + 1)^2 = y + 3$$

This is a parabola of the form $(x - h)^2 = 4p(y - k)$. Therefore, $h = -1$, $k = -3$, and $4p = 1$ or $p = \frac{1}{4}$. The curve $y = x^2 + 2x - 2$ is a parabola with vertex $(-1, -3)$, focus $\left(-1, -\frac{11}{4}\right)$, and the directrix is the line $y = -p + k = -\left(\frac{1}{4}\right) + (-3) = -\frac{13}{4}$. Also, since $p = \frac{1}{4} > 0$, the parabola opens upward.

Remark: The graph of the equation $y = ax^2 + bx + c$, $a \neq 0$, is always a parabola.

If we remove the parentheses from the standard form $(x - h)^2 = 4p(y - k)$, then we get an equation of the form

$$Ax^2 + Dx + Ey + F = 0 \tag{I}$$

Conversely, by completing the square on x in equation (I) we can show that (I) can be written in the standard form $(x - h)^2 = 4p(y - k)$.

Similarly, removing the parentheses from the standard form $(y - k)^2 = 4p(x - h)$ we see that the equation

$$Cy^2 + Dx + Ey + F = 0 \tag{II}$$

represents a parabola. We can now state these results as a theorem.

> **Theorem 9:** The graph of the equation $Ax^2 + Cy^2 + Dx + Ey + F = 0$ is a parabola if either A or C equals zero, but not both.

Parabolas play a prominent role in many situations in the real world. In many scientific applications it can be shown that either an arc of a parabola is used or the variable quantities are related in the same manner as the coordinates of a point on a parabola. The reader is referred to Chapter 3, Section 3.5, in which some of the applications of parabolas are given. In Chapter 3 we showed that the path of a projectile, if air resistance is neglected, is a parabola. Also, the equation expressing the relation between the distance a freely falling body travels and the time required is the equation of a parabola. Other applications are in the design of suspension bridges and arches; large reflectors in a reflecting telescope are usually parabolic.

For example, if equal weights are placed in a line, equally spaced, and then suspended from a thin wire cable, the cable will assume, approximately, a parabolic shape (see Fig. 199). Since the parabolic shape of the thin wire cable is natural for supporting a uniform load from above, by symmetry it follows that a parabolic arch will give uniform support to a load from below. The arch of the human foot is such an example.

Another occurrence of a parabola in nature is the flight formation of a group of migrating birds. It is believed that such a parabolic flight formation helps the birds support each other flying against horizontal resistance of the air (see Fig. 200).

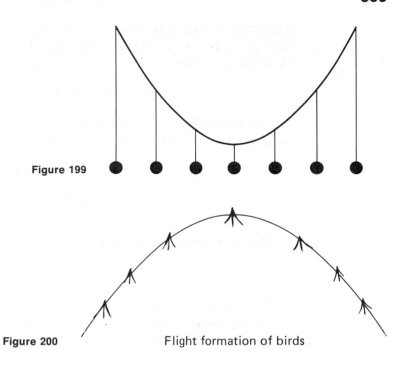

Figure 199

Figure 200   Flight formation of birds

PROBLEM 6

A parabolic arch has a span of 20 feet and is 10 feet high at its vertex. What will be the height of the arch at a distance of 5 feet from the center of the span?

Solution: We choose a coordinate system which has the base of the arch along the x-axis and the center of the arch at the origin (see Fig. 201).

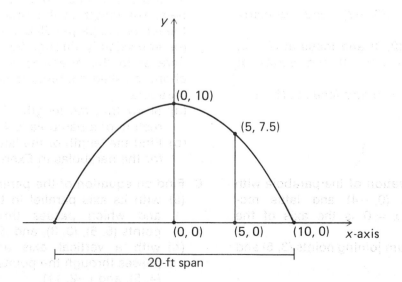

Figure 201   20-ft span

Since the vertex is along the y-axis, the equation of the parabola takes the form $(x - h)^2 = 4p(y - k)$. The vertex is the point $(0, 10)$; therefore, the equation becomes

$$x^2 = 4p(y - 10)$$

We also know that the point $(10, 0)$ lies on the parabola. Substituting these values for x and y, we obtain a value for p:

$$(10)^2 = 4p(0 - 10)$$

$$100 = -40\, p \text{ or } p = -\frac{5}{2}$$

Thus, the equation representing the arch is

$$x^2 = -10(y - 10)$$

To find the height of the arch 5 feet from the center we substitute $x = 5$ and solve for y: $25 = -10(y - 10)$, or $y = 7.5$.

7.3 EXERCISES

1. Determine the vertex, focus, and equation of the directrix for each of the following parabolas:
   (a) $y^2 = 6x$
   (b) $x^2 = 8y$
   (c) $4y^2 = -3x$
   (d) $(y - 1)^2 = -8(x - 5)$
   (e) $(y - 2)^2 = 8(x - 3)$
   (f) $(x - 2)^2 = 3(y + 3)$

2. Put the following parabolas in standard form and determine the vertex, focus, and directrix. Graph.
   (a) $y^2 - 2y - 8x + 25 = 0$
   (b) $x^2 - 4x - 2y - 8 = 0$
   (c) $3y^2 - 8x + 16 = 0$
   (d) $5x^2 + 4y - 12 = 0$

3. Write equations of the following parabolas:
   (a) focus at $(6, -2)$ and directrix $x - 2 = 0$
   (b) vertex at $(2, 3)$ and focus at $(2, -2)$
   (c) directrix $x + 2 = 0$ and vertex at $(1, 3)$
   (d) directrix $y = 0$ and focus at $(3, 1)$

4. A chord of a parabola is a line segment joining two distinct points of the parabola. The length of the chord through the focus and perpendicular to the axis (or, equivalently, through the focus and parallel to the directrix) is $4|p|$. This chord is called the **latus rectum** of the parabola.
   (a) Show that the length of the latus rectum of a parabola is $4 \cdot |p|$.
   (b) Find the length of the latus rectum for the parabolas in Exercise 3.

5. Derive an equation of the parabola with
   (a) vertex at $(0, -4)$ and latus rectum $= 9$; $x = 0$ is the axis of the parabola.
   (b) latus rectum joining points $(3, 5)$ and $(3, -3)$.

6. Find an equation of the parabola
   (a) with its axis parallel to the x-axis, and which passes through the points $(6, 5)$, $(3, 3)$, and $(6, -3)$.
   (b) with a vertical axis and which passes through the points $(-4, 21)$, $(4, 5)$, and $(-2, 11)$.

7. The two supporting towers of a parabolic suspension bridge are 70 feet high above the ground. They are 200 feet apart. The lowest point of the cable is 20 feet above the ground. How high above the ground is the point on the cable 50 feet from the center of the bridge?

8. The ends of a parabolic arch are 40 feet apart and the arch is 25 feet high. Find the height of the arch 8 feet from one end of the arch.

9. Show that the determinant equations (a) and (b) are parabolas with vertex (h, k).

(a) $\begin{vmatrix} (x - h) & 4p \\ (y - k) & (x - h) \end{vmatrix} = 0$

(b) $\begin{vmatrix} (y - k) & 4p \\ (x - h) & (y - k) \end{vmatrix} = 0$

10. The path of a projectile thrown horizontally from a point y feet above the ground with a velocity of v feet/second is a parabola whose equation is

$$x^2 = \frac{-2v^2}{9} y$$

where x is the distance measured horizontally from the point of the projection and the origin being the starting point. If a ball is thrown horizontally from the top of a building 30 feet tall with a horizontal velocity of 192 feet/second, then how far will the ball travel horizontally before it hits the ground?

7.4 ELLIPSE

If we cut a right-circular cone with a plane which is parallel to no generator, then the curve we obtain is an ellipse (see Fig. 202).

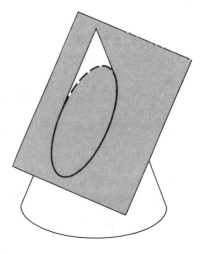

Figure 202

Definition: An **ellipse** is the collection of all points in the plane, the sum of whose distances from two fixed points is a constant. The two fixed points are called the **foci.**

To derive an equation of an ellipse we choose our coordinate system such that the foci are the points $F_1\,(c, 0)$ and $F_2\,(-c, 0)$, for $c > 0$. Thus, the line F_1F_2 lies along the x-axis and the midpoint of the line segment F_1F_2, called the **center** of the ellipse, is located at the origin (see Figure 203). Let the constant sum in

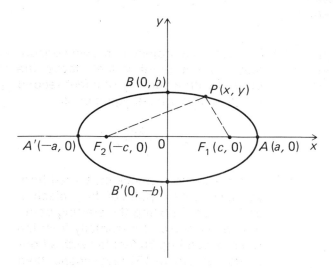

Figure 203

the definition be $2a$, $a > c$, and let $P\,(x, y)$ be any point on the ellipse. Then from the definition of an ellipse and the distance formula we have

$$\overline{F_2P} + \overline{PF_1} = 2a$$

or

$$\sqrt{(x + c)^2 + (y - 0)^2} + \sqrt{(x - c)^2 + (y - 0)^2} = 2a$$

or

$$\sqrt{(x + c)^2 + y^2} = 2a - \sqrt{(x - c)^2 + y^2}$$

Squaring both sides, we get

$$x^2 + 2cx + c^2 + y^2 = 4a^2 - 4a\sqrt{(x - c)^2 + y^2} + x^2 - 2cx + c^2 + y^2$$

or

$$-4a^2 + 4cx = -4a\sqrt{(x - c)^2 + y^2}$$

or

$$-a^2 + cx = -a\sqrt{(x - c)^2 + y^2}$$

Squaring both sides again,

$$a^4 - 2a^2cx + c^2x^2 = a^2x^2 - 2ca^2x + a^2c^2 + a^2y^2$$

or

$$x^2(a^2 - c^2) + a^2y^2 = a^2(a^2 - c^2)$$

Since $a > c$, then we have $a^2 > c^2$ or $a^2 - c^2 > 0$.

Dividing through by $a^2(a^2 - c^2)$, the equation becomes

$$\frac{x^2}{a^2} + \frac{y^2}{a^2 - c^2} = 1$$

Since $a^2 - c^2$ is positive, we can write $a^2 - c^2 = b^2$ and we have the standard form of the equation of an ellipse centered at the origin.

$$\frac{x^2}{a^2} + \frac{y^2}{b^2} = 1, a > b \tag{I}$$

Since the equation contains only even powers of x and y, the graph is symmetric with respect to the x-axis, y-axis, and the origin. The longer axis of symmetry, $\overline{A'A}$, is called the **major axis** and the shorter axis of symmetry, $\overline{BB'}$, is called the **minor axis.** The points A, A', B, and B' are called the **vertices.**

If the foci had been at $(0, c)$ and $(0, -c)$, then the major axis would have been on the y-axis and the standard form would be

$$\frac{x^2}{b^2} + \frac{y^2}{a^2} = 1, a > b \tag{II}$$

PROBLEM 1

Discuss the graph of $4x^2 + y^2 = 16$ and find its vertices and foci.

Solution: Dividing by 16, we obtain

$$\frac{x^2}{4} + \frac{y^2}{16} = 1$$

which is the standard form for an ellipse whose major axis lies along the y-axis; $a^2 = 16$ so $a = 4$, $b^2 = 4$ so $b = 2$. Therefore, $c^2 = a^2 - b^2 = 16 - 4 = 12$ and $c = \sqrt{12}$. The foci are located at $(0, \sqrt{12})$ and $(0, -\sqrt{12})$. The four vertices are $(0, 4)$, $(0, -4)$, $(2, 0)$, and $(-2, 0)$. The graph is given in Figure 204.

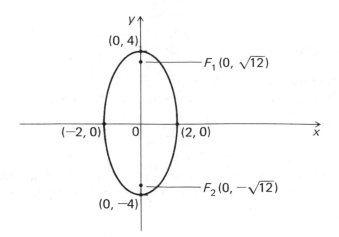

Figure 204

PROBLEM 2

Find an equation of the ellipse with foci (± 3, 0) and vertices (± 5, 0).

Solution: The ellipse has its center at the origin and its foci are on the x-axis. Thus, its equation is of the form $\dfrac{x^2}{a^2} + \dfrac{y^2}{b^2} = 1$. Since $c = 3$ and $a = 5$ we have $b^2 = a^2 - c^2 = 25 - 9 = 16$, and $b = 4$. Therefore, an equation of the ellipse is

$$\frac{x^2}{25} + \frac{y^2}{16} = 1$$

PROBLEM 3

A point $P(x, y)$ moves so that the sum of its distances from the points (0, 5) and (0, -5) is 20. Derive the equation of the curve traced by such points.

Solution: Let F_1 be (0, 5) and F_2 be (0, -5). Then

$$\overline{F_1P} + \overline{PF_2} = 20$$

or

$$\sqrt{(x - 0)^2 + (y - 5)^2} + \sqrt{(x - 0)^2 + (y + 5)^2} = 20$$

or

$$\sqrt{x^2 + (y - 5)^2} = 20 - \sqrt{x^2 + (y + 5)^2}$$

Squaring both sides and collecting terms, we get

$$20y + 400 = 40\sqrt{x^2 + (y + 5)^2}$$

or

$$y + 20 = 2\sqrt{x^2 + (y + 5)^2}$$

Squaring again, we have

$$4x^2 + 3y^2 = 300$$

or

$$\frac{x^2}{75} + \frac{y^2}{100} = 1$$

Thus, the locus of the points is an ellipse.

Center Not at Origin

If the center of the ellipse is at a point (h, k) and its major axis is parallel to either the x-axis or y-axis, then we can determine an equation for the ellipse by translating the axes.

Suppose the center is at (h, k) and the major axis is parallel to the x-axis. Also, let the foci be $(h - c, k)$ and $(h + c, k)$. By translating to the new axes $x' = x - h$ and $y' = y - k$, then an equation of the ellipse in the new coordinate system is $\frac{(x')^2}{a^2} + \frac{(y')^2}{b^2} = 1$. In terms of the original coordinate system we have

$$\frac{(x - h)^2}{a^2} + \frac{(y - k)^2}{b^2} = 1, a > b \qquad \text{(III)}$$

A similar argument holds for an ellipse with center at (h, k) and foci at $(h, k - c)$ and $(h, k + c)$. Thus, if the major axis is parallel to the y-axis with foci $(h, k - c)$ and $(h, k + c)$ and center (h, k), then an equation is

$$\frac{(x - h)^2}{b^2} + \frac{(y - k)^2}{a^2} = 1, a > b \qquad \text{(IV)}$$

PROBLEM 4

Determine the center, foci, major axis, and vertices for the ellipse $\frac{(x - 6)^2}{36} + \frac{(y + 4)^2}{16} = 1$.

Solution: The equation is of the form of equation (III). Thus, the center is $(6, -4)$ and the major axis is parallel to the x-axis. Since $a = 6$ and

$b = 4$, we have $c^2 = a^2 - b^2 = 36 - 16 = 20$ or $c = \sqrt{20} = 2\sqrt{5}$. Hence, the foci are $(h - c, k) = (6 - 2\sqrt{5}, -4)$ and $(h + c, k) = (6 + 2\sqrt{5}, -4)$, and the vertices are $(6, 0)$, $(6, -8)$, $(0, -4)$, and $(12, -4)$ (see Fig. 205).

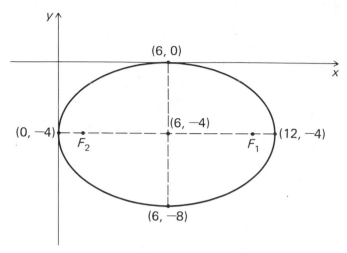

Figure 205

PROBLEM 5

Discuss the graph of $16x^2 + 25y^2 - 64x + 50y - 311 = 0$.

Solution: The equation can be put in standard form by completing the square.

$$16x^2 + 25y^2 - 64x + 50y - 311 = 0$$
$$16x^2 - 64x + 25y^2 + 50y = 311$$
$$16(x^2 - 4x) + 25(y^2 + 2y) = 311$$

Completing the square for x and y we obtain

$$16(x^2 - 4x + 4) + 25(y^2 + 2y + 1) = 311 + 64 + 25$$
$$16(x - 2)^2 + 25(y + 1)^2 = 400$$

Dividing by 400, we get

$$\frac{(x - 2)^2}{25} + \frac{(y + 1)^2}{16} = 1$$

This equation shows that the graph is an ellipse with center $(2, -1)$, with the major axis parallel to the x-axis. The foci are at $(-1, -1)$ and $(5, -1)$.

PROBLEM 6

Find an equation of the ellipse with its center at $(4, -1)$, focus at $(1, -1)$, and passing through $(8, 0)$.

Solution: Since the center is (4, −1) and the focus is (1, −1), the major axis is parallel to the *x*-axis and the standard form is

$$\frac{(x - 4)^2}{a^2} + \frac{(y + 1)^2}{b^2} = 1$$

Using the fact that (8, 0) lies on the ellipse, we have $\frac{(8 - 4)^2}{a^2} + \frac{(0 + 1)^2}{b^2} = 1$ or $\frac{16}{a^2} + \frac{1}{b^2} = 1$. However, since the coordinates of the focus are $(h - c, k)$, we have $(4 - c, -1) = (1, -1)$ or $c = 3$. Recalling that $b^2 = a^2 - c^2 = a^2 - 9$ and substituting this value for b^2 we obtain $\frac{16}{a^2} + \frac{1}{b^2} = \frac{16}{a^2} + \frac{1}{a^2 - 9} = 1$. Solving this equation, we find $a^2 = 18$.

Thus, an equation of the ellipse is $\frac{(x - 4)^2}{18} + \frac{(y + 1)^2}{9} = 1$.

Since a translation of axes will change equations (III) and (IV) into equations (I) and (II), respectively, any equation of the form of equation (III) or (IV) represents an ellipse. In any case, the **general form** of the equation of an ellipse is $Ax^2 + Cy^2 + Dx + Ey + F = 0$ if A and C agree in sign.

Theorem 10: The equation

$$Ax^2 + Cy^2 + Dx + Ey + F = 0$$

represents an ellipse if A and C are of the same sign (that is, if $AC > 0$).

Proof. If A and C are of the same sign, then the equation may be written so that both are positive. Completing the squares, we can write

$$A\left(x^2 + \frac{D}{A} + \frac{D^2}{4A^2}\right) + C\left(y^2 + \frac{E}{C}y + \frac{E^2}{4C^2}\right) = \frac{D^2}{4A} + \frac{E^2}{4C} - F$$

$$A\left(x + \frac{D}{2A}\right)^2 + C\left(y + \frac{E}{2C}\right)^2 = \frac{D^2}{4A} + \frac{E^2}{4C} - F$$

Dividing by AC we obtain

$$\frac{\left(x + \frac{D}{2A}\right)^2}{C} + \frac{\left(y + \frac{E}{2C}\right)^2}{A} = \frac{CD^2 + AE^2 - 4ACF}{4A^2C^2}$$

which is an equation of the form (III) or (IV). Therefore, the equation $Ax^2 + Cy^2 + Dx + Ey + F = 0$ represents an ellipse if A and C are of the same sign.

Among the many applications of the ellipse are architecture, map projections which are designed to preserve relative areas, and astronomy. In particular the planets move in elliptical orbits about the sun, with the sun at a focus.

Before giving an example, we define the **eccentricity (e)** of an ellipse. The eccentricity of an ellipse is the ratio

$$\frac{\text{distance between foci}}{\text{length of major axis}}$$

For the ellipse in standard position $\dfrac{x^2}{a^2} + \dfrac{y^2}{b^2} = \dfrac{x^2}{a^2} + \dfrac{y^2}{a^2 - c^2} = 1$, the eccentricity is

$$e = \frac{c}{a}$$

PROBLEM 7

The earth's orbit around the sun is an ellipse with the sun at one of the foci. If the major axis of the ellipse is 93 million miles and the eccentricity is approximately $\dfrac{1}{62}$, find the closest and furthest distances of the earth from the sun.

Solution: Consider the graph in Figure 206. We choose our coordinate

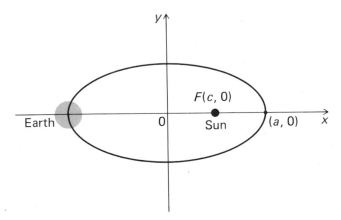

Figure 206

system so that the center of the elliptical orbit is at $(0, 0)$. Since the eccentricity $e = \dfrac{c}{a}$, we have $\dfrac{1}{62} = \dfrac{c}{93,000,000}$ or $c = 1,500,000$. Thus, the closest the earth gets to the sun is $a - c = 91,500,000$ miles, while the furthest it gets from the sun is $a + c = 94,500,000$ miles.

7.4 EXERCISES

1. Determine the center, foci, vertices, and major axis for each of the following ellipses. Graph.
   (a) $9x^2 + 16y^2 = 576$
   (b) $16(x - 6)^2 + 36(y + 4)^2 = 576$
   (c) $9x^2 + 4y^2 - 18x + 16y - 11 = 0$
   (d) $16x^2 + 4y^2 + 96x - 8y + 84 = 0$

2. Write an equation of the ellipse satisfying the following conditions:
   (a) center $(0, 0)$, vertex $(0, 13)$, and focus $(0, -5)$
   (b) foci $(\pm 4, 0)$, vertices $(\pm 5, 0)$
   (c) foci $(\pm 5, 0)$, eccentricity $= \dfrac{5}{8}$

3. Find an equation of the ellipse with center $(1, 2)$, focus $(6, 2)$, and passing through the point $(4, 6)$.

4. Using Theorem 10, write an equation of the ellipse which passes through $(-6, 4)$, $(-8, 1)$, $(2, -4)$, and $(8, -3)$, and which has axes parallel to the x-axis and y-axis.

5. Describe the graph of the ellipse $\dfrac{x^2}{a^2} + \dfrac{y^2}{b^2} = 1$ if $a = b$.

6. (a) Does the graph of $4x^2 + y^2 = 4$ represent a function?

   (b) Graph $y = 2\sqrt{1 - x^2}$. Is it a function? If so, find its domain and range.

   (c) Do the same as in (b) for $y = -2\sqrt{1 - x^2}$.

7. Show that the determinant equation

$$\begin{vmatrix} \dfrac{(x - h)}{a} & \dfrac{(y - k)}{b} \\[2ex] \dfrac{-(y - k)}{b} & \dfrac{(x - h)}{a} \end{vmatrix} = 1, \text{ where } a > b$$

is an equation of an ellipse with center at (h, k).

7.5 HYPERBOLA

A hyperbola is a curve formed if the cutting plane is parallel to two generators and intersects both nappes of a right-circular cone (see Fig. 207). The two

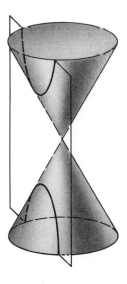

Figure 207

parts of a hyperbola are called **branches.** A more formal geometric definition of a hyperbola is given as follows.

> **Definition:** A **hyperbola** is the set of all points $P(x, y)$ in the plane such that the absolute value of the difference between the distances from P to two fixed points is constant. The two fixed points are called the **foci** of the hyperbola.

To derive an equation of a hyperbola we consider a hyperbola such that the distance between the foci is $2c$, and the constant difference is $2a$, $a < c$. The graph of such a hyperbola is given in Figure 208. We choose the line through

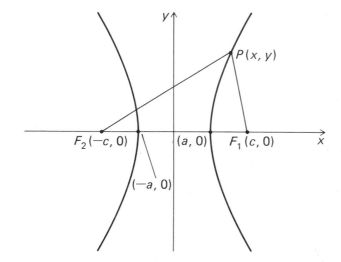

Figure 208

the foci as the x-axis and the midpoint between the foci as the origin. Thus, the coordinates of the foci are $(c, 0)$ and $(-c, 0)$. By the definition of a hyperbola, we have

$$\overline{F_2P} - \overline{F_1P} = \pm 2a$$

where the positive sign is used for a point P on the right of the y-axis, and the negative sign is used for a point P on the left of the y-axis. In either case, we have $\overline{F_2P} = \sqrt{(x + c)^2 + y^2}$ and $\overline{F_1P} = \sqrt{(x - c)^2 + y^2}$. Therefore,

$$\sqrt{(x + c)^2 + y^2} - \sqrt{(x - c)^2 + y^2} = \pm 2a$$

or

$$\sqrt{(x + c)^2 + y^2} = \pm 2a + \sqrt{(x - c)^2 + y^2}$$

Squaring, we get

$$x^2 + 2cx + c^2 + y^2 = 4a^2 \pm 4a\sqrt{(x - c)^2 + y^2} + x^2 - 2cx + c^2 + y^2$$

Simplifying,

$$4cx - 4a^2 = \pm 4a\sqrt{(x - c)^2 + y^2}$$

or

$$cx - a^2 = \pm a \sqrt{(x - c)^2 + y^2}$$

Squaring again, we obtain

$$c^2x^2 - 2a^2cx + a^4 = a^2x^2 - 2a^2cx + a^2c^2 + a^2y^2$$

or

$$(c^2 - a^2)x^2 - a^2y^2 = a^2(c^2 - a^2)$$

Since $c > a$, $c^2 > a^2$, and $c^2 - a^2 > 0$, so we can let $b^2 = c^2 - a^2$, and the equation becomes

$$b^2x^2 - a^2y^2 = a^2b^2$$

or

$$\frac{x^2}{a^2} - \frac{y^2}{b^2} = 1 \tag{I}$$

Therefore, we have the following theorem.

Theorem 11: An equation of a hyperbola with center C (0, 0), foci $(c, 0)$ and $(-c, 0)$, and vertices $(a, 0)$ and $(-a, 0)$ is given by

$$\frac{x^2}{a^2} - \frac{y^2}{b^2} = 1, \text{ where } c^2 = a^2 + b^2$$

Conversely, any equation of the form $\frac{x^2}{a^2} - \frac{y^2}{b^2} = 1$ represents a hyperbola. The proof consists in showing that the steps taken in obtaining equation (I) can be reversed.

PROBLEM 1

Discuss the graph of $\frac{x^2}{16} - \frac{y^2}{9} = 1$.

Solution: The equation takes the form of equation (I) where $a^2 = 16$, $a = 4$, and $b^2 = 9$, $b = 3$. Since $a^2 = 16$ and $b^2 = 9$, $c^2 = a^2 + b^2 = 16 + 9 = 25$ and $c = 5$. Thus, the equation $\frac{x^2}{16} - \frac{y^2}{9} = 1$ is a hyperbola with center (0, 0), foci (5, 0) and $(-5, 0)$, and vertices (4, 0) and $(-4, 0)$. The graph is given in Figure 209. Note that each branch of the

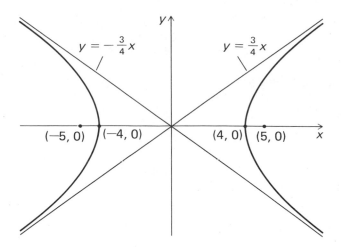

Figure 209

hyperbola gets closer to the two lines $y = \dfrac{b}{a}x = \dfrac{3}{4}x$ and $y = -\dfrac{b}{a}x = -\dfrac{3}{4}x$. These lines are called the **asymptotes** of the hyperbola.

Definition: The **asymptotes** of the hyperbola $\dfrac{x^2}{a^2} - \dfrac{y^2}{b^2} = 1$ are the lines given by $y = \dfrac{b}{a}x$ and $y = -\dfrac{b}{a}x$.

If the foci of the hyperbola lie on the y-axis, say at points $(0, c)$ and $(0, -c)$, then the equation of the hyperbola may be derived as for equation (I). The equation is

$$\frac{y^2}{a^2} - \frac{x^2}{b^2} = 1 \qquad\qquad \textbf{(II)}$$

and we have the following theorem.

Theorem 12: An equation of the hyperbola with center $C(0, 0)$, foci $(0, c)$ and $(0, -c)$, and vertices $(0, a)$ and $(0, -a)$, is

$$\frac{y^2}{a^2} - \frac{x^2}{b^2} = 1, \text{ where } c^2 = a^2 + b^2$$

and equations of the asymptotes are

$$y = \frac{a}{b}x \text{ and } y = -\frac{a}{b}x$$

PROBLEM 2

Analyze the equation $4x^2 - 16y^2 + 25 = 0$.

Solution: Dividing both sides by -25, we get

$$\frac{16y^2}{25} - \frac{4x^2}{25} = 1$$

The equation can be written

$$\frac{y^2}{\frac{25}{16}} - \frac{x^2}{\frac{25}{4}} = 1$$

and this is a hyperbola of the form $\dfrac{y^2}{a^2} - \dfrac{x^2}{b^2} = 1$, where $a^2 = \dfrac{25}{16}$ or $a = \dfrac{5}{4}$ and $b^2 = \dfrac{25}{4}$ or $b = \dfrac{5}{2}$. Using the fact that $c^2 = a^2 + b^2 = \dfrac{25}{16} + \dfrac{25}{4} = \dfrac{125}{16}$, we get $c = \sqrt{\dfrac{125}{16}} = \dfrac{5}{4}\sqrt{5}$. Thus, the equation $4x^2 - 16y^2 + 25 = 0$ is a hyperbola with center at $(0, 0)$, foci $F_1 \left(0, \dfrac{5\sqrt{5}}{4}\right)$ and $F_2 \left(0, \dfrac{-5\sqrt{5}}{4}\right)$, vertices $V_1 \left(0, \dfrac{5}{4}\right)$ and $V_2 \left(0, -\dfrac{5}{4}\right)$, and asymptotes $y = \dfrac{1}{2}x$ and $y = -\dfrac{1}{2}x$ (see Fig. 210).

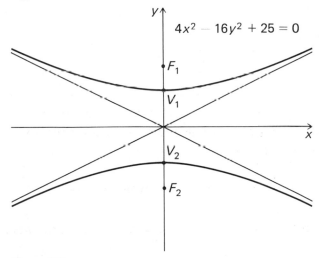

Figure 210

Center Not at Origin

By an argument similar to that used for the other conic sections one may show that for the center at (h, k) instead of at the origin an equation of a hyperbola can be obtained by using the translation of axes equations.

Theorem 13: An equation of a hyperbola with center (h, k) and foci $F_1(h + c, k)$ and $F_2(h - c, k)$ is

$$\frac{(x - h)^2}{a^2} - \frac{(y - k)^2}{b^2} = 1, \text{ where } c^2 = a^2 + b^2$$

The vertices are $V_1(h + a, k)$ and $V_2(h - a, k)$. Equations of the asymptotes are

$$y = \frac{b}{a}(x - h) + k \quad \text{and} \quad y = -\frac{b}{a}(x - h) + k$$

PROBLEM 3

Find the center, vertices, foci, and equations of the asymptotes for $9(x - 3)^2 - 4(y + 1)^2 = 144$.

Solution: Putting $9(x - 3)^2 - 4(y + 1)^2 = 144$ in standard form, we get

$$\frac{(x - 3)^2}{16} - \frac{(y + 1)^2}{36} = 1$$

Thus, $h = 3$, $k = -1$, $a = 4$, $b = 6$, and $c = \sqrt{52}$. The center $C(h, k) = (3, -1)$, the vertices are $V_1(h + a, k) = (3 + 4, -1) = (7, -1)$ and $V_2(h - a, k) = (3 - 4, -1) = (-1, -1)$. The foci are $F_1(h + c, k) = (3 + \sqrt{52}, -1)$ and $F_2(h - c, k) = (3 - \sqrt{52}, -1)$. Finally, the equations of the asymptotes are $y = \frac{b}{a}(x - h) + k = \frac{6}{4}(x - 3) - 1 = \frac{3}{2}x - \frac{11}{2}$ and $y = -\frac{b}{a}(x - h) + k = -\frac{6}{4}(x - 3) - 1 = -\frac{3}{2}x + \frac{7}{2}$.

Theorem 14: An equation of a hyperbola with center (h, k) and foci $F_1(h, k + c)$ and $F_2(h, k - c)$ is

$$\frac{(y - k)^2}{a^2} - \frac{(x - h)^2}{b^2} = 1, \text{ where } c^2 = a^2 + b^2$$

The vertices are $V_1(h, k + a)$ and $V_2(h, k - a)$. Equations of the asymptotes are

$$y = \frac{a}{b}(x - h) + k \quad \text{and} \quad y = -\frac{a}{b}(x - h) + k$$

PROBLEM 4

Find the center, foci, vertices, and equations of the asymptotes for the hyperbola $4y^2 - 9x^2 - 36x - 8y - 68 = 0$.

Solution: To put the equation in standard form we complete the square on x and y:

$$4y^2 - 9x^2 - 36x - 8y - 68 = 0$$
$$4(y^2 - 2y) - 9(x^2 + 4x) = 68$$
$$4(y^2 - 2y + 1) - 9(x^2 + 4x + 4) = 68 + 4 - 36$$
$$4(y - 1)^2 - 9(x + 2)^2 = 36$$

$$\frac{(y - 1)^2}{9} - \frac{(x + 2)^2}{4} = 1$$

This equation shows that the center of the hyperbola is at $(-2, 1)$. We have $a = 3$, $b = 2$, and $c = \sqrt{13}$. Therefore, the foci are $(-2, 1 + \sqrt{13})$ and $(-2, 1 - \sqrt{13})$ and the vertices are $(-2, 4)$ and $(-2, -2)$. Equations of the asymptotes are $y = \frac{3}{2}x + 4$ and $y = -\frac{3}{2}x - 2$.

PROBLEM 5

Find an equation of the hyperbola having foci $(8, 17)$ and $(8, -9)$ and vertices at $(8, 9)$ and $(8, -1)$.

Solution: Since the center is the midpoint of the line connecting the foci, we have the center at $(8, 4)$. Thus, $h = 8$, $k = 4$, and computation shows that $a = 5$ and $c = 13$. Using $b = \sqrt{c^2 - a^2} = 12$, we obtain the equation $\frac{(y - 4)^2}{25} - \frac{(x - 8)^2}{144} = 1$.

We now state a theorem which states those conditions on the general quadratic equation which give a hyperbola.

Theorem 15: The equation $Ax^2 + Cy^2 + Dx + Ey + F = 0$ represents a hyperbola if A and C are opposite in sign (that is, if $AC < 0$).

Proof. Completing the square, we can write

$$A\left(x^2 + \frac{D}{A}x + \frac{D^2}{4A^2}\right) + C\left(y^2 + \frac{E}{C}y + \frac{E^2}{4C^2}\right) = \frac{D^2}{4A} + \frac{E^2}{4C} - F$$

$$A\left(x + \frac{D}{2A}\right)^2 + C\left(y + \frac{E}{2C}\right)^2 = \frac{D^2}{4A} + \frac{E^2}{4C} - F$$

Dividing by $\frac{D^2}{4A} + \frac{E^2}{4C} - F$ we get an equation in standard form, since A and C are opposite in sign.

7.5 EXERCISES

1. Find the coordinates of the foci, vertices, and equations of the asymptotes of the following hyperbolas. Graph.
   (a) $x^2 - y^2 = 25$
   (b) $4x^2 - 45y^2 = 180$
   (c) $y^2 - x^2 = 9$
   (d) $\dfrac{x^2}{16} - \dfrac{y^2}{9} = 1$

2. Do the same as in Exercise 1 for the following, and also find the center.
   (a) $\dfrac{(x - 1)^2}{9} - \dfrac{(y - 2)^2}{16} = 1$
   (b) $\dfrac{(y + 1)^2}{4} - \dfrac{(x - 1)^2}{1} = 1$
   (c) $x^2 - 3y^2 - 4x + 18y - 50 = 0$
   (d) $9x^2 - 4y^2 + 36x + 24y + 36 = 0$
   (e) $16x^2 - 9y^2 + 90y - 81 = 0$
   (f) $4x^2 - y^2 - 8x - 12 = 0$

3. Find an equation of the hyperbola with foci at $(-4, -2)$ and $(6, -2)$, and vertices $(-3, -2)$ and $(5, -2)$.

4. Find an equation of a hyperbola with center $(5, -4)$, vertex $(5, 2)$, and passing through the point $(4, 8)$.

5. Graph the following hyperbolas:
   (a) $xy = 1$ (b) $xy = -2$

6. Find an equation of the hyperbola with foci at $(10, 0)$ and $(-10, 0)$, asymptotes $4y \pm 3x = 0$.

7. Show that the determinant equation

$$\begin{vmatrix} \dfrac{(x - h)}{a} & \dfrac{(y - k)}{b} \\ \dfrac{(y - k)}{b} & \dfrac{(x - h)}{a} \end{vmatrix} = 1$$

is an equation of a hyperbola with center (h, k).

8. A hyperbola whose asymptotes are perpendicular is known as an **equilateral hyperbola.** Show that the equilateral hyperbola having foci at $\left(\dfrac{c}{\sqrt{2}}, \dfrac{c}{\sqrt{2}}\right)$ and $\left(\dfrac{-c}{\sqrt{2}}, \dfrac{-c}{\sqrt{2}}\right)$ has an equation $xy = \dfrac{c^2}{4}$.

SUMMARY

Circle

1. A **circle** is the set of all points in a plane at a given distance from a fixed point. The fixed point is called the **center** and the fixed distance is called the **radius.**
2. The equation in standard form of the circle whose center is (h, k) and whose radius equals r is

$$(x - h)^2 + (y - k)^2 = r^2$$

Translation of Axes

1. If (x, y) represents a point P with respect to a given set of axes, and (x', y') is a representation of P after the axes are translated to a new origin having coordinates (h, k) with respect to the given axes, then

$$x = x' + h \quad \text{and} \quad y = y' + k$$

or

$$x' = x - h \quad \text{and} \quad y' = y - k$$

Parabola

1. A **parabola** is the set of all points in a plane equidistant from a fixed point **(focus)** and a fixed line **(directrix).**
2. An equation of the parabola having its focus at $(p, 0)$ and as its directrix the line $x = -p$ is

$$y^2 = 4px$$

3. An equation of the parabola having its focus at $(0, p)$ and as its directrix the line $y = -p$ is

$$x^2 = 4py$$

4. An equation of a parabola with focus $F(h + p, k)$ and directrix $x = -p + h$ is

$$(y - k)^2 = 4p(x - h)$$

   (i) The vertex is $V(h, k)$.
   (ii) If $p > 0$, then the parabola opens to the right.
   (iii) If $p < 0$, then the parabola opens to the left.
5. An equation of a parabola with focus $F(h, k + p)$ and directrix $y = -p + k$ is

$$(x - h)^2 = 4p(y - k)$$

   (i) The vertex is $V(h, k)$.
   (ii) If $p > 0$, then the parabola opens upward.
   (iii) If $p < 0$, then the parabola opens downward.

Ellipse

1. An **ellipse** is the collection of all points in the plane, the sum of whose distances from two fixed points **(foci)** is a constant.
2. An equation of an ellipse with center (h, k) and the major axis parallel to the x-axis is

$$\frac{(x - h)^2}{a^2} + \frac{(y - k)^2}{b^2} = 1, a > b$$

3. Similarly, an equation of an ellipse with center (h, k) and the major axis parallel to the y-axis is

$$\frac{(x - h)^2}{b^2} + \frac{(y - k)^2}{a^2} = 1, a > b$$

Hyperbola

1. A **hyperbola** is the set of all points $P(x, y)$ in the plane such that the absolute value of the difference between the distance from P to two fixed points (**foci**) is constant.
2. An equation of a hyperbola with center (h, k) and foci $F_1(h + c, k)$ and F_2 $(h - c, k)$ is

$$\frac{(x - h)^2}{a^2} - \frac{(y - k)^2}{b^2} = 1, \text{ where } c^2 = a^2 + b^2$$

The vertices are $(h + a, k)$ and $(h - a, k)$. Equations of the asymptotes are

$$y = \frac{b}{a}(x - h) + k \text{ and } y = -\frac{b}{a}(x - h) + k$$

3. An equation of a hyperbola with center (h, k) and foci $F_1(h, k + c)$ and $F_2(h, k - c)$ is

$$\frac{(y - k)^2}{a^2} - \frac{(x - h)^2}{b^2} = 1, \text{ where } c^2 = a^2 + b^2$$

The vertices are $V_1(h, k + a)$ and $V_2(h, k - a)$. Equations of the asymptotes are

$$y = \frac{a}{b}(x - h) + k \text{ and } y = -\frac{a}{b}(x - h) + k$$

General Equation

The conic sections with lines of symmetry parallel to the coordinate axes have second degree equations of the form $Ax^2 + Cy^2 + Dx + Ey + F = 0$, where either A or C must be nonzero. Then the graph is a(n)

(i) **circle**, if $A = C \neq 0$.
(ii) **parabola**, if either $A = 0$ or $C = 0$, but not both.
(iii) **ellipse**, if $A \neq C$ and $AC > 0$ (same sign).
(iv) **hyperbola**, if $AC < 0$ (opposite signs).

CHAPTER 7 EXERCISES _____

1. Determine the center and radius of the following circles:
   (a) $x^2 + y^2 + 4x - 2y - 4 = 0$
   (b) $x^2 + y^2 - x - 2y = 0$
   (c) $4x^2 + 4y^2 + 16x + 15 = 0$
   (d) $2x^2 + 2y^2 + 3x + 5y + 2 = 0$

2. Find an equation of the circle satisfying the following conditions:
   (a) center $(-1, -3)$ and radius $r = 3$
   (b) passing through $(0, 1)$, $(2, 1)$, and $(-2, 5)$
   (c) diameter is the segment from $(-2, 3)$ to $(4, 3)$

3. Determine a translation of axes that will transform the following into equations without first degree terms.
   (a) $3x^2 - 4y^2 + 6x + 24y - 135 = 0$
   (b) $3x^2 + 4y^2 - 12x + 4y + 13 = 0$
   (c) $2x^2 + 5y^2 - 12x + 10y - 17 = 0$

4. Find the focus and directrix of the following parabolas:
   (a) $y^2 = 16x$
   (b) $(x - 2)^2 = 24(y + 3)$
   (c) $x^2 + 2x + 4y + 5 = 0$

5. Find an equation of the parabola with
   (a) vertex $(0, 0)$ and focus $(2, 0)$
   (b) focus $(1, 1)$ and directrix $y = -1$
   (c) focus $(2, -2)$ and directrix $x = 5$

6. Find the vertices and foci of the following ellipses:
   (a) $4x^2 + 25y^2 = 25$
   (b) $3x^2 + 5y^2 + 15x - 15y = 0$

7. Find an equation of the ellipse that satisfies the following conditions:
   (a) vertices $(1, 5)$, $(-7, 5)$, $(-3, 7)$, and $(-3, 3)$
   (b) foci at $(2, 9)$ and $(2, -3)$ and minor axis of length 10

8. Determine the vertices, foci, and equations of the asymptotes of the following hyperbolas:
   (a) $9x^2 - 16y^2 + 144 = 0$
   (b) $x^2 - y^2 - 4 = 0$
   (c) $4x^2 - 9y^2 + 16 = 0$

9. Find an equation of the hyperbola satisfying the given conditions:
   (a) foci at $(\pm 4, 0)$ and vertices $(\pm 2, 0)$
   (b) vertices at $(\pm 2, 0)$ and asymptotes $y = \pm 2x$
   (c) vertices at $(0, \pm 2)$ and passing through $(3, 4)$
   (d) asymptotes $2x - y - 3 = 0$ and $2x + y - 5 = 0$, and passing through $(4, 7)$

10. Determine whether the following equations represent an ellipse, circle, parabola, or hyperbola:
    (a) $2x^2 + 5y^2 - 7x + 9y - 4 = 0$
    (b) $3x^2 - 7y^2 - 6x + 12y - 1 = 0$
    (c) $-8x^2 - 8y^2 + 4x - 5y + 7 = 0$
    (d) $4y^2 - 7x + y + 10 = 0$

APPENDIX I

BINOMIAL THEOREM

One of the procedures we learn in algebra is the process of raising a binomial like $(a + b)$ to an integer power. In this section, we obtain a simplified method of expanding $(a + b)^n$, where n is any positive integer, which avoids the tiresomeness of repeated multiplication and additions. We begin by calculating $(a + b)^n$ for several values of n.

$$(a + b)^1 = a + b$$
$$(a + b)^2 = a^2 + 2ab + b^2$$
$$(a + b)^3 = a^3 + 3a^2b + 3ab^2 + b^3$$
$$(a + b)^4 = a^4 + 4a^3b + 6a^2b^2 + 4ab^3 + b^4$$

We now investigate this triangular array to discover any pattern. First, we note that the exponents of a decrease starting with the value of n and descending to zero. The exponents of b do just the opposite, and increase from zero to the value of n. Also, we note that the sum of the exponents of a and b in any term should equal n. Following this pattern we can write

$$(a + b)^5 = \underline{\hspace{1cm}}a^5 + \underline{\hspace{1cm}}a^4b + \underline{\hspace{1cm}}a^3b^2 + \underline{\hspace{1cm}}a^2b^3 + \underline{\hspace{1cm}}ab^4 + \underline{\hspace{1cm}}b^5$$

We have yet to determine the coefficients associated with each term.

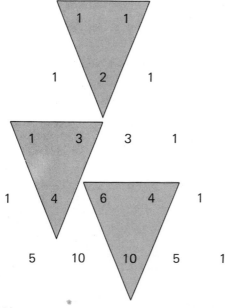

The triangular array composed of the coefficients of $(a + b)^n$ is called **Pascal's triangle.** Note that each row of this array begins and ends with a "1." Any other number in the array can be obtained by adding the two numbers to the right and left of it in the preceding row. For example, the 4 in the fourth row is obtained by adding the 3 to the right of it and the 1 to the left of it in the third row. This and other examples are illustrated above by the little triangular outlines. That is, $2 = 1 + 1$, $10 = 6 + 4$. Thus, we can fill in the coefficients of $(a + b)^5$:

$$(a + b)^5 = a^5 + 5a^4b + 10a^3b^2 + 10a^2b^3 + 5ab^4 + b^5$$

PROBLEM 1

Compute $(a + b)^6$.

Solution: The coefficients of $(a + b)^6$ are given by the sixth row of Pascal's triangle. Thus,

$$(a + b)^6 = a^6 + 6a^5b + 15a^4b^2 + 20a^3b^3 + 15a^2b^4 + 6ab^5 + b^6$$

PROBLEM 2

Compute $(2x^3 + 3y)^5$.

Solution: Let $a = 2x^3$ and $b = 3y$. Then $(2x^3 + 3y)^5 = (a + b)^5$. Pascal's triangle gives

$$(a + b)^5 = a^5 + 5a^4b + 10a^3b^2 + 10a^2b^3 + 5ab^4 + b^5$$

Substituting for a and b, we obtain

$$\begin{aligned}
(2x^3 + 3y)^5 &= (2x^3)^5 + 5(2x^3)^4(3y) + 10(2x^3)^3(3y)^2 \\
&\quad + 10(2x^3)^2(3y)^3 + 5(2x^3)(3y)^4 + (3y)^5 \\
&= 32x^{15} + 240x^{12}y + 720x^9y^2 + 1080x^6y^3 \\
&\quad + 810x^3y^4 + 243y^5
\end{aligned}$$

While the method given above works well when the powers of the binomial are small, for large values the method can be very tedious. For example, $(a + b)^{35}$ would require that we compute the first 35 rows of Pascal's triangle. However, Pascal's triangle can be written using **binomial coefficients** $\binom{n}{r}$. To achieve this, we define the meaning of **factorial n.**

Definition: Factorial n, written $n!$, is defined by the identity $n! = n \cdot (n - 1) \cdot (n - 2) \ldots 3 \cdot 2 \cdot 1$, where $n > 0$, and n is an integer.

EXAMPLE 1

$$1! = 1, 2! = 2 \cdot 1 = 2, 3! = 3 \cdot 2 \cdot 1 = 6, 4! = 4 \cdot 3 \cdot 2 \cdot 1 = 24$$

$$17! = 17 \cdot 16 \cdot 15 \ldots 4 \cdot 3 \cdot 2 \cdot 1$$
$$(r - 1)! = (r - 1) \cdot (r - 2) \ldots 3 \cdot 2 \cdot 1$$
$$(n + 1)! = (n + 1) \cdot n \cdot (n - 1) \ldots 3 \cdot 2 \cdot 1 = (n + 1) \cdot n!$$

Definition: $0! = 1$

We can now define the binomial coefficient using the factorial notation.

Definition: If r and n are integers, where $0 \leq r \leq n$, then the **binomial coefficient**, written $\binom{n}{r}$, is given by

$$\binom{n}{r} = \frac{n!}{r!(n - r)!}$$

PROBLEM 3

Compute $\binom{4}{2}$, $\binom{13}{10}$, and $\binom{7}{5}$.

Solution: (a) $\binom{4}{2} = \dfrac{4!}{2!(4 - 2)!} = \dfrac{4!}{2!2!} = \dfrac{4 \cdot 3 \cdot 2 \cdot 1}{(2 \cdot 1)(2 \cdot 1)} = 6$

(b) $\binom{13}{10} = \dfrac{13!}{10!(13 - 10)!} = \dfrac{13!}{10!3!}$

$$= \frac{13 \cdot 12 \cdot 11 \cdot 10 \ldots 3 \cdot 2 \cdot 1}{(10 \cdot 9 \ldots 1)(3 \cdot 2 \cdot 1)} = 286$$

(c) $\binom{7}{5} = \dfrac{7!}{5!(7 - 5)!} = \dfrac{7!}{5!2!}$

$$= \frac{7 \cdot 6 \cdot 5 \cdot 4 \cdot 3 \cdot 2 \cdot 1}{(5 \cdot 4 \cdot 3 \cdot 2 \cdot 1)(2 \cdot 1)} = 21$$

PROBLEM 4

Prove that $\binom{n}{r - 1} + \binom{n}{r} = \binom{n + 1}{r}$.

Solution: $\binom{n}{r - 1} + \binom{n}{r} = \dfrac{n!}{(r - 1)!(n - r + 1)!} + \dfrac{n!}{r!(n - r)!}$

$$= \frac{(n!)r + (n!)(n - r + 1)}{r!(n - r + 1)!}$$

$$= \frac{n!(n + 1)}{r!(n + 1 - r)!}$$

$$= \frac{(n + 1)!}{r!(n + 1 - r)!}$$

$$= \binom{n + 1}{r}$$

We can now write the entries of Pascal's triangle using binomial coefficients

$$
\begin{array}{ccccccc}
1 & & 1 & & & & \\
1 & & 2 & & 1 & & \\
1 & & 3 & & 3 & & 1 \\
1 & & 4 & & 6 & & 4 & & 1 \\
1 & & 5 & & 10 & & 10 & & 5 & & 1
\end{array}
$$

$$\binom{1}{0} \quad \binom{1}{1}$$

$$\binom{2}{0} \quad \binom{2}{1} \quad \binom{2}{2}$$

$$\binom{3}{0} \quad \binom{3}{1} \quad \binom{3}{2} \quad \binom{3}{3}$$

$$\binom{4}{0} \quad \binom{4}{1} \quad \binom{4}{2} \quad \binom{4}{3} \quad \binom{4}{4}$$

$$\binom{5}{0} \quad \binom{5}{1} \quad \binom{5}{2} \quad \binom{5}{3} \quad \binom{5}{4} \quad \binom{5}{5}$$

Thus,

$$(a + b)^5 = \binom{5}{0} a^5 + \binom{5}{1} a^4 b + \binom{5}{2} a^3 b^2 + \binom{5}{3} a^2 b^3 + \binom{5}{4} ab^4 + \binom{5}{5} b^5$$

If we generalize the results from above, we have an equation referred to as the **Binomial Theorem.**

Binomial Theorem: For positive integral values of n,

$$(a + b)^n = a^n + n \cdot a^{n-1} \cdot b + \frac{n(n - 1)}{2!} \cdot a^{n-2} \cdot b^2$$

$$+ \frac{n(n - 1)(n - 2)}{3!} \cdot a^{n-3} \cdot b^3$$

$$+ \cdots + \frac{n(n - 1)(n - 2) \cdots (n - r + 1)}{r!} a^{n-r} b^r$$

$$+ \cdots + b^n$$

$$= \binom{n}{0} a^n + \binom{n}{1} a^{n-1} \cdot b + \binom{n}{2} a^{n-2} \cdot b^2 + \binom{n}{3} a^{n-3} b^3$$

$$+ \cdots + \binom{n}{r} a^{n-r} \cdot b^r + \cdots + \binom{n}{n} b^n$$

EXERCISES

1. Evaluate:

   (a) $\dfrac{8!}{5!}$

   (b) $\dfrac{3!}{0!}$

   (c) $\dfrac{4!3!}{5!}$

   (d) $\dfrac{19!}{18!}$

   (e) $\dfrac{n!}{(n-1)!}$

   (f) $\dfrac{n!}{(n-2)!}$

   (g) $\dfrac{(2n+1)!}{(2n-1)!}$

2. Prove $\dbinom{n}{k} + \dbinom{n}{k+1} = \dbinom{n+1}{k+1}$.

3. Expand by using the binomial formula.

   (a) $\left(a + \dfrac{1}{3}\right)^5$

   (b) $(x - 2y)^6$

   (c) $(\sqrt{a} + \sqrt{b})^6$

   (d) $(x^{-1} + b^{1/2})^4$

   (e) $\left(\dfrac{t^2}{2} + \dfrac{3}{t}\right)^4$

   (f) $\left(\dfrac{1}{\sqrt{x}} + \sqrt{x}\right)^6$

APPENDIX II

MATHEMATICAL INDUCTION

Consider the problem of adding consecutive odd integers beginning with 1 and finding a formula for the sum in terms of the number added. For example,

$$1 = 1 = 1^2$$
$$1 + 3 = 4 = 2^2$$
$$1 + 3 + 5 = 9 = 3^2$$
$$1 + 3 + 5 + 7 = 16 = 4^2$$

Thus, from these four examples we might conjecture that the formula

$$1 + 3 + 5 + \cdots + (2n - 1) = n^2, n \geq 1$$

Is true.

However, what if we wanted to show that the formula is true for all natural numbers? The Principle of Mathematical Induction Is a method of proving a statement involving n, where n is a positive integer. There are two distinct steps in the proof.

Step 1: Show that the statement is true for $n = 1$.

Step 2: Show that if the statement is true for $n = k$, then it is true for $n = k + 1$.

The principle of mathematical induction implies that if the statement is true for $k = 1$, then it is true for $k + 1 = 1 + 1 = 2$. Now, $k = 2$ implies the statement is true for $k + 1 = 2 + 1 = 3$, and so on.

We now prove the formula

$$1 + 3 + 5 + 7 + \cdots + (2n - 1) = n^2$$

Step 1: For $n = 1$, we have

$$1 = 1^2 = 1$$

Thus, the formula is true for $n = 1$.

Step 2: Now, assume the formula is true for $n = k$. We must show that it is true for $n = k + 1$. Assume that

$$1 + 3 + 5 + 7 + \cdots + (2k - 1) = k^2 \tag{A}$$

We want to show that

$$1 + 3 + 5 + 7 + \cdots + (2k - 1) + [2(k + 1) - 1] = (k + 1)^2$$

or,

$$1 + 3 + 5 + 7 + \cdots + (2k - 1) + (2k + 1) \qquad = (k + 1)^2$$

Adding $(2k + 1)$ to both sides of (**A**) we get

$$\begin{aligned} 1 + 3 + 5 + 7 + \cdots + (2k - 1) + (2k + 1) &= k^2 + (2k + 1) \\ &= k^2 + 2k + 1 \\ &= (k + 1)^2 \end{aligned}$$

PROBLEM 1

Prove $1 + 2 + 3 + \cdots + n = \dfrac{n(n + 1)}{2}$.

Solution:

Step 1: For $n = 1$, we have $1 = \dfrac{1(1 + 1)}{2} = \dfrac{1 \cdot 2}{2} = 1$.

Step 2: Assume that the formula is true for $n = k$:

$$1 + 2 + 3 + \cdots + k = \frac{k(k + 1)}{2} \tag{B}$$

We want to show that $1 + 2 + 3 + \cdots + k + (k + 1) = \dfrac{(k + 1)(k + 2)}{2}$.

Adding $(k + 1)$ to both sides of (**B**) we have

$$\begin{aligned} 1 + 2 + 3 + \cdots + k + (k + 1) &= \frac{k(k + 1)}{2} + (k + 1) \\ &= \frac{k(k + 1) + 2(k + 1)}{2} \\ &= \frac{k^2 + 3k + 2}{2} \\ &= \frac{(k + 1)(k + 2)}{2} \end{aligned}$$

We now make use of the principle of mathematical induction to prove the Binomial Theorem. Recall that the Binomial Theorem says that if a and b are real numbers and n is a positive integer, then

$$(a + b)^n = \binom{n}{0} a^n + \binom{n}{1} a^{n-1}b + \cdots + \binom{n}{r} a^{n-r}b^r + \cdots + \binom{n}{n} b^n$$

Proof. By mathematical induction on *n*.
Step 1: For $n = 1$, we have $(a + b)^1 = a + b$ and the formula is true.
Step 2: Assume the formula to be true for $n = k$.

$$(a + b)^k = \binom{k}{0} a^k + \binom{k}{1} a^{k-1}b + \cdots + \binom{k}{r} a^{k-r}b^r + \cdots + \binom{k}{k} b^k \quad \text{(C)}$$

We want to show it to be true for $n = k + 1$; that is,

$$(a + b)^{k+1} = \binom{k+1}{0} a^{k+1} + \binom{k+1}{1} a^k b + \cdots + \binom{k+1}{r} a^{k+1-r}b^r$$
$$+ \cdots + \binom{k+1}{k+1} b^{k+1}$$

Multiplying both sides of **(C)** by $(a + b)$ and collecting terms, we obtain

$$(a + b)^{k+1} = \binom{k}{0} a^{k+1} + \left[\binom{k}{0} + \binom{k}{1} \right] a^k b$$
$$+ \cdots + \left[\binom{k}{r-1} + \binom{k}{r} \right] a^{k-r+1}b^r$$
$$+ \cdots + \left[\binom{k}{k-1} + \binom{k}{k} \right] ab^k + \binom{k}{k} b^{k+1}$$

From Problem 4 in Appendix I, we know that

$$\binom{k}{r-1} + \binom{k}{r} = \binom{k+1}{r} \quad \text{and} \quad \binom{k}{0} = \binom{k+1}{0} = \binom{k}{k} = \binom{k+1}{k+1}$$

We have

$$(a + b)^{k+1} = \binom{k+1}{0} a^{k+1} + \binom{k+1}{1} a^k b + \cdots + \binom{k+1}{r} a^{k+1-r}b^r$$
$$+ \cdots + \binom{k+1}{k} ab^k + \binom{k+1}{k+1} b^{k+1}$$

Thus, we have proved the Binomial Theorem.

EXERCISES

Prove:

1. $2 + 4 + 6 + \cdots + 2n = n^2 + n$

2. $1^2 + 2^2 + 3^2 + \cdots + n^2 = \dfrac{n(n + 1)(2n + 1)}{6}$

3. $\dfrac{1}{1 \cdot 2} + \dfrac{1}{2 \cdot 3} + \dfrac{1}{3 \cdot 4} + \cdots + \dfrac{1}{n(n + 1)} = \dfrac{n}{n + 1}$

4. $\dfrac{1}{1 \cdot 3} + \dfrac{1}{3 \cdot 5} + \dfrac{1}{5 \cdot 7} + \cdots + \dfrac{1}{(2n - 1)(2n + 1)} = \dfrac{n}{2n + 1}$

APPENDIX III

TABLES

TABLE A. Common Logarithms

	0	1	2	3	4	5	6	7	8	9
1.0	.0000	.0043	.0086	.0128	.0170	.0212	.0253	.0294	.0334	.0374
1.1	.0414	.0453	.0492	.0531	.0569	.0607	.0645	.0682	.0719	.0755
1.2	.0792	.0828	.0864	.0899	.0934	.0969	.1004	.1038	.1072	.1106
1.3	.1139	.1173	.1206	.1239	.1271	.1303	.1335	.1367	.1399	.1430
1.4	.1461	.1492	.1523	.1553	.1584	.1614	.1644	.1673	.1703	.1732
1.5	.1761	.1790	.1818	.1847	.1875	.1903	.1931	.1959	.1987	.2014
1.6	.2041	.2068	.2095	.2122	.2148	.2175	.2201	.2227	.2253	.2279
1.7	.2304	.2330	.2355	.2380	.2405	.2430	.2455	.2480	.2504	.2529
1.8	.2553	.2577	.2601	.2625	.2648	.2672	.2695	.2718	.2742	.2765
1.9	.2788	.2810	.2833	.2856	.2878	.2900	.2923	.2945	.2967	.2989
2.0	.3010	.3032	.3054	.3075	.3096	.3118	.3139	.3160	.3181	.3201
2.1	.3222	.3243	.3263	.3284	.3304	.3324	.3345	.3365	.3385	.3404
2.2	.3424	.3444	.3464	.3483	.3502	.3522	.3541	.3560	.3579	.3598
2.3	.3617	.3636	.3655	.3674	.3692	.3711	.3729	.3747	.3766	.3784
2.4	.3802	.3820	.3838	.3856	.3874	.3892	.3909	.3927	.3945	.3962
2.5	.3979	.3997	.4014	.4031	.4048	.4065	.4082	.4099	.4116	.4133
2.6	.4150	.4166	.4183	.4200	.4216	.4232	.4249	.4265	.4281	.4298
2.7	.4314	.4330	.4346	.4362	.4378	.4393	.4409	.4425	.4440	.4456
2.8	.4472	.4487	.4502	.4518	.4533	.4548	.4564	.4579	.4594	.4609
2.9	.4624	.4639	.4654	.4669	.4683	.4698	.4713	.4728	.4742	.4757
3.0	.4771	.4786	.4800	.4814	.4829	.4843	.4857	.4871	.4886	.4900
3.1	.4914	.4928	.4942	.4955	.4969	.4983	.4997	.5011	.5024	.5038
3.2	.5051	.5065	.5079	.5092	.5105	.5119	.5132	.5145	.5159	.5172
3.3	.5185	.5198	.5211	.5224	.5237	.5250	.5263	.5276	.5289	.5302
3.4	.5315	.5328	.5340	.5353	.5366	.5378	.5391	.5403	.5416	.5428
3.5	.5441	.5453	.5465	.5478	.5490	.5502	.5514	.5527	.5539	.5551
3.6	.5563	.5575	.5587	.5599	.5611	.5623	.5635	.5647	.5658	.5670
3.7	.5682	.5694	.5705	.5717	.5729	.5740	.5752	.5763	.5775	.5786
3.8	.5798	.5809	.5821	.5832	.5843	.5855	.5866	.5877	.5888	.5899
3.9	.5911	.5922	.5933	.5944	.5955	.5966	.5977	.5988	.5999	.6010
4.0	.6021	.6031	.6042	.6053	.6064	.6075	.6085	.6096	.6107	.6117
4.1	.6128	.6138	.6149	.6160	.6170	.6180	.6191	.6201	.6212	.6222
4.2	.6232	.6243	.6253	.6263	.6274	.6284	.6294	.6304	.6314	.6325
4.3	.6335	.6345	.6355	.6365	.6375	.6385	.6395	.6405	.6415	.6425
4.4	.6435	.6444	.6454	.6464	.6474	.6484	.6493	.6503	.6513	.6522
4.5	.6532	.6542	.6551	.6561	.6571	.6580	.6590	.6599	.6609	.6618
4.6	.6628	.6637	.6646	.6656	.6665	.6675	.6684	.6693	.6702	.6712
4.7	.6721	.6730	.6739	.6749	.6758	.6767	.6776	.6785	.6794	.6803
4.8	.6812	.6821	.6830	.6839	.6848	.6857	.6866	.6875	.6884	.6893
4.9	.6902	.6911	.6920	.6928	.6937	.6946	.6955	.6964	.6972	.6981
5.0	.6990	.6998	.7007	.7016	.7024	.7033	.7042	.7050	.7059	.7067
5.1	.7076	.7084	.7093	.7101	.7110	.7118	.7126	.7135	.7143	.7152
5.2	.7160	.7168	.7177	.7185	.7193	.7202	.7210	.7218	.7226	.7235
5.3	.7243	.7251	.7259	.7267	.7275	.7284	.7292	.7300	.7308	.7316
5.4	.7324	.7332	.7340	.7348	.7356	.7364	.7372	.7380	.7388	.7396
5.5	.7404	.7412	.7419	.7427	.7435	.7443	.7451	.7459	.7466	.7474
5.6	.7482	.7490	.7497	.7505	.7513	.7520	.7528	.7536	.7543	.7551
5.7	.7559	.7566	.7574	.7582	.7589	.7597	.7604	.7612	.7619	.7627
5.8	.7634	.7642	.7649	.7657	.7664	.7672	.7679	.7686	.7694	.7701
5.9	.7709	.7716	.7723	.7731	.7738	.7745	.7752	.7760	.7767	.7774

TABLE A. Common Logarithms (Continued)

	0	1	2	3	4	5	6	7	8	9
6.0	.7782	.7789	.7796	.7803	.7810	.7818	.7825	.7832	.7839	.7846
6.1	.7853	.7860	.7868	.7875	.7882	.7889	.7896	.7903	.7910	.7917
6.2	.7924	.7931	.7938	.7945	.7952	.7959	.7966	.7973	.7980	.7987
6.3	.7993	.8000	.8007	.8014	.8021	.8028	.8035	.8041	.8048	.8055
6.4	.8062	.8069	.8075	.8082	.8089	.8096	.8102	.8109	.8116	.8122
6.5	.8129	.8136	.8142	.8149	.8156	.8162	.8169	.8176	.8182	.8189
6.6	.8195	.8202	.8209	.8215	.8222	.8228	.8235	.8241	.8248	.8254
6.7	.8261	.8267	.8274	.8280	.8287	.8293	.8299	.8306	.8312	.8319
6.8	.8325	.8331	.8338	.8344	.8351	.8357	.8363	.8370	.8376	.8382
6.9	.8388	.8395	.8401	.8407	.8414	.8420	.8426	.8432	.8439	.8445
7.0	.8451	.8457	.8463	.8470	.8476	.8482	.8488	.8494	.8500	.8506
7.1	.8513	.8519	.8525	.8531	.8537	.8543	.8549	.8555	.8561	.8567
7.2	.8573	.8579	.8585	.8591	.8597	.8603	.8609	.8615	.8621	.8627
7.3	.8633	.8639	.8645	.8651	.8657	.8663	.8669	.8675	.8681	.8686
7.4	.8692	.8698	.8704	.8710	.8716	.8722	.8727	.8733	.8739	.8745
7.5	.8751	.8756	.8762	.8768	.8774	.8779	.8785	.8791	.8797	.8802
7.6	.8808	.8814	.8820	.8825	.8831	.8837	.8842	.8848	.8854	.8859
7.7	.8865	.8871	.8876	.8882	.8887	.8893	.8899	.8904	.8910	.8915
7.8	.8921	.8927	.8932	.8938	.8943	.8949	.8954	.8960	.8965	.8971
7.9	.8976	.8982	.8987	.8993	.8998	.9004	.9009	.9015	.9020	.9026
8.0	.9031	.9036	.9042	.9047	.9053	.9058	.9063	.9069	.9074	.9079
8.1	.9085	.9090	.9096	.9101	.9106	.9112	.9117	.9122	.9128	.9133
8.2	.9138	.9143	.9149	.9154	.9159	.9165	.9170	.9175	.9180	.9186
8.3	.9191	.9196	.9201	.9206	.9212	.9217	.9222	.9227	.9232	.9238
8.4	.9243	.9248	.9253	.9258	.9263	.9269	.9274	.9279	.9284	.9289
8.5	.9294	.9299	.9304	.9309	.9315	.9320	.9325	.9330	.9335	.9340
8.6	.9345	.9350	.9355	.9360	.9365	.9370	.9375	.9380	.9385	.9390
8.7	.9395	.9400	.9405	.9410	.9415	.9420	.9425	.9430	.9435	.9440
8.8	.9445	.9450	.9455	.9460	.9465	.9469	.9474	.9479	.9484	.9489
8.9	.9494	.9499	.9504	.9509	.9513	.9518	.9523	.9528	.9533	.9538
9.0	.9542	.9547	.9552	.9557	.9562	.9566	.9571	.9576	.9581	.9586
9.1	.9590	.9595	.9600	.9605	.9609	.9614	.9619	.9624	.9628	.9633
9.2	.9638	.9643	.9647	.9652	.9657	.9661	.9666	.9671	.9675	.9680
9.3	.9685	.9689	.9694	.9699	.9703	.9708	.9713	.9717	.9722	.9727
9.4	.9731	.9736	.9741	.9745	.9750	.9754	.9759	.9763	.9768	.9773
9.5	.9777	.9782	.9786	.9791	.9795	.9800	.9805	.9809	.9814	.9818
9.6	.9823	.9827	.9832	.9836	.9841	.9845	.9850	.9854	.9859	.9863
9.7	.9868	.9872	.9877	.9881	.9886	.9890	.9894	.9899	.9903	.9908
9.8	.9912	.9917	.9921	.9926	.9930	.9934	.9939	.9943	.9948	.9952
9.9	.9956	.9961	.9965	.9969	.9974	.9978	.9983	.9987	.9991	.9996

TABLE B.   Natural Logarithms

$$\ln x$$

x	0	0.01	0.02	0.03	0.04	0.05	0.06	0.07	0.08	0.09
1.0	0.0000	0.0100	0.0198	0.0296	0.0392	0.0488	0.0583	0.0677	0.0	
1.1	0.0953	0.1044	0.1133	0.1222	0.1310	0.1398	0.1484	0.1570	0.1	
1.2	0.1823	0.1906	0.1989	0.2070	0.2151	0.2231	0.2311	0.2390	0.2469	0.2547
1.3	0.2624	0.2700	0.2776	0.2852	0.2927	0.3001	0.3075	0.3148	0.3221	0.3293
1.4	0.3365	0.3436	0.3507	0.3577	0.3646	0.3716	0.3784	0.3853	0.3920	0.3988
1.5	0.4055	0.4121	0.4187	0.4253	0.4318	0.4383	0.4447	0.4511	0.4574	0.4637
1.6	0.4700	0.4762	0.4824	0.4886	0.4947	0.5008	0.5068	0.5128	0.5188	0.5247
1.7	0.5306	0.5365	0.5423	0.5481	0.5539	0.5596	0.5653	0.5710	0.5766	0.5822
1.8	0.5878	0.5933	0.5988	0.6043	0.6098	0.6152	0.6206	0.6259	0.6313	0.6366
1.9	0.6419	0.6471	0.6523	0.6575	0.6627	0.6678	0.6729	0.6780	0.6831	0.6881
2.0	0.6931	0.6981	0.7031	0.7080	0.7129	0.7178	0.7227	0.7275	0.7324	0.7372
2.1	0.7419	0.7467	0.7514	0.7561	0.7608	0.7655	0.7701	0.7747	0.7793	0.7839
2.2	0.7885	0.7930	0.7975	0.8020	0.8065	0.8109	0.8154	0.8198	0.8242	0.8286
2.3	0.8329	0.8372	0.8416	0.8459	0.8502	0.8544	0.8587	0.8629	0.8671	0.8713
2.4	0.8755	0.8796	0.8838	0.8879	0.8920	0.8961	0.9002	0.9042	0.9083	0.9123
2.5	0.9163	0.9203	0.9243	0.9282	0.9322	0.9361	0.9400	0.9439	0.9478	0.9517
2.6	0.9555	0.9594	0.9632	0.9670	0.9708	0.9746	0.9783	0.9821	0.9858	0.9895
2.7	0.9933	0.9969	1.0006	1.0043	1.0080	1.0116	1.0152	1.0188	1.0225	1.0260
2.8	1.0296	1.0332	1.0367	1.0403	1.0438	1.0473	1.0508	1.0543	1.0578	1.0613
2.9	1.0647	1.0682	1.0716	1.0750	1.0784	1.0818	1.0852	1.0886	1.0919	1.0953
3.0	1.0986	1.1019	1.1053	1.1086	1.1119	1.1151	1.1184	1.1217	1.1249	1.1282
3.1	1.1314	1.1346	1.1378	1.1410	1.1442	1.1474	1.1506	1.1537	1.1569	1.1600
3.2	1.1632	1.1663	1.1694	1.1725	1.1756	1.1787	1.1817	1.1848	1.1878	1.1909
3.3	1.1939	1.1969	1.2000	1.2030	1.2060	1.2090	1.2119	1.2149	1.2179	1.2208
3.4	1.2238	1.2267	1.2296	1.2326	1.2355	1.2384	1.2413	1.2442	1.2470	1.2499
3.5	1.2528	1.2556	1.2585	1.2613	1.2641	1.2669	1.2698	1.2726	1.2754	1.2782
3.6	1.2809	1.2837	1.2865	1.2892	1.2920	1.2947	1.2975	1.3002	1.3029	1.3056
3.7	1.3083	1.3110	1.3137	1.3164	1.3191	1.3218	1.3244	1.3271	1.3297	1.3324
3.8	1.3350	1.3376	1.3403	1.3429	1.3455	1.3481	1.3507	1.3533	1.3558	1.3584
3.9	1.3610	1.3635	1.3661	1.3686	1.3712	1.3737	1.3762	1.3788	1.3813	1.3838
4.0	1.3863	1.3888	1.3913	1.3938	1.3962	1.3987	1.4012	1.4036	1.4061	1.4085
4.1	1.4110	1.4134	1.4159	1.4183	1.4207	1.4231	1.4255	1.4279	1.4303	1.4327
4.2	1.4351	1.4375	1.4398	1.4422	1.4446	1.4469	1.4493	1.4516	1.4540	1.4563
4.3	1.4586	1.4609	1.4633	1.4656	1.4679	1.4702	1.4725	1.4748	1.4770	1.4793
4.4	1.4816	1.4839	1.4861	1.4884	1.4907	1.4929	1.4951	1.4974	1.4996	1.5019
4.5	1.5041	1.5063	1.5085	1.5107	1.5129	1.5151	1.5173	1.5195	1.5217	1.5239
4.6	1.5261	1.5282	1.5304	1.5326	1.5347	1.5369	1.5390	1.5412	1.5433	1.5454
4.7	1.5476	1.5497	1.5518	1.5539	1.5560	1.5581	1.5602	1.5623	1.5644	1.5665
4.8	1.5686	1.5707	1.5728	1.5748	1.5769	1.5790	1.5810	1.5831	1.5851	1.5872
4.9	1.5892	1.5913	1.5933	1.5953	1.5974	1.5994	1.6014	1.6034	1.6054	1.6074
5.0	1.6094	1.6114	1.6134	1.6154	1.6174	1.6194	1.6214	1.6233	1.6253	1.6273
5.1	1.6292	1.6312	1.6332	1.6351	1.6371	1.6390	1.6409	1.6429	1.6448	1.6467
5.2	1.6487	1.6506	1.6525	1.6544	1.6563	1.6582	1.6601	1.6620	1.6639	1.6658
5.3	1.6677	1.6696	1.6715	1.6734	1.6752	1.6771	1.6790	1.6808	1.6827	1.6845
5.4	1.6864	1.6882	1.6901	1.6919	1.6938	1.6956	1.6974	1.6993	1.7011	1.7029
x	0	0.01	0.02	0.03	0.04	0.05	0.06	0.07	0.08	0.09

TABLE B. Natural Logarithms (*Continued*)
ln x

x	0	0.01	0.02	0.03	0.04	0.05	0.06	0.07	0.08	0.09
5.5	1.7047	1.7066	1.7084	1.7102	1.7120	1.7138	1.7156	1.7174	1.7192	1.7210
5.6	1.7228	1.7246	1.7263	1.7281	1.7299	1.7317	1.7334	1.7352	1.7370	1.7387
5.7	1.7405	1.7422	1.7440	1.7457	1.7475	1.7492	1.7509	1.7527	1.7544	1.7561
5.8	1.7579	1.7596	1.7613	1.7630	1.7647	1.7664	1.7681	1.7699	1.7716	1.7733
5.9	1.7750	1.7766	1.7783	1.7800	1.7817	1.7834	1.7851	1.7867	1.7884	1.7901
6.0	1.7918	1.7934	1.7951	1.7967	1.7984	1.8001	1.8017	1.8034	1.8050	1.8066
6.1	1.8083	1.8099	1.8116	1.8132	1.8148	1.8165	1.8181	1.8197	1.8213	1.8229
6.2	1.8245	1.8262	1.8278	1.8294	1.8310	1.8326	1.8342	1.8358	1.8374	1.8390
6.3	1.8405	1.8421	1.8437	1.8453	1.8469	1.8485	1.8500	1.8516	1.8532	1.8547
6.4	1.8563	1.8579	1.8594	1.8610	1.8625	1.8641	1.8656	1.8672	1.8687	1.8703
6.5	1.8718	1.8733	1.8749	1.8764	1.8779	1.8795	1.8810	1.8825	1.8840	1.8856
6.6	1.8871	1.8886	1.8901	1.8916	1.8931	1.8946	1.8961	1.8976	1.8991	1.9006
6.7	1.9021	1.9036	1.9051	1.9066	1.9081	1.9095	1.9110	1.9125	1.9140	1.9155
6.8	1.9169	1.9184	1.9199	1.9213	1.9228	1.9242	1.9257	1.9272	1.9286	1.9301
6.9	1.9315	1.9330	1.9344	1.9359	1.9373	1.9387	1.9402	1.9416	1.9430	1.9445
7.0	1.9459	1.9473	1.9488	1.9502	1.9516	1.9530	1.9544	1.9559	1.9573	1.9587
7.1	1.9601	1.9615	1.9629	1.9643	1.9657	1.9671	1.9685	1.9699	1.9713	1.9727
7.2	1.9741	1.9755	1.9769	1.9782	1.9796	1.9810	1.9824	1.9838	1.9851	1.9865
7.3	1.9879	1.9892	1.9906	1.9920	1.9933	1.9947	1.9961	1.9974	1.9988	2.0001
7.4	2.0015	2.0028	2.0042	2.0055	2.0069	2.0082	2.0096	2.0109	2.0122	2.0136
7.5	2.0149	2.0162	2.0176	2.0189	2.0202	2.0215	2.0229	2.0242	2.0255	2.0268
7.6	2.0281	2.0295	2.0308	2.0321	2.0334	2.0347	2.0360	2.0373	2.0386	2.0399
7.7	2.0412	2.0425	2.0438	2.0451	2.0464	2.0477	2.0490	2.0503	2.0516	2.0528
7.8	2.0541	2.0554	2.0567	2.0580	2.0592	2.0605	2.0618	2.0631	2.0643	2.0656
7.9	2.0669	2.0681	2.0694	2.0707	2.0719	2.0732	2.0744	2.0757	2.0769	2.0782
8.0	2.0794	2.0807	2.0819	2.0832	2.0844	2.0857	2.0869	2.0882	2.0894	2.0906
8.1	2.0919	2.0931	2.0943	2.0956	2.0968	2.0980	2.0992	2.1005	2.1017	2.1029
8.2	2.1041	2.1054	2.1066	2.1078	2.1090	2.1102	2.1114	2.1126	2.1138	2.1150
8.3	2.1163	2.1175	2.1187	2.1199	2.1211	2.1223	2.1235	2.1247	2.1258	2.1270
8.4	2.1282	2.1294	2.1306	2.1318	2.1330	2.1342	2.1353	2.1365	2.1377	2.1389
8.5	2.1401	2.1412	2.1424	2.1436	2.1448	2.1459	2.1471	2.1483	2.1494	2.1506
8.6	2.1518	2.1529	2.1541	2.1552	2.1564	2.1576	2.1587	2.1599	2.1610	2.1622
8.7	2.1633	2.1645	2.1656	2.1668	2.1679	2.1691	2.1702	2.1713	2.1725	2.1736
8.8	2.1748	2.1759	2.1770	2.1782	2.1793	2.1804	2.1815	2.1827	2.1838	2.1849
8.9	2.1861	2.1872	2.1883	2.1894	2.1905	2.1917	2.1928	2.1939	2.1950	2.1961
9.0	2.1972	2.1983	2.1994	2.2006	2.2017	2.2028	2.2039	2.2050	2.2061	2.2072
9.1	2.2083	2.2094	2.2105	2.2116	2.2127	2.2138	2.2148	2.2159	2.2170	2.2181
9.2	2.2192	2.2203	2.2214	2.2225	2.2235	2.2246	2.2257	2.2268	2.2279	2.2289
9.3	2.2300	2.2311	2.2322	2.2332	2.2343	2.2354	2.2364	2.2375	2.2386	2.2396
9.4	2.2407	2.2418	2.2428	2.2439	2.2450	2.2460	2.2471	2.2481	2.2492	2.2502
9.5	2.2513	2.2523	2.2534	2.2544	2.2555	2.2565	2.2576	2.2586	2.2597	2.2607
9.6	2.2618	2.2628	2.2638	2.2649	2.2659	2.2670	2.2680	2.2690	2.2701	2.2711
9.7	2.2721	2.2732	2.2742	2.2752	2.2762	2.2773	2.2783	2.2793	2.2803	2.2814
9.8	2.2824	2.2834	2.2844	2.2854	2.2865	2.2875	2.2885	2.2895	2.2905	2.2915
9.9	2.2925	2.2935	2.2946	2.2956	2.2966	2.2976	2.2986	2.2996	2.3006	2.3016
x	0	0.01	0.02	0.03	0.04	0.05	0.06	0.07	0.08	0.09

TABLE C. Exponential Functions
e^x

x	0	0.01	0.02	0.03	0.04	0.05	0.06	0.07	0.08	0.09
0.0	0.1000×10^1	0.1010×10^1	0.1020×10^1	0.1030×10^1	0.1041×10^1	0.1051×10^1	0.1062×10^1	0.1073×10^1	0.1083×10^1	0.1094×10^1
0.1	0.1105	0.1116	0.1127	0.1139	0.1150	0.1162	0.1174	0.1185	0.1197	0.1209
0.2	0.1221	0.1234	0.1246	0.1259	0.1271	0.1284	0.1297	0.1310	0.1323	0.1336
0.3	0.1350	0.1363	0.1377	0.1391	0.1405	0.1419	0.1433	0.1448	0.1462	0.1477
0.4	0.1492	0.1507	0.1522	0.1537	0.1553	0.1568	0.1584	0.1600	0.1616	0.1632
0.5	0.1649	0.1665	0.1682	0.1699	0.1716	0.1733	0.1751	0.1768	0.1786	0.1804
0.6	0.1822	0.1840	0.1859	0.1878	0.1896	0.1916	0.1935	0.1954	0.1974	0.1994
0.7	0.2014	0.2034	0.2054	0.2075	0.2096	0.2117	0.2138	0.2160	0.2181	0.2203
0.8	0.2226	0.2248	0.2270	0.2293	0.2316	0.2340	0.2363	0.2387	0.2411	0.2435
0.9	0.2460	0.2484	0.2509	0.2535	0.2560	0.2586	0.2612	0.2638	0.2664	0.2691
1.0	0.2718	0.2746	0.2773	0.2801	0.2829	0.2858	0.2886	0.2915	0.2945	0.2974
1.1	0.3004	0.3034	0.3065	0.3096	0.3127	0.3158	0.3190	0.3222	0.3254	0.3287
1.2	0.3320	0.3353	0.3387	0.3421	0.3456	0.3490	0.3525	0.3561	0.3597	0.3633
1.3	0.3669	0.3706	0.3743	0.3781	0.3819	0.3857	0.3896	0.3935	0.3975	0.4015
1.4	0.4055	0.4096	0.4137	0.4179	0.4221	0.4263	0.4306	0.4349	0.4393	0.4437
1.5	0.4482	0.4527	0.4572	0.4618	0.4665	0.4711	0.4759	0.4807	0.4855	0.4904
1.6	0.4953	0.5003	0.5053	0.5104	0.5155	0.5207	0.5259	0.5312	0.5366	0.5419
1.7	0.5474	0.5529	0.5585	0.5641	0.5697	0.5755	0.5812	0.5871	0.5930	0.5989
1.8	0.6050	0.6110	0.6172	0.6234	0.6297	0.6360	0.6424	0.6488	0.6554	0.6619
1.9	0.6686	0.6753	0.6821	0.6890	0.6959	0.7029	0.7099	0.7171	0.7243	0.7316
2.0	0.7389	0.7463	0.7538	0.7614	0.7691	0.7768	0.7845	0.7925	0.8004	0.8085
2.1	0.8166	0.8248	0.8331	0.8415	0.8499	0.8585	0.8671	0.8758	0.8846	0.8935
2.2	0.9025	0.9116	0.9207	0.9300	0.9393	0.9488	0.9583	0.9679	0.9777	0.9875
2.3	0.9974	0.1007×10^2	0.1018×10^2	0.1028×10^2	0.1038×10^2	0.1049×10^2	0.1059×10^2	0.1070×10^2	0.1080×10^2	0.1091×10^2
2.4	0.1102×10^2	0.1113	0.1125	0.1136	0.1147	0.1159	0.1170	0.1182	0.1194	0.1206

TABLE C. Exponential Functions (*Continued*)
$$e^x$$

x	0	0.01	0.02	0.03	0.04	0.05	0.06	0.07	0.08	0.09
2.5	0.1218×10^2	0.1230×10^2	0.1243×10^2	0.1255×10^2	0.1268×10^2	0.1281×10^2	0.1294×10^2	0.1307×10^2	0.1320×10^2	0.1333×10^2
2.6	0.1346	0.1360	0.1374	0.1387	0.1401	0.1415	0.1430	0.1444	0.1459	0.1473
2.7	0.1488	0.1503	0.1518	0.1533	0.1549	0.1564	0.1580	0.1596	0.1612	0.1628
2.8	0.1644	0.1661	0.1678	0.1695	0.1712	0.1729	0.1746	0.1764	0.1781	0.1799
2.9	0.1817	0.1836	0.1854	0.1873	0.1892	0.1911	0.1930	0.1949	0.1969	0.1989
3.0	0.2009	0.2029	0.2049	0.2070	0.2091	0.2112	0.2133	0.2154	0.2176	0.2198
3.1	0.2220	0.2242	0.2265	0.2287	0.2310	0.2334	0.2357	0.2381	0.2405	0.2429
3.2	0.2453	0.2478	0.2503	0.2528	0.2553	0.2579	0.2605	0.2631	0.2658	0.2684
3.3	0.2711	0.2739	0.2766	0.2794	0.2822	0.2850	0.2879	0.2908	0.2937	0.2967
3.4	0.2996	0.3027	0.3057	0.3088	0.3119	0.3150	0.3182	0.3214	0.3246	0.3279
3.5	0.3312	0.3345	0.3378	0.3412	0.3447	0.3481	0.3516	0.3552	0.3587	0.3623
3.6	0.3660	0.3697	0.3734	0.3771	0.3809	0.3847	0.3886	0.3925	0.3965	0.4004
3.7	0.4045	0.4085	0.4126	0.4168	0.4210	0.4252	0.4295	0.4338	0.4382	0.4426
3.8	0.4470	0.4515	0.4560	0.4606	0.4653	0.4699	0.4747	0.4794	0.4842	0.4891
3.9	0.4940	0.4990	0.5040	0.5091	0.5142	0.5194	0.5246	0.5298	0.5352	0.5405
4.0	0.5460	0.5515	0.5570	0.5626	0.5683	0.5740	0.5797	0.5856	0.5915	0.5974
4.1	0.6034	0.6095	0.6156	0.6218	0.6280	0.6343	0.6407	0.6472	0.6537	0.6602
4.2	0.6669	0.6736	0.6803	0.6872	0.6941	0.7011	0.7081	0.7152	0.7224	0.7297
4.3	0.7370	0.7444	0.7519	0.7594	0.7671	0.7748	0.7826	0.7904	0.7984	0.8064
4.4	0.8145	0.8227	0.8310	0.8393	0.8477	0.8563	0.8649	0.8736	0.8823	0.8912
4.5	0.9002	0.9092	0.9184	0.9276	0.9369	0.9463	0.9558	0.9654	0.9751	0.9849
4.6	0.9948	0.1005×10^3	0.1015×10^3	0.1025×10^3	0.1035×10^3	0.1046×10^3	0.1056×10^3	0.1067×10^3	0.1078×10^3	0.1089×10^3
4.7	0.1099×10^3	0.1111	0.1122	0.1133	0.1144	0.1156	0.1167	0.1179	0.1191	0.1203
4.8	0.1215	0.1227	0.1240	0.1252	0.1265	0.1277	0.1290	0.1303	0.1316	0.1330
4.9	0.1343	0.1356	0.1370	0.1384	0.1398	0.1412	0.1426	0.1440	0.1455	0.1469
5.0	0.1484	0.1499	0.1514	0.1529	0.1545	0.1560	0.1576	0.1592	0.1608	0.1624

TABLE D. Four-Place Values of Trigonometric Functions and Radians

Degrees	Radians	Sin θ	Cos θ	Tan θ	Cot θ	Sec θ	Csc θ		
0° 00′	.0000	.0000	1.0000	.0000	—	1.000	—	1.5708	90° 00′
10	.0029	.0029	1.0000	.0029	343.8	1.000	343.8	1.5679	50
20	.0058	.0058	1.0000	.0058	171.9	1.000	171.9	1.5650	40
30	.0087	.0087	1.0000	.0087	114.6	1.000	114.6	1.5621	30
40	.0116	.0116	.9999	.0116	85.94	1.000	85.95	1.5592	20
50	.0145	.0145	.9999	.0145	68.75	1.000	68.76	1.5563	10
1° 00′	.0175	.0175	.9998	.0175	57.29	1.000	57.30	1.5533	89° 00′
10	.0204	.0204	.9998	.0204	49.10	1.000	49.11	1.5504	50
20	.0233	.0233	.9997	.0233	42.96	1.000	42.98	1.5475	40
30	.0262	.0262	.9997	.0262	38.19	1.000	38.20	1.5446	30
40	.0291	.0291	.9996	.0291	34.37	1.000	34.38	1.5417	20
50	.0320	.0320	.9995	.0320	31.24	1.001	31.26	1.5388	10
2° 00′	.0349	.0349	.9994	.0349	28.64	1.001	28.65	1.5359	88° 00′
10	.0378	.0378	.9993	.0378	26.43	1.001	26.45	1.5330	50
20	.0407	.0407	.9992	.0407	24.54	1.001	24.56	1.5301	40
30	.0436	.0436	.9990	.0437	22.90	1.001	22.93	1.5272	30
40	.0465	.0465	.9989	.0466	21.47	1.001	21.49	1.5243	20
50	.0495	.0494	.9988	.0495	20.21	1.001	20.23	1.5213	10
3° 00′	.0524	.0523	.9986	.0524	19.08	1.001	19.11	1.5184	87° 00′
10	.0553	.0552	.9985	.0553	18.07	1.002	18.10	1.5155	50
20	.0582	.0581	.9983	.0582	17.17	1.002	17.20	1.5126	40
30	.0611	.0610	.9981	.0612	16.35	1.002	16.38	1.5097	30
40	.0640	.0640	.9980	.0641	15.60	1.002	15.64	1.5068	20
50	.0669	.0669	.9978	.0670	14.92	1.002	14.96	1.5039	10
4° 00′	.0698	.0698	.9976	.0699	14.30	1.002	14.34	1.5010	86° 00′
10	.0727	.0727	.9974	.0729	13.73	1.003	13.76	1.4981	50
20	.0756	.0756	.9971	.0758	13.20	1.003	13.23	1.4952	40
30	.0785	.0785	.9969	.0787	12.71	1.003	12.75	1.4923	30
40	.0814	.0814	.9967	.0816	12.25	1.003	12.29	1.4893	20
50	.0844	.0843	.9964	.0846	11.83	1.004	11.87	1.4864	10
5° 00′	.0873	.0872	.9962	.0875	11.43	1.004	11.47	1.4835	85° 00′
10	.0902	.0901	.9959	.0904	11.06	1.004	11.10	1.4806	50
20	.0931	.0929	.9957	.0934	10.71	1.004	10.76	1.4777	40
30	.0960	.0958	.9954	.0963	10.39	1.005	10.43	1.4748	30
40	.0989	.0987	.9951	.0992	10.08	1.005	10.13	1.4719	20
50	.1018	.1016	.9948	.1022	9.788	1.005	9.839	1.4690	10
6° 00′	.1047	.1045	.9945	.1051	9.514	1.006	9.567	1.4661	84° 00′
		Cos θ	Sin θ	Cot θ	Tan θ	Csc θ	Sec θ	Radians	Degrees

Source: Washington, Allyn J.: *Basic Technical Mathematics with Calculus*, 2nd ed. Menlo Park, Calif., Cummings Publishing Co., 1970, pp. 737–744.

TABLE D.   Four-Place Values of Trigonometric Functions and Radians
(Continued)

Degrees	Radians	Sin θ	Cos θ	Tan θ	Cot θ	Sec θ	Csc θ		
6° 00′	.1047	.1045	.9945	.1051	9.514	1.006	9.567	1.4661	84° 00′
10	.1076	.1074	.9942	.1080	9.255	1.006	9.309	1.4632	50
20	.1105	.1103	.9939	.1110	9.010	1.006	9.065	1.4603	40
30	.1134	.1132	.9936	.1139	8.777	1.006	8.834	1.4573	30
40	.1164	.1161	.9932	.1169	8.556	1.007	8.614	1.4544	20
50	.1193	.1190	.9929	.1198	8.345	1.007	8.405	1.4515	10
7° 00′	.1222	.1219	.9925	.1228	8.144	1.008	8.206	1.4486	83° 00′
10	.1251	.1248	.9922	.1257	7.953	1.008	8.016	1.4457	50
20	.1280	.1276	.9918	.1287	7.770	1.008	7.834	1.4428	40
30	.1309	.1305	.9914	.1317	7.596	1.009	7.661	1.4399	30
40	.1338	.1334	.9911	.1346	7.429	1.009	7.496	1.4370	20
50	.1367	.1363	.9907	.1376	7.269	1.009	7.337	1.4341	10
8° 00′	.1396	.1392	.9903	.1405	7.115	1.010	7.185	1.4312	82° 00′
10	.1425	.1421	.9899	.1435	6.968	1.010	7.040	1.4283	50
20	.1454	.1449	.9894	.1465	6.827	1.011	6.900	1.4254	40
30	.1484	.1478	.9890	.1495	6.691	1.011	6.765	1.4224	30
40	.1513	.1507	.9886	.1524	6.561	1.012	6.636	1.4195	20
50	.1542	.1536	.9881	.1554	6.435	1.012	6.512	1.4166	10
9° 00′	.1571	.1564	.9877	.1584	6.314	1.012	6.392	1.4137	81° 00′
10	.1600	.1593	.9872	.1614	6.197	1.013	6.277	1.4108	50
20	.1629	.1622	.9868	.1644	6.084	1.013	6.166	1.4079	40
30	.1658	.1650	.9863	.1673	5.976	1.014	6.059	1.4050	30
40	.1687	.1679	.9858	.1703	5.871	1.014	5.955	1.4021	20
50	.1716	.1708	.9853	.1733	5.769	1.015	6.855	1.3992	10
10° 00′	.1745	.1736	.9848	.1763	5.671	1.015	5.759	1.3963	80° 00′
10	.1774	.1765	.9843	.1793	5.576	1.016	5.665	1.3934	50
20	.1804	.1794	.9838	.1823	5.485	1.016	5.575	1.3904	40
30	.1833	.1822	.9833	.1853	5.396	1.017	5.487	1.3875	30
40	.1862	.1851	.9827	.1883	5.309	1.018	5.403	1.3846	20
50	.1891	.1880	.9822	.1914	5.226	1.018	5.320	1.3817	10
11° 00′	.1920	.1908	.9816	.1944	5.145	1.019	5.241	1.3788	79° 00′
10	.1949	.1937	.9811	.1974	5.066	1.019	5.164	1.3759	50
20	.1978	.1965	.9805	.2004	4.989	1.020	5.089	1.3730	40
30	.2007	.1994	.9799	.2035	4.915	1.020	5.016	1.3701	30
40	.2036	.2022	.9793	.2065	4.843	1.021	4.945	1.3672	20
50	.2065	.2051	.9787	.2095	4.773	1.022	4.876	1.3643	10
12° 00′	.2094	.2079	.9781	.2126	4.705	1.022	4.810	1.3614	78° 00′
		Cos θ	Sin θ	Cot θ	Tan θ	Csc θ	Sec θ	Radians	Degrees

TABLE D.   Four-Place Values of Trigonometric Functions and Radians
(Continued)

Degrees	Radians	Sin θ	Cos θ	Tan θ	Cot θ	Sec θ	Csc θ		
12° 00′	.2094	.2079	.9781	.2126	4.705	1.022	4.810	1.3614	78° 00′
10	.2123	.2108	.9775	.2156	4.638	1.023	4.745	1.3584	50
20	.2153	.2136	.9769	.2186	4.574	1.024	4.682	1.3555	40
30	.2182	.2164	.9763	.2217	4.511	1.024	4.620	1.3526	30
40	.2211	.2193	.9757	.2247	4.449	1.025	4.560	1.3497	20
50	.2240	.2221	.9750	.2278	4.390	1.026	4.502	1 3468	10
13° 00′	.2269	.2250	.9744	.2309	4.331	1.026	4.445	1.3439	77° 00′
10	.2298	.2278	.9737	.2339	4.275	1.027	4.390	1.3410	50
20	.2327	.2306	.9730	.2370	4.219	1.028	4.336	1.3381	40
30	.2356	.2334	.9724	.2401	4.165	1.028	4.284	1.3352	30
40	.2385	.2363	.9717	.2432	4.113	1.029	4.232	1.3323	20
50	.2414	.2391	.9710	.2462	4.061	1.030	4.182	1.3294	10
14° 00′	.2443	.2419	.9703	.2493	4.011	1.031	4.134	1.3265	76° 00′
10	.2473	.2447	.9696	.2524	3.962	1.031	4.086	1.3235	50
20	.2502	.2476	.9689	.2555	3.914	1.032	4.039	1.3206	40
30	.2531	.2504	.9681	.2586	3.867	1.033	3.994	1.3177	30
40	.2560	.2532	.9674	.2617	3.821	1.034	3.950	1.3148	20
50	.2589	.2560	.9667	.2648	3.776	1.034	3.906	1.3119	10
15° 00′	.2618	.2588	.9659	.2679	3.732	1.035	3.864	1.3090	75° 00′
10	.2647	.2616	.9652	.2711	3.689	1.036	3.822	1.3061	50
20	.2676	.2644	.9644	.2742	3.647	1.037	3.782	1.3032	40
30	.2705	.2672	.9636	.2773	3.606	1.038	3.742	1.3003	30
40	.2734	.2700	.9628	.2805	3.566	1.039	3.703	1.3974	20
50	.2763	.2728	.9621	.2836	3.526	1.039	3.665	1.3945	10
16° 00′	.2793	.2756	.9613	.2867	3.487	1.040	3.628	1.2915	74° 00′
10	.2822	.2784	.9605	.2899	3.450	1.041	3.592	1.2886	50
20	.2851	.2812	.9596	.2931	3.412	1.042	3.556	1.2857	40
30	.2880	.2840	.9588	.2962	3.376	1.043	3.521	1.2828	30
40	.2909	.2868	.9580	.2994	3.340	1.044	3.487	1.2799	20
50	.2938	.2896	.9572	.3026	3.305	1.045	3.453	1.2770	10
17° 00′	.2967	.2924	.9563	.3057	3.271	1.046	3.420	1.2741	73° 00′
10	.2996	.2952	.9555	.3089	3.237	1.047	3.388	1.2712	50
20	.3025	.2979	.9546	.3121	3.204	1.048	3.356	1.2683	40
30	.3054	.3007	.9537	.3153	3.172	1.049	3.326	1.2654	30
40	.3083	.3035	.9528	.3185	3.140	1.049	3.295	1.2625	20
50	.3113	.3062	.9520	.3217	3.108	1.050	3.265	1.2595	10
18° 00′	.3142	.3090	.9511	.3249	3.078	1.051	3.236	1.2566	72° 00′
		Cos θ	Sin θ	Cot θ	Tan θ	Csc θ	Sec θ	Radians	Degrees

TABLE D.   Four-Place Values of Trigonometric Functions and Radians
(*Continued*)

Degrees	Radians	Sin θ	Cos θ	Tan θ	Cot θ	Sec θ	Csc θ		
18° 00′	.3142	.3090	.9511	.3249	3.078	1.051	3.236	1.2566	72° 00′
10	.3171	.3118	.9502	.3281	3.047	1.052	3.207	1.2537	50
20	.3200	.3145	.9492	.3314	3.018	1.053	3.179	1.2508	40
30	.3229	.3173	.9483	.3346	2.989	1.054	3.152	1.2479	30
40	.3258	.3201	.9474	.3378	2.960	1.056	3.124	1.2450	20
50	.3287	.3228	.9465	.3411	2.932	1.057	3.098	1.2421	10
19° 00′	.3316	.3256	.9455	.3443	2.904	1.058	3.072	1.2392	71° 00′
10	.3345	.3283	.9446	.3476	2.877	1.059	3.046	1.2363	50
20	.3374	.3311	.9436	.3508	2.850	1.060	3.021	1.2334	40
30	.3403	.3338	.9426	.3541	2.824	1.061	2.996	1.2305	30
40	.3432	.3365	.9417	.3574	2.798	1.062	2.971	1.2275	20
50	.3462	.3393	.9407	.3607	2.773	1.063	2.947	1.2246	10
20° 00′	.3491	.3420	.9397	.3640	2.747	1.064	2.924	1.2217	70° 00′
10	.3520	.3448	.9387	.3673	2.723	1.065	2.901	1.2188	50
20	.3549	.3475	.9377	.3706	2.699	1.066	2.878	1.2159	40
30	.3578	.3502	.9367	.3739	2.675	1.068	2.855	1.2130	30
40	.3607	.3529	.9356	.3772	2.651	1.069	2.833	1.2101	20
50	.3636	.3557	.9346	.3805	2.628	1.070	2.812	1.2072	10
21° 00′	.3665	.3584	.9336	.3839	2.605	1.071	2.790	1.2043	69° 00′
10	.3694	.3611	.9325	.3872	2.583	1.072	2.769	1.2014	50
20	.3723	.3638	.9315	.3906	2.560	1.074	2.749	1.1985	40
30	.3752	.3665	.9304	.3939	2.539	1.075	2.729	1.1956	30
40	.3782	.3692	.9293	.3973	2.517	1.076	2.709	1.1926	20
50	.3811	.3719	.9283	.4006	2.496	1.077	2.689	1.1897	10
22° 00′	.3840	.3746	.9272	.4040	2.475	1.079	2.669	1.1868	68° 00′
10	.3869	.3773	.9261	.4074	2.455	1.080	2.650	1.1839	50
20	.3898	.3800	.9250	.4108	2.434	1.081	2.632	1.1810	40
30	.3927	.3827	.9239	.4142	2.414	1.082	2.613	1.1781	30
40	.3956	.3854	.9228	.4176	2.394	1.084	2.595	1.1752	20
50	.3985	.3881	.9216	.4210	2.375	1.085	2.577	1.1723	10
23° 00′	.4014	.3907	.9205	.4245	2.356	1.086	2.559	1.1694	67° 00′
10	.4043	.3934	.9194	.4279	2.337	1.088	2.542	1.1665	50
20	.4072	.3961	.9182	.4314	2.318	1.089	2.525	1.1636	40
30	.4102	.3987	.9171	.4348	2.300	1.090	2.508	1.1606	30
40	.4131	.4014	.9159	.4383	2.282	1.092	2.491	1.1577	20
50	.4160	.4041	.9147	.4417	2.264	1.093	2.475	1.1548	10
24° 00′	.4189	.4067	.9135	.4452	2.246	1.095	2.459	1.1519	66° 00′
		Cos θ	Sin θ	Cot θ	Tan θ	Csc θ	Sec θ	Radians	Degrees

TABLE D. Four-Place Values of Trigonometric Functions and Radians
(Continued)

Degrees	Radians	Sin θ	Cos θ	Tan θ	Cot θ	Sec θ	Csc θ		
24° 00′	.4189	.4067	.9135	.4452	2.246	1.095	2.459	1.1519	66° 00′
10	.4218	.4094	.9124	.4487	2.229	1.096	2.443	1.1490	50
20	.4247	.4120	.9112	.4522	2.211	1.097	2.427	1.1461	40
30	.4276	.4147	.9100	.4557	2.194	1.099	2.411	1.1432	30
40	.4305	.4173	.9088	.4592	2.177	1.100	2.396	1.1403	20
50	.4334	.4200	.9075	.4628	2.161	1.102	2.381	1.1374	10
25° 00′	.4363	.4226	.9063	.4663	2.145	1.103	2.366	1.1345	65° 00′
10	.4392	.4253	.9051	.4699	2.128	1.105	2.352	1.1316	50
20	.4422	.4279	.9038	.4734	2.112	1.106	2.337	1.1286	40
30	.4451	.4305	.9026	.4770	2.097	1.108	2.323	1.1257	30
40	.4480	.4331	.9013	.4806	2.081	1.109	2.309	1.1228	20
50	.4509	.4358	.9001	.4841	2.066	1.111	2.295	1.1199	10
26° 00′	.4538	.4384	.8988	.4877	2.050	1.113	2.281	1.1170	64° 00′
10	.4567	.4410	.8975	.4913	2.035	1.114	2.268	1.1141	50
20	.4596	.4436	.8962	.4950	2.020	1.116	2.254	1.1112	40
30	.4025	.4402	.8949	.4986	2.006	1.117	2.241	1.1083	30
40	.4654	.4488	.8936	.5022	1.991	1.119	2.228	1.1054	20
50	.4683	.4514	.8923	.5059	1.977	1.121	2.215	1.1025	10
27° 00′	.4712	.4540	.8910	.5095	1.963	1.122	2.203	1.0996	63° 00′
10	.4741	.4566	.8897	.5132	1.949	1.124	2.190	1.0966	50
20	.4771	.4592	.8884	.5169	1.935	1.126	2.178	1.0937	40
30	.4800	.4617	.8870	.5206	1.921	1.127	2.166	1.0908	30
40	.4829	.4643	.8857	.5243	1.907	1.129	2.154	1.0879	20
50	.4858	.4669	.8843	.5280	1.894	1.131	2.142	1.0850	10
28° 00′	.4887	.4695	.8829	.5317	1.881	1.133	2.130	1.0821	62° 00′
10	.4916	.4720	.8816	.5354	1.868	1.134	2.118	1.0792	50
20	.4945	.4746	.8802	.5392	1.855	1.136	2.107	1.0763	40
30	.4974	.4772	.8788	.5430	1.842	1.138	2.096	1.0734	30
40	.5003	.4797	.8774	.5467	1.829	1.140	2.085	1.0705	20
50	.5032	.4823	.8760	.5505	1.816	1.142	2.074	1.0676	10
29° 00′	.5061	.4848	.8746	.5543	1.804	1.143	2.063	1.0647	61° 00′
10	.5091	.4874	.8732	.5581	1.792	1.145	2.052	1.0617	50
20	.5120	.4899	.8718	.5619	1.780	1.147	2.041	1.0588	40
30	.5149	.4924	.8704	.5658	1.767	1.149	2.031	1.0559	30
40	.5178	.4950	.8689	.5696	1.756	1.151	2.020	1.0530	20
50	.5207	.4975	.8675	.5735	1.744	1.153	2.010	1.0501	10
30° 00′	.5236	.5000	.8660	.5774	1.732	1.155	2.000	1.0472	60° 00′
		Cos θ	Sin θ	Cot θ	Tan θ	Csc θ	Sec θ	Radians	Degrees

TABLE D.   Four-Place Values of Trigonometric Functions and Radians
(*Continued*)

Degrees	Radians	Sin θ	Cos θ	Tan θ	Cot θ	Sec θ	Csc θ		
30° 00′	.5236	.5000	.8660	.5774	1.732	1.155	2.000	1.0472	60° 00′
10	.5265	.5025	.8646	.5812	1.720	1.157	1.990	1.0443	50
20	.5294	.5050	.8631	.5851	1.709	1.159	1.980	1.0414	40
30	.5323	.5075	.8616	.5890	1.698	1.161	1.970	1.0385	30
40	.5352	.5100	.8601	.5930	1.686	1.163	1.961	1.0356	20
50	.5381	.5125	.8587	.5969	1.675	1.165	1.951	1.0327	10
31° 00′	.5411	.5150	.8572	.6009	1.664	1.167	1.942	1.0297	59° 00′
10	.5440	.5175	.8557	.6048	1.653	1.169	1.932	1.0268	50
20	.5469	.5200	.8542	.6088	1.643	1.171	1.923	1.0239	40
30	.5498	.5225	.8526	.6128	1.632	1.173	1.914	1.0210	30
40	.5527	.5250	.8511	.6168	1.621	1.175	1.905	1.0181	20
50	.5556	.5275	.8496	.6208	1.611	1.177	1.896	1.0152	10
32° 00′	.5585	.5299	.8480	.6249	1.600	1.179	1.887	1.0123	58° 00′
10	.5614	.5324	.8465	.6289	1.590	1.181	1.878	1.0094	50
20	.5643	.5348	.8450	.6330	1.580	1.184	1.870	1.0065	40
30	.5672	.5373	.8434	.6371	1.570	1.186	1.861	1.0036	30
40	.5701	.5398	.8418	.6412	1.560	1.188	1.853	1.0007	20
50	.5730	.5422	.8403	.6453	1.550	1.190	1.844	.9977	10
33° 00′	.5760	.5446	.8387	.6494	1.540	1.192	1.836	.9948	57° 00′
10	.5789	.5471	.8371	.6536	1.530	1.195	1.828	.9919	50
20	.5818	.5495	.8355	.6577	1.520	1.197	1.820	.9890	40
30	.5847	.5519	.8339	.6619	1.511	1.199	1.812	.9861	30
40	.5876	.5544	.8323	.6661	1.501	1.202	1.804	.9832	20
50	.5905	.5568	.8307	.6703	1.492	1.204	1.796	.9803	10
34° 00′	.5934	.5592	.8290	.6745	1.483	1.206	1.788	.9774	56° 00′
10	.5963	.5616	.8274	.6787	1.473	1.209	1.781	.9745	50
20	.5992	.5640	.8258	.6830	1.464	1.211	1.773	.9716	40
30	.6021	.5664	.8241	.6873	1.455	1.213	1.766	.9687	30
40	.6050	.5688	.8225	.6916	1.446	1.216	1.758	.9657	20
50	.6080	.5712	.8208	.6959	1.437	1.218	1.751	.9628	10
35° 00′	.6109	.5736	.8192	.7002	1.428	1.221	1.743	.9599	55° 00′
10	.6138	.5760	.8175	.7046	1.419	1.223	1.736	.9570	50
20	.6167	.5783	.8158	.7089	1.411	1.226	1.729	.9541	40
30	.6196	.5807	.8141	.7133	1.402	1.228	1.722	.9512	30
40	.6225	.5831	.8124	.7177	1.393	1.231	1.715	.9483	20
50	.6254	.5854	.8107	.7221	1.385	1.233	1.708	.9454	10
36° 00′	.6283	.5878	.8900	.7265	1.376	1.236	1.701	.9425	54° 00′
		Cos θ	Sin θ	Cot θ	Tan θ	Csc θ	Sec θ	Radians	Degrees

TABLE D.   Four-Place Values of Trigonometric Functions and Radians
(*Continued*)

Degrees	Radians	Sin θ	Cos θ	Tan θ	Cot θ	Sec θ	Csc θ		
36° 00′	.6283	.5878	.8090	.7265	1.376	1.236	1.701	.9425	54° 00′
10	.6312	.5901	.8073	.7310	1.368	1.239	1.695	.9396	50
20	.6341	.5925	.8056	.7355	1.360	1.241	1.688	.9367	40
30	.6370	.5948	.8039	.7400	1.351	1.244	1.681	.9338	30
40	.6400	.5972	.8021	.7445	1.343	1.247	1.675	.9308	20
50	.6429	.5995	.8004	.7490	1.335	1.249	1.668	.9279	10
37° 00′	.6458	.6018	.7986	.7536	1.327	1.252	1.662	.9250	53° 00′
10	.6487	.6041	.7969	.7581	1.319	1.255	1.655	.9221	50
20	.6516	.6065	.7951	.7627	1.311	1.258	1.649	.9192	40
30	.6545	.6088	.7934	.7673	1.303	1.260	1.643	.9163	30
40	.6574	.6111	.7916	.7720	1.295	1.263	1.636	.9134	20
50	.6603	.6134	.7898	.7766	1.288	1.266	1.630	.9105	10
38° 00′	.6632	.6157	.7880	.7813	1.280	1.269	1.624	.9076	52° 00′
10	.6661	.6180	.7862	.7860	1.272	1.272	1.618	.9047	50
20	.6690	.6202	.7844	.7907	1.265	1.275	1.612	.9018	40
30	.6720	.6225	.7826	.7954	1.257	1.278	1.606	.8988	30
40	.6749	.6248	.7808	.8002	1.250	1.281	1.601	.8959	20
50	.6778	.6271	.7790	.8050	1.242	1.284	1.595	.8930	10
39° 00′	.6807	.6293	.7771	.8098	1.235	1.287	1.589	.8901	51° 00′
10	.6836	.6316	.7753	.8146	1.228	1.290	1.583	.8872	50
20	.6865	.6338	.7735	.8195	1.220	1.293	1.578	.8843	40
30	.6894	.6361	.7716	.8243	1.213	1.296	1.572	.8814	30
40	.6923	.6383	.7698	.8292	1.206	1.299	1.567	.8785	20
50	.6952	.6406	.7679	.8342	1.199	1.302	1.561	.8756	10
40° 00′	.6981	.6428	.7660	.8391	1.192	1.305	1.556	.8727	50° 00′
10	.7010	.6450	.7642	.8441	1.185	1.309	1.550	.8698	50
20	.7039	.6472	.7623	.8491	1.178	1.312	1.545	.8668	40
30	.7069	.6494	.7604	.8541	1.171	1.315	1.540	.8639	30
40	.7098	.6517	.7585	.8591	1.164	1.318	1.535	.8610	20
50	.7127	.6539	.7566	.8642	1.157	1.322	1.529	.8581	10
41° 00′	.7156	.6561	.7547	.8693	1.150	1.325	1.524	.8552	49° 00′
10	.7185	.6583	.7528	.8744	1.144	1.328	1.519	.8523	50
20	.7214	.6604	.7509	.8796	1.137	1.332	1.514	.8494	40
30	.7243	.6626	.7490	.8847	1.130	1.335	1.509	.8465	30
40	.7272	.6648	.7470	.8899	1.124	1.339	1.504	.8436	20
50	.7301	.6670	.7451	.8952	1.117	1.342	1.499	.8407	10
42° 00′	.7330	.6691	.7431	.9004	1.111	1.346	1.494	.8378	48° 00′
		Cos θ	Sin θ	Cot θ	Tan θ	Csc θ	Sec θ	Radians	Degrees

TABLE D.   Four-Place Values of Trigonometric Functions and Radians
(*Continued*)

Degrees	Radians	Sin θ	Cos θ	Tan θ	Cot θ	Sec θ	Csc θ		
42° 00′	.7330	.6691	.7431	.9004	1.111	1.346	1.494	.8378	48° 00′
10	.7359	.6713	.7412	.9057	1.104	1.349	1.490	.8348	50
20	.7389	.6734	.7392	.9110	1.098	1.353	1.485	.8319	40
30	.7418	.6756	.7373	.9163	1.091	1.356	1.480	.8290	30
40	.7447	.6777	.7353	.9217	1.085	1.360	1.476	.8261	20
50	.7476	.6799	.7333	.9271	1.079	1.364	1.471	.8232	10
43° 00′	.7505	.6820	.7314	.9325	1.072	1.367	1.466	.8203	47° 00′
10	.7534	.6841	.7294	.9380	1.066	1.371	1.462	.8174	50
20	.7563	.6862	.7274	.9435	1.060	1.375	1.457	.8145	40
30	.7592	.6884	.7254	.9490	1.054	1.379	1.453	.8116	30
40	.7621	.6905	.7234	.9545	1.048	1.382	1.448	.8087	20
50	.7650	.6926	.7214	.9601	1.042	1.386	1.444	.8058	10
44° 00′	.7679	.6947	.7193	.9657	1.036	1.390	1.440	.8029	46° 00′
10	.7709	.6967	.7173	.9713	1.030	1.394	1.435	.7999	50
20	.7738	.6988	.7153	.9770	1.024	1.398	1.431	.7970	40
30	.7767	.7009	.7133	.9827	1.018	1.402	1.427	.7941	30
40	.7796	.7030	.7112	.9884	1.012	1.406	1.423	.7912	20
50	.7825	.7050	.7092	.9942	1.006	1.410	1.418	.7883	10
45° 00′	.7854	.7071	.7071	1.000	1.000	1.414	1.414	.7854	45° 00′
		Cos θ	Sin θ	Cot θ	Tan θ	Csc θ	Sec θ	Radians	Degrees

ANSWERS TO SELECTED ODD-NUMBERED PROBLEMS

CHAPTER 1

1.1 Exercises

1. $\frac{1}{7} = .\overline{142857}\ldots$

3. $\frac{3}{4} = .75$

5. $\frac{9}{7} = 1.\overline{285714}\ldots$

7. $\frac{10}{8} = 1.25$

9. $\frac{2}{9} = .\overline{2}$

11. $\frac{6}{11}$

13. $\frac{501}{500}$

15. $\frac{30,007}{10,000}$

17. $\frac{37}{99}$

19. $\frac{5}{8}$

21. Rational

23. Irrational

25. Rational

27. Rational

29. Irrational

1.2 Exercises

11. (a) True   (b) False   (c) False   (d) True   (e) False

13. (a) Addition Law   (b) Multiplication Law   (c) Multiplication Law   (d) Transitive Law

1.3 Exercises

1. $(0, +\infty)$

3. $[0, +\infty)$

5. $(-\infty, -1]$

15. $x > -5$

17. $x \leq -\frac{17}{8}$ 19. $\frac{5}{2} < x < \frac{7}{2}$

21. $-1 < x < 1$

23. $x \leq -4$ or $x \geq 2$

25. $0 < x < 2$ or $x > 4$

27. True

1.4 Exercises

1. (a) 1   (b) 5   (c) 5   (d) 5   (e) -1

    (f) 6   (g) $\frac{2}{3}$ (h) -11

3. (a)

-2 4

(b)

-1 $\dfrac{5}{3}$

(c)

-5 5

(d)

-6 10

(e)

$x = 1$ or $x = 2$

(f)

All the reals

5. (a) $|x| = 7$ (b) $|x| < 2$ (c) $|x| > 4$

    (d) $|x - 3| < 8$ (e) $|x - 1| < 19$ (f) $|3x + 2| < 2$

1.5 Exercises

1. (a) Quadrant IV   (b) Quadrant I   (c) Quadrant III
   (d) y-axis   (e) Origin   (f) Quadrant II

3. (a)

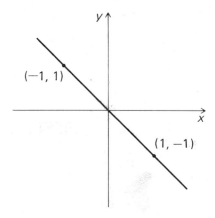

(c)

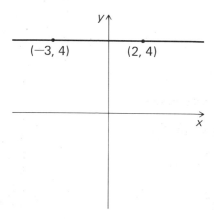

(b)

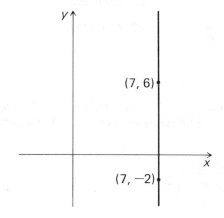

(d)

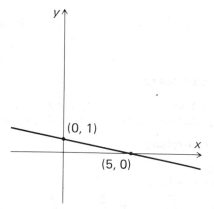

5. $(x - 1)^2 + (y + 3)^2 = 25$

Chapter 1 Exercises

1. (a) $\dfrac{2}{7} = 0.\overline{285714}$ (b) $0.75\overline{0}$

   (c) $0.\overline{923076}$ (d) $1.1\overline{45}$

3. (a) $-2 > -2.5$ (b) $|-9| > |3|$ (c) $0 \le x^2$

(d) $-b < -a$ (e) $\left(\frac{1}{2}\right)^2 > \left(\frac{1}{2}\right)^3$ (f) $-(3x^2 + 1) < 0$

5. (a) No x's   (b) $\{x| - 2 \le x \le 1|$ or $x \ge 3\}$
   (c) All x's   (d) All x's

7. (a) $-4 < y < -\frac{2}{3}$ (b) $y < 1$ or $y > 8$

9. (a) $3\sqrt{10}$ (b) 5   (c) $\sqrt{34}$

(d) $5\sqrt{2}$ (e) $\sqrt{13}$ (f) $2\sqrt{13}$

CHAPTER 2

2.1(A) Exercises

1. (a) $\{1, 2, 3, 4, 5\}$ (b) $\{-2\}$
   (c) $\{25, 26, 27, 28, 29\}$ (d) $\{x|x = 2n + 1, n = 0, 1, 2, \ldots\}$
   (e) $\{-1, 2\}$
3. R is a subset of C, T is not a subset of C.
5. Empty
7. (a) False   (b) False   (c) True   (d) True   (e) False
   (f) False   (g) True   (h) True
9. (a) False   (b) False   (c) False   (d) False   (e) True

2.1(B) Exercises

1. $A' = \{d, e, f, g, h\}$, $B' = \{a, b, e, f, h\}$, $C' = \{b, c, d, g\}$, $D' = \varnothing$
7. $D' \subset C'$
9. $D = \{1, -3, 5\}$

2.2 Exercises

1. $A \times B = \{(1,0), (1, 1), (2, 0), (2, 1), (3, 0), (3, 1)\}$
   Domain of $A \times B$ is A and range of $A \times B$ is B.
   $B \times A = \{(0, 1), (1, 1), (0, 2), (1, 2), (0, 3), (1, 3)\}$
   Domain of $B \times A$ is B and range of $B \times A$ is A.
3. $A \times B = \{(5, -1), (5, -2)\}$; $B \times A = \{(-1, 5), (-2, 5)\}$
5. $A \times B = \{(a, a),\ (a, b),\ (a, c),\ (a, d),\ (a, e),\ (b, a),\ (b, b),\ (b, c),\ (b, d),\ (b, e),\ (c, a),\ (c, b),$
          $(c, c),\ (c, d),\ (c, e),\ (d, a),\ (d, b),\ (d, c),\ (d, d),\ (d, e),\ (e, a),\ (e, b),\ (e, c),\ (e, d),\ (e, e)\}$
7. $\{(-2, -2), (1, -2), (1, 0), (1, 1), (0, -2), (0, 0), (4, -2), (4, 0), (4, 1), (4, 4), (6, -2), (6, 0), (6, 1),$
   $(6, 4), (6, 6)\}$
9. $\{(-2, -2), (1, 1), (0, 0), (4, 4), (6, 6)\}$
11. $\{(-2, 0), (-2, 1), (-2, 4), (-2, 6), (0, -2), (0, 1), (0, 4), (0, 6), (1, -2), (1, 0), (1, 4), (1, 6), (4, -2),$
   $(4, 0), (4, 1), (4, 6), (6, -2), (6, 0), (6, 1), (6, 4)\}$
15.

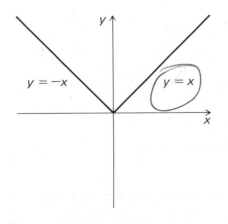

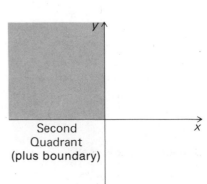

Second
Quadrant
(plus boundary)

17. (a)

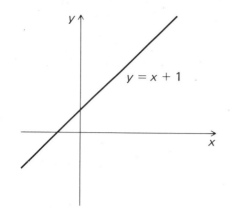

Domain = {x|x is real}

(b)

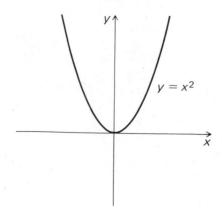

Domain = {x|x is real}

(c)

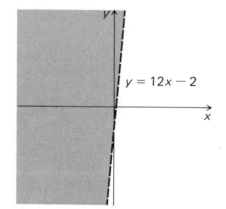

Domain = {x|x is real}

(d)

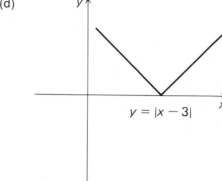

Domain = {x|x is real}

(e)

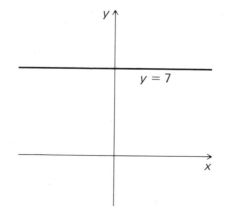

Domain = {x| x is real}

(f)

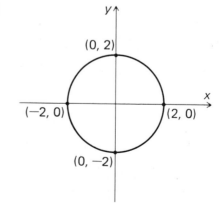

Domain = {x| −2 ≤ x ≤ 2}

(g) Domain = {x| x ≥ 3}

(h) Domain = {x| x is real and x ≠ 3}

19. $A \times B = \{4\} \times \{reals\}$ is the set of ordered pairs with first coordinate equal to 4; that is, the vertical line $x = 4$.

2.3(A) Exercises

1. f is a function   3. h is a function   5. q is a function

7. $g(-2) = 4$; $g(2) = 4$; $g(b) = b^2$; $g(a - b) = (a - b)^2 = a^2 - 2ab + b^2$

9. $p(3) = 0$; $p(-3) = 0$; $p(a) = \sqrt{a^2 - 9}$; $p(x + 1) = \sqrt{(x + 1)^2 - 9}$

11. $s(0) = 0$; $s(1) = 0$; $s(-1) = 2$; $s\left(-\dfrac{1}{3}\right) = \dfrac{2}{3}$

13. $v(0) = 7$; $v(7) = 0$; $v(1) = 3$; $v(t) = \left|\dfrac{t - 7}{t + 1}\right|$

15. $y(0) = 0$; $y(1) = -1$; $y(a) = a^2(a - 2)$; $y(a - b) = (a - b)^2(a - b - 2)$

17. $\dfrac{f(x + h) - f(x)}{h} = \dfrac{[(x + h)^2 - 1] - [x^2 - 1]}{h} = \dfrac{x^2 + 2hx + h^2 - 1 - x^2 + 1}{h} = 2x + h$

19. $\dfrac{f(x + h) - f(x)}{h} = \dfrac{\dfrac{1}{x + h} - \dfrac{1}{x}}{h} = \dfrac{\dfrac{x - (x + h)}{x(x + h)}}{h} = \dfrac{-1}{x(x + h)}$

21. Function of x 23. Function of x

25. Domain $= \{x|\ x$ is real$\}$; range $= \{f(x)|\ f(x) \geq 0\}$

27. Domain $= \{x|\ x$ is real$\}$; range $= \{7\}$

29. Domain $= \{x|\ x^2 - 2x - 15 \geq 0\}$; that is, $\{x|\ x \leq -3$ or $x \geq 5\}$
    Range $= \{q(x)|\ q(x) \geq 0\}$

31. $A = \pi r^2$

2.3(B) Exercises

1. $g(x) = x^3 + 2$

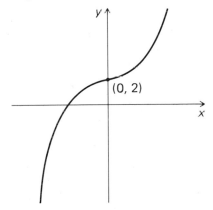

(0, 2)

3. $g(x) = -2x^3$

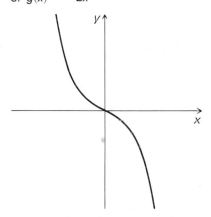

5. $g(x) = (x + 2)^3$

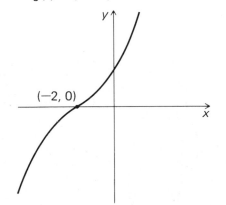

(−2, 0)

7. $g(x) = |x + 3|$

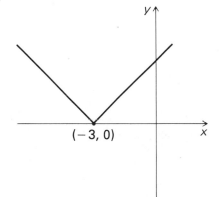

(−3, 0)

9. $g(x) = \dfrac{1}{2}|x| + 1$

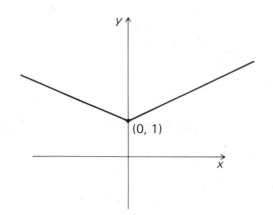

11. $y = -x^2 + 2$

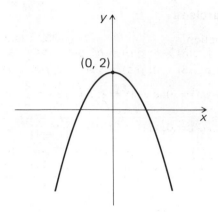

13. $y = 2x^2 - 3$

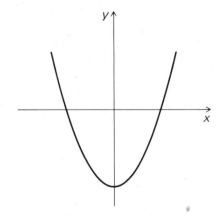

15. $y = -(x - 4)^2 - 2$

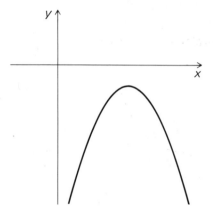

17.

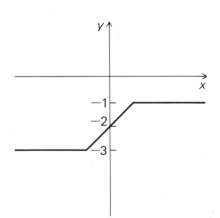

19.

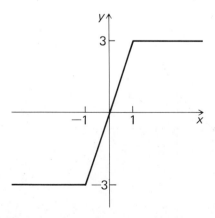

23. $y = \left(x - \dfrac{3}{4}\right)^2 + \dfrac{27}{16}$

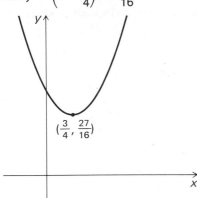

$\left(\dfrac{3}{4}, \dfrac{27}{16}\right)$

25. $y = -3(x + 3)^2 - 2$

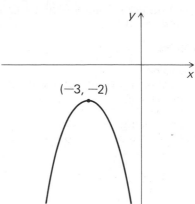

$(-3, -2)$

2.4 Exercises

1. (a) $(f + g)(x) = f(x) + g(x)$
$= 3x^2 + 2x - 4$

   (b) $(f - g)(x) = f(x) - g(x)$
$= -3x^2 + 2x - 2$

   (c) $(fg)(x) = f(x)g(x)$
$= 6x^3 - 9x^2 - 2x + 3$

   (d) $\left(\dfrac{f}{g}\right)(x) = \dfrac{f(x)}{g(x)} = \dfrac{2x - 3}{3x^2 - 1}$

3. Domain $f + g, f - g, fg$ is $\{x \mid x$ is real and $x \geq 0\}$
   Domain $\dfrac{f}{g} = \left\{x \mid x$ is real and $x \geq 0$ and $x \neq -\dfrac{7}{3}\right\}$

5. Domain $f + g, f - g, fg$ is $\{x \mid x$ is real$\}$
   Domain $\dfrac{f}{g}$ is $\{x \mid x$ is real and $x \neq 1\}$

7. Domain $f + g, f - g, fg$ is $\{x \mid x$ is real$\}$
   Domain $\dfrac{f}{g}$ is $\{x \mid x$ is real$\}$

9. Domain $f + g, f - g, fg, \dfrac{f}{g}$ is $\{x \mid x$ is real$\}$

13. False. Let $f(x) = x^2$; then $f(x - 1) = (x - 1)^2$, but $f(x) - f(1) = x^2 - 1$.

15. False. Let $f(x) = x^3$; then $f(3x) = (3x)^3 = 27x^3$, but $3f(x) = 3x^3$.

17. False. Let $f(x) = x$ and $g(x) = 1$; then $(f - g)^2(x) = (f^2 - 2fg + g^2)(x) = [f(x)]^2 - 2f(x)g(x) + [g(x)]^2 = x^2 - 2x + 1$, but $(f^2 - g^2)(x) = [f(x)]^2 - [g(x)]^2 = x^2 - 1$.

19. True   21. True   23. True

2.5(A) Exercises

1. (a) Symmetric with respect to origin
   (b) Symmetric with respect to y-axis
   (c) Symmetric with respect to x-axis, y-axis, and origin
   (d) Symmetric with respect to origin
   (e) Symmetric with respect to x-axis
   (f) No symmetry

3. All are symmetric with respect to y-axis.

5. If n is odd, f_n is symmetric with respect to the origin; if n is even, f_n is symmetric with respect to the y-axis.

2.5(B) Exercises

1. Even function   3. Odd function   5. Odd function
7. Neither even nor odd function   9. Neither even nor odd function
11. (a) If f is odd, then $f(-x) = -f(x)$ and $f(x) + f(-x) = f(x) - f(x) = 0$.
    (b) If f is even, then $f(-x) = f(x)$ and $f(x) - f(-x) = f(x) - f(x) = 0$.
13. (a) and (d) represent even functions, (b) represents odd functions.
15. If $a = 0$, then f is an even function; if $b = 0$, then f is an odd function.

17. $\dfrac{1}{f}$ is an even function provided f is not the zero function.

19. If g is an even function, then $b = 0$; if g is an odd function, then $ax^2 + c = 0$ for all x, which implies $a = 0$ and $c = 0$.

2.5(C)   Exercises

1. Strictly increasing   3. Strictly decreasing
5. (a) Strictly decreasing for $x > 0$; strictly decreasing for $x < 0$
   (b) Strictly increasing
   (c) Neither
   (d) Strictly decreasing
7. No, because $f(x) = c = f(-x)$, but $x \neq -x$ for all $x \neq 0$.
11. (a) Either $x \geq 0$ or $x \leq 0$
   (b) Either $x \geq 0$ or $x \leq 0$
   (c) Either $x > 0$ or $x < 0$
   (d) Either $-1 \leq x \leq 0$ or $0 \leq x \leq 1$

2.6   Exercises

1. (a) $(f \circ g)(x) = f[g(x)] = f\left[\dfrac{4x + 1}{3}\right] = 3\left(\dfrac{4x + 1}{3}\right)^2 - 1$

   (b) $(g \circ f)(x) = g[f(x)] = g[3x^2 - 1] = \dfrac{4(3x^2 - 1) + 1}{3}$

   (c) $(f \circ f)(x) = f[f(x)] = f(3x^2 - 1) = 3(3x^2 - 1)^2 - 1$

   (d) $(g \circ g)(x) = g[g(x)] = g\left[\dfrac{4x + 1}{3}\right] = \dfrac{4\left(\dfrac{4x + 1}{3}\right) + 1}{3}$

3. (a) $(f \circ g)(x) = f[g(x)] = f[x + 2] = (x + 2)^3$
   (b) $(g \circ f)(x) = g[f(x)] = g[x^3] = x^3 + 2$
   (c) Domain $f \circ g = \{x \mid x$ is real$\}$; range $= \{$reals$\}$
   (d) Domain $g \circ f = \{x \mid x$ is real$\}$; range $= \{$reals$\}$
5. (a) $f(x) = x^2 - 3$ and $g(x) = x + 1$

   (b) $f(x) = \sqrt{x^3} + \dfrac{1}{\sqrt[3]{x}}$ and $g(x) = x - 1$

   (c) $f(x) = |x| - \dfrac{1}{|2x|^2}$ and $g(x) = 3x$

7. (a) Even function   (b) Even function   (c) Even function   (d) Odd function
9. (a) Let $x =$ number of gallons; then the number of quarts is defined by $f(x) = 4x$.
   (b) Let $y =$ number of quarts; then the number of pints is defined by $g(y) = 2y$.
   (c) The function which converts gallons to pints is defined by $(g \circ f)(x) = g[f(x)] = g[4x] = 8x$.
11. $(f \circ g)(x) = f[g(x)] = f[\sqrt[3]{x - 1}] = (\sqrt[3]{x - 1})^3 + 1 = x$
   $(g \circ f)(x) = g[f(x)] = g[x^3 + 1] = \sqrt[3]{x^3 + 1 - 1} = x$
   Domain of $f \circ g$ and $g \circ f = \{x \mid x$ is real$\}$
13. $(f \circ g)(x) = f[g(x)] = f[\sqrt{x}] = \sqrt{\dfrac{\sqrt{x} + 1}{\sqrt{x} - 1}}$

   $(g \circ f)(x) = g[f(x)] = g\left[\sqrt{\dfrac{x + 1}{x - 1}}\right] = \sqrt{\sqrt{\dfrac{x + 1}{x - 1}}}$

   Domain $f \circ g$ and $g \circ f = \{x \mid x$ is real and $x > 1\}$
15. $(h \circ (f \circ g))(x) = h\{f[g(x)]\} = (h \circ f)(g(x)) = ((h \circ f) \circ g)(x)$

2.7   Exercises

1. $f^{-1}(x) = x$ 3. $f^{-1}(x) = \sqrt[5]{x}$

5. No inverse, not $1 - 1$; $f(2) = f(3) = 5$, but $2 \neq 3$

7. For $x > 1$, $f^{-1}(x) = -\dfrac{(1 + x^2)}{(1 - x^2)} = \dfrac{x^2 + 1}{x^2 - 1}$

9. $f^{-1}(x) = \dfrac{1 - x}{x}$ 11. $f^{-1}(x) = \dfrac{x - b}{m}$

13. $f^{-1}(x) = \dfrac{x + 5}{3}$ 15. No, not $1 - 1$

17. Either $x \geq 0$ or $x \leq 0$

19. If f is an even function, then f is symmetric with respect to the y-axis and, thus, f is not $1 - 1$. Hence, f has no inverse.

Chapter 2   Exercises

1. (a) $A \times B = \{(1, -1), (\sqrt{2}, -1), (-3, -1), (1, -5), (\sqrt{2}, -5), (-3, -5)\}$

(b) $B \times A = \{(-1, 1), (-1, \sqrt{2}), (-1, -3), (-5, 1), (-5, \sqrt{2}), (-5, -3)\}$

(c) $B \times C = \left\{\left(-1, \dfrac{1}{2}\right), (-1, 0), (-1, 4), (-1, -7), \left(-5, \dfrac{1}{2}\right), (-5, 0), (-5, 4), (-5, -7)\right\}$

(d) $C \times B = \left\{\left(\dfrac{1}{2}, -1\right), (0, -1), (4, -1), (-7, -1), \left(\dfrac{1}{2}, -5\right), (0, -5), (4, -5), (-7, -5)\right\}$

(e) $C \times A = \left\{(0, 1), \left(\dfrac{1}{2}, 1\right), (4, 1), (-7, 1), \left(\dfrac{1}{2}, \sqrt{2}\right), (0, \sqrt{2}), (4, \sqrt{2}), (-7, \sqrt{2}), \left(\dfrac{1}{2}, -3\right), (0, -3), (4, -3), (-7, -3)\right\}$

(f) $A \times A = \{(1, 1), (1, \sqrt{2}), (1, -3), (\sqrt{2}, 1), (\sqrt{2}, \sqrt{2}), (\sqrt{2}, -3), (-3, 1), (-3, \sqrt{2}), (-3, -3)\}$

(g) $B \times B = \{(-1, -1), (-1, -5), (-5, -1), (-5, -5)\}$

(h) $C \times C = \left\{\left(\dfrac{1}{2}, \dfrac{1}{2}\right), \left(\dfrac{1}{2}, 0\right), \left(\dfrac{1}{2}, 4\right), \left(\dfrac{1}{2}, -7\right), \left(0, \dfrac{1}{2}\right), (0, 0), (0, 4), (0, -7), \left(4, \dfrac{1}{2}\right), (4, 0), (4, 4), (4, -7), \left(-7, \dfrac{1}{2}\right), (-7, 0), (-7, 4), (-7, -7)\right\}$

3. (a) and (c) represent functions.

5. (a) $g(0) = -2$ (b) $g(-3) = 22$

(c) $g(a + b) = 3(a + b)^2 + 3(a + b) - 2$

(d) $\dfrac{g(x + h) - g(x)}{h} = \dfrac{[3(x + h)^2 + 3(x + h) - 2] - [3x^2 + 3x - 2]}{h} = 6x + 3h + 3$

7. (a) Symmetric with respect to y-axis

(b) Symmetric with respect to x-axis, y-axis, and origin

(c) Symmetric with respect to origin

(d) No symmetry

9. (a) Even function   (b) Neither

(c) Neither   (d) Neither

11. Area remains the same.

13. (a) $(f \circ g)(x) = f[g(x)] = 2(4x^2 - 2x + 3) - 1 = 8x^2 - 4x + 5$

(b) $(g \circ f)(x) = g[f(x)] = 4(2x - 1)^2 - 2(2x - 1) + 3 = 16x^2 - 20x + 9$

(c) $(f \circ f)(x) = f[f(x)] = 2(2x - 1) - 1 = 4x - 3$

(d) $(g \circ g)(x) = g[g(x)] = 4(4x^2 - 2x + 3)^2 - 2(4x^2 - 2x + 3) + 3$

17. $g^{-1}(x) = \dfrac{x^3 - 2}{5}$

19. $a = -1$

CHAPTER 3

3.1   Exercises

1. (a) $a_1 = 2, a_0 = 7$ (b) $a_0 = -5$

(c) $a_2 = -3, a_1 = 1, a_0 = 0$ (d) $a_3 = 7, a_2 = 4, a_1 = 0, a_0 = \sqrt{2}$

(e) $a_1 = m, a_0 = b$ (f) $a_3 = -1, a_2 = 0, a_1 = 0, a_0 = 5$

(g) $a_4 = \sqrt{3}, a_3 = 0, a_2 = 0, a_1 = 7, a_0 = -5$ (h) $a_{10} = 1, a_9 = a_8 = \ldots = a_1 = a_0 = 0$

(i) $a_2 = a, a_1 = b, a_0 = c$ (j) $a_4 = 17, a_3 = -12, a_2 = 5, a_1 = 3, a_0 = -1$

3. (a) $p(0) = -1, p(-1) = -5, p(1) = -1$

   (b) $p(3) = 3\sqrt{2} - 3, p(\sqrt{2}) = 2 - \sqrt{2}, p(a) = a\sqrt{2} - a$

   (c) $p(0) = -6, p(3) = 0, p(-2) = 0$

   (d) $p(0) = b, p\left(-\dfrac{b}{m}\right) = 0, p(x + h) = mx + mh + b$

   (e) $p(1) = a + b + c, p\left(\dfrac{1}{2}\right) = \dfrac{1}{4}a + \dfrac{1}{2}b + c, p(x + h) = a(x + h)^2 + b(x + h) + c$

5. (a) $r(x) = -x^2 + 8x + 12$ (b) $r(x) = 1$

   (c) $r(x) = x^4 - 3x^3 + 7x^2 - 5x - 1$

7. (a) $q(x) = x^2 - \dfrac{5}{4}x + \dfrac{7}{4}; r(x) = x + 3$ (b) $q(x) = 4x^2 - 4x - 3; r(x) = -5$

   (c) $q(x) = 3x^2 - 4x + 13; r(x) = 0$ (d) $q(x) = x^2 - bx + b^2; r(x) = 0$

   (e) $q(x) = 4x^6 + 2x^4 + x^2 + \dfrac{1}{2}; r(x) = -\dfrac{1}{2}$

9. $c = -5$ or 1

11. $k = 3$

13. (a) $q(x) = x^3 - 3x^2 + 4$; rem $= 0$

   (b) $q(x) = 6x^2 - 9x + 3$; rem $= 0$

   (c) $q(x) = +4x^3 - 12x^2 + 19x - 57$; rem $= 169$

   (d) $q(x) = x^4 + \sqrt[5]{2}x^3 + \sqrt[5]{4}x^2 + \sqrt[5]{8}x + \sqrt[5]{16}$; rem $= 0$

   (e) $q(x) = x^5 - ax^4 + a^2x^3 - a^3x^2 + a^4x - a^5$; rem $= 2a^6$

15. (a) 2 is a root.   (b) -5 is a root.

   (c) $\sqrt{6}$ is a root.   (d) Neither is a root.

   (e) Both are roots.

3.2 Exercises

1. (a) $y = x + 3$

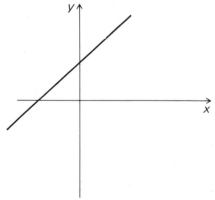

                                        (b) $f(x) = -x$

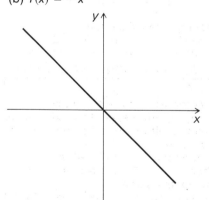

(c) $y = 7$

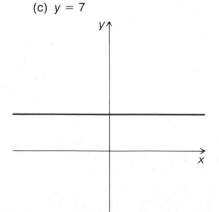

                                        (d) $3y = 2x + 4$

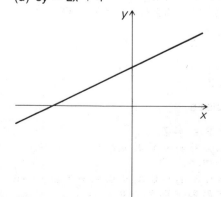

(e) $4x - y + 5 = 0$

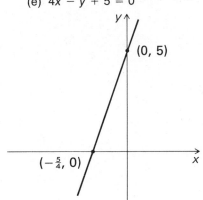

(0, 5)

$(-\frac{5}{4}, 0)$

(f) $\frac{2}{3}x + \frac{4}{9}y = \frac{1}{9}$

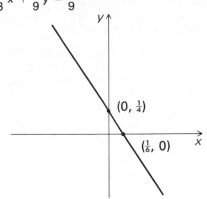

$(0, \frac{1}{4})$

$(\frac{1}{6}, 0)$

3. (a), (b), (c), and (f) slant downward from left to right; (d) slants upward from left to right; (e) is horizontal.

5. (a) $x = 0$　　　(b) $x = -1$　　　(c) $x = 4$

　(d) $x = \pi$　　　(e) $x = a$　　　(f) $x = \frac{2}{3}$

7. (a) slope = 4, y-intercept = -5　　　　(b) slope = 3, y-intercept = -2
　(c) slope = -1, y-intercept = 0　　　　(d) slope = 2, y-intercept = 5

　(e) slope = $-\frac{8}{15}$, y-intercept = $\frac{4}{5}$　(f) slope = 0, y-intercept = 3

　(g) slope = $-\frac{a}{b}$, y-intercept = 0　(h) slope = $-\frac{P}{Q}$, y-intercept = $-\frac{R}{Q}$

9. $y = \frac{4}{3}x + 4$ and $y = \frac{a}{b}x + \frac{c}{b}$

11. $y + 3 = -\frac{3}{4}(x - 1)$　　　13. $k = \frac{1}{3}$

15. (a) Collinear　　　(b) Not collinear

17. If $f(x) = ax + b$ and $f(0) = 0$, then $0 = a(0) + b$ or $b = 0$. Thus, $f(x) = ax$, and $f(x + y) = a(x + y) = ax + ay = f(x) + f(y)$.

19. (a) Distance = 5　　　(b) Distance = $\frac{57}{10}$

3.3 Exercises

1. (a) Consistent and independent　　　　(b) Inconsistent and independent
　(c) Consistent and dependent　　　　　(d) Consistent and independent
　(e) Consistent and independent　　　　(f) Inconsistent and independent

3. (a) $x = 1, y = 0$　　　(b) $x = \frac{ac - bd}{a^2 - b^2}, y = \frac{bc - ad}{b^2 - a^2}$

7. $x = \frac{1}{2}, y = -\frac{1}{3}, z = \frac{1}{6}$

9. Number = 817

3.4 Exercises

1. (a) det = 7　　　(b) det = -2　　　(c) det = 2.07　　　(d) det = 0

3. (a) Consistent and independent   (b) Consistent and dependent
   (c) Inconsistent and independent
7. (a) det $= -12$ (b) det $= 0$ (c) det $= 60$
   (d) det $= 825$ (e) det $= 1$
9. $x = 2$
11. (a) $6x + 7y + 17 = 0$ (b) $3x + 5y = 0$
    (c) $x - 5 = 0$ (d) $y - 2 = 0$
13. (a) Area $= 24.5$ (b) Area $= 30$ (c) Area $= 7\sqrt{2} - 2$

15. Volume $= \left| -\dfrac{1}{3} \right| = \dfrac{1}{3}$ cubic unit

3.5 Exercises

1. $h = \dfrac{-b}{2a}$ and $k = \dfrac{4ac - b^2}{4a}$

   (a) $h = \dfrac{6}{2} = 3, k = \dfrac{-48}{4} = -12; p(x) = (x - 3)^2 - 12$

   (b) $h = \dfrac{4}{-4} = -1, k = \dfrac{-24}{-8} = 3; p(x) = -2(x + 1)^2 + 3$

   (c) $h = \dfrac{-5}{6}, k = \dfrac{-73}{12}; p(x) = 3\left(x + \dfrac{5}{6}\right)^2 - \dfrac{73}{12}$

   (d) $h = \dfrac{-2}{16} = \dfrac{-1}{8}, k = \dfrac{-100}{32} = \dfrac{-25}{8}; p(x) = 8\left(x + \dfrac{1}{8}\right)^2 - \dfrac{25}{8}$

3. (a) $b^2 - 4ac = 25 + 24 = 49$; two real roots
   (b) $b^2 - 4ac = 25 - 16 = 9$; two real roots
   (c) $b^2 - 4ac = 4 - 8 = -4$; no real roots
   (d) $b^2 - 4ac = 48 - 48 = 0$; one real root

5. (a) Opens downward; maximum at $\left(\dfrac{3}{8}, \dfrac{25}{16}\right)$ (b) Opens upward; minimum at $\left(\dfrac{1}{6}, \dfrac{-25}{12}\right)$

   (c) Opens upward; minimum at $(0, 0)$ (d) Opens downward; maximum at $(1, 4)$

7. $s = 48t - 16t^2$
   (a) $t = 1.5$ sec   (b) 36 ft   (c) $t = 3$ sec
9. 146 boats

3.6 Exercises

1. (a) $\pm 1, \pm \dfrac{1}{3}$ (b) $\pm 1, \pm 3, \pm 9$

   (c) $\pm 1, \pm 2, \pm 3, \pm 4, \pm 6, \pm 12, \pm \dfrac{1}{3}, \pm \dfrac{2}{3}, \pm \dfrac{4}{3}$

   (d) $\pm 1, \pm 2, \pm 3, \pm 4, \pm 6, \pm 8, \pm 12, \pm 24$

   (e) $\pm 1, \pm \dfrac{1}{2}, \pm \dfrac{1}{4}$ (f) $\pm 1, \pm 5, \pm \dfrac{1}{5}$

   (g) $\pm 1, \pm 2, \pm 5, \pm 10$ (h) $\pm 1, \pm 5, \pm \dfrac{1}{3}, \pm \dfrac{5}{3}$

   (i) $\pm 1, \pm 2, \pm \dfrac{1}{3}, \pm \dfrac{2}{3}$

   (j) $\pm 1, \pm 2, \pm 4, \pm 7, \pm 14, \pm 28, \pm \dfrac{1}{3}, \pm \dfrac{2}{3}, \pm \dfrac{4}{3}, \pm \dfrac{7}{3}, \pm \dfrac{14}{3}, \pm \dfrac{28}{3}, \pm \dfrac{1}{9}, \pm \dfrac{2}{9}, \pm \dfrac{4}{9}, \pm \dfrac{7}{9}, \pm \dfrac{14}{9}, \pm \dfrac{28}{9}$

3. (a) $1, 2, 3$ (b) $3, -2, 4$ (c) $\dfrac{1}{2}, 2 \pm \sqrt{2}$

   (d) $0, -3, -4$ (e) $2, -2, 2 \pm \sqrt{3}$

5. (a)

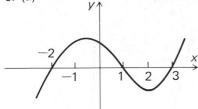

(b)

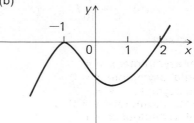

(c)

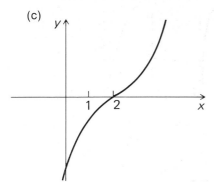

(d)

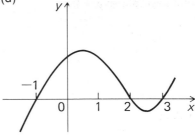

(e)

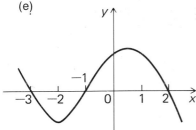

7. (a) $p(x) = x^2 - 7$ (b) $p(x) = x^3 - 4$
   (c) $p(x) = x^4 - 5$ (d) $p(x) = x^5 - 6$

3.7 Exercises

1. (a) $3i$ (b) $5i$ (c) $6i$ (d) $9i$
   (e) $i\sqrt{2}$ (f) $i\sqrt{5}$ (g) $2i\sqrt{3}$ (h) $2i\sqrt{7}$
   (i) $\dfrac{i}{3}$ (j) $\dfrac{4i}{7}$ (k) $\dfrac{i}{2\sqrt{5}}$ (l) $\dfrac{i}{\sqrt{2}}$

3. (a) $-2i$ (b) $1 + 4i$ (c) $\sqrt{2} - 6i$ (d) $-3 + 8i$
   (e) 14 (f) $\sqrt{3}i$ (g) $\dfrac{1}{3} + \dfrac{1}{5}i$ (h) $-4i$

5. (a) $R = 1$, multiplicity 1   (b) $R = 2$, multiplicity 2
       $R = 7$, multiplicity 1 $R = -1$, multiplicity 1
   (c) $R = 3$, multiplicity 1   (d) $R = 0$, multiplicity 1
       $R = 5$, multiplicity 3 $R = -9$, multiplicity 2
   (e) $R = 1$, multiplicity 1 $R = 8$, multiplicity 1
       $R = -1 + i$, multiplicity 1 $R = -1$, multiplicity 3
       $R = -1 - i$, multiplicity 1   (f) $R = \sqrt{2}$, multiplicity 2
                                     $R = 3i$, multiplicity 1
                                     $R = -3i$, multiplicity 1

7. (a) $-1 + 4i$ (b) $-2i$ (c) $-\dfrac{1}{3} + i$ (d) $5\sqrt{2} - 4i$
   (e) $2a$ (f) $2bi$

9. (a) $-2i$ (b) $\dfrac{1}{10} + \dfrac{7}{10}i$ (c) $\dfrac{1}{4} + \dfrac{\sqrt{3}}{4}i$ (d) $\dfrac{1}{4} - \dfrac{3\sqrt{15}}{4}i$

3.8(A) Exercises

1. (a) Vertical asymptote: $x = 4$
   Horizontal asymptote: $y = 0$
   (b) Vertical asymptote: $x = -7$
   Horizontal asymptote: $y = 0$
   (c) Vertical asymptote: $x = 1$
   Horizontal asymptote: $y = 0$
   (d) Vertical asymptotes: $x = -3$, $x = 2$
   Horizontal asymptote: $y = 2$

3. $f(x) = \dfrac{x}{x^2 - 1}$

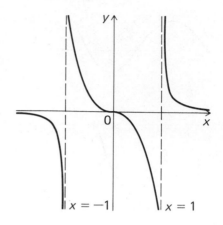

3.8(B) Exercises

1. $\dfrac{1}{x^2 - 4} = \dfrac{1}{4(x - 2)} - \dfrac{1}{4(x + 2)}$

3. $\dfrac{2x^2 + 3}{x(x - 1)^2} = \dfrac{3}{x} - \dfrac{1}{x - 1} + \dfrac{5}{(x - 1)^2}$

5. $\dfrac{1}{x(x^2 + x + 1)} = \dfrac{1}{x} - \dfrac{x + 1}{x^2 + x + 1}$

7. $\dfrac{2x^2 + 1}{(x^2 - x + 1)^2} = \dfrac{2}{x^2 - x + 1} + \dfrac{2x - 1}{(x^2 - x + 1)^2}$

9. $\dfrac{x^3 + 5x^2 + 2x - 4}{x(x^2 + 4)^2} = \dfrac{-1}{4x} + \dfrac{x + 4}{4(x^2 + 4)} + \dfrac{6x - 2}{(x^2 + 4)^2}$

Chapter 3 Exercises

1. $-1\,|\ 1 \quad\ 1 \ \ 1 \quad\ 1 \ \ 1 \quad\ 1 \ \ 1 \qquad 1$
   $\underline{\ \ -1 \ \ 0 \ -1 \ \ 0 \ -1 \ \ 0 \ -1 \ }$
   $\ 1 \quad\ 0 \ \ 1 \quad\ 0 \ \ 1 \quad\ 0 \ \ 1 \ \ \underline{|\,0}$

   Quotient $q(x) = x^6 + x^4 + x^2 + 1$

3. (a) $y - 1 = \dfrac{1}{9}(x - 7)$ (b) $y = -3$ (c) $y = \dfrac{1}{2}x - 4$

   (d) $x = 4$ (e) $y = -\dfrac{3}{5}x$

5. $x_1 = 3$, $x_2 = 0$, $x_3 = -5$

7. (a) $x = \dfrac{3 \pm \sqrt{41}}{4}$ (b) $x = \dfrac{1}{2}$ (double root)   (c) $x = \dfrac{2 \pm i\sqrt{2}}{3}$

9.   The vertex $(-1, -3)$ is a minimum.

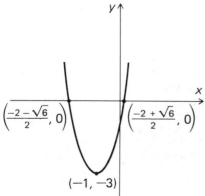

11. $p(x) = (x + 4i)(x - 4i)(x - 7)(x + 5) = (x^2 + 16)(x - 7)(x + 5)$

13. (a) $f(x) = \dfrac{3}{x + 2}$

(b) $f(x) = \dfrac{x}{x^2 + 1}$

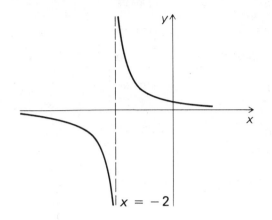

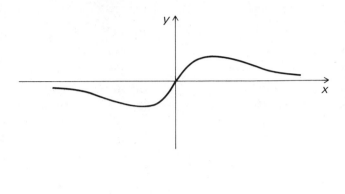

Vertical asymptote: $x = -2$
Horizontal asymptote: $y = 0$

No vertical asymptote
Horizontal asymptote: $y = 0$

(c) $f(x) = \dfrac{2x^2 + 8x + 6}{x + 3}$

(d) $f(x) = \dfrac{2x^2}{x^2 + x - 2}$

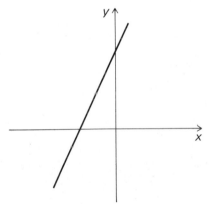

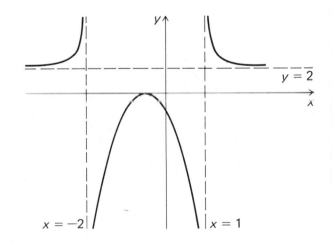

The point $(-3, -4)$ is not on the line.

No asymptote

Vertical asymptotes: $x = -2$ and $x = 1$
Horizontal asymptote: $y = 2$

15. (a) $\dfrac{x^2 + x + 1}{(2x + 1)(x^2 + 1)} = \dfrac{3}{5(2x + 1)} + \dfrac{x + 2}{5(x^2 + 1)}$

(b) $\dfrac{2x^2 + 1}{(x^2 - x + 1)^2} = \dfrac{2}{x^2 - x + 1} + \dfrac{2x - 1}{(x^2 - x + 1)^2}$

CHAPTER 4

4.1 Exercises

1. (a) 49   (b) $\dfrac{125}{16}$ (c) a^5 (d) x^3

   (e) 17^6 (f) $a^5b^5c^{10}$ (g) ab^2 (h) 10^8

   (i) $-\dfrac{8a^3}{27b^6}$ (j) $\dfrac{81y^4}{16x^8}$ (k) $-2401x^4$ (l) $9x^3y^{10}$

   (m) a^{2m} (n) 10^{n+2} (o) y^{3n} (p) am^2

3. (a) 16   (b) -8 (c) 9   (d) 25

   (e) $\dfrac{1}{3}$ (f) $\dfrac{1}{16}$ (g) $a^{5/6}$ (h) $x^{1/3}$

   (i) a^3 (j) $x^{1/3}y^{1/2}$ (k) $\dfrac{a^6}{b^4}$ (l) $a^{5/6}+1$

   (m) $y^{24/5}+y^{-1/5}$ (n) $a^{n+4/2}b^{\frac{-(16n+1)}{2}}$ (o) $a^{1/6}y^{-1/4}$

5. (a) $x=0$ or $x=-1$ (b) $x=2$ or $x=-1$

4.2 Exercises

1. (a) $f(x)=2^{-x}$ (b) $f(x)=3^x$ (c) $f(x)=4^x$

   (d) $f(x)=\left(\dfrac{1}{4}\right)^x$ (e) $f(x)=\left(\dfrac{1}{7}\right)^x$ (f) $f(x)=a^x,\ a>0$

3. (a) and (b)   (c) and (d)

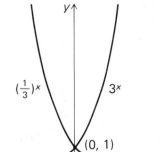 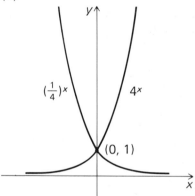

(e) and (f)

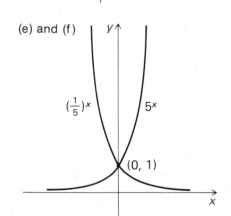

5. (a) $f(x+y)=3^{2(x+y)+1}=3^{2x+2y+1}\ne f(x)\cdot f(y)=3^{2x+1}\cdot 3^{2y+1}=3^{2x+2y+2}$
    $\therefore f$ is NOT an exponential function.
   (b) $f(x+y)=2^{x+y}\cdot 2^{3(x+y)}=2^x\cdot 2^{3x}\cdot 2^y\cdot 2^{3y}=f(x)\cdot f(y)$
    $\therefore f$ is an exponential function.
   (c) $f(x+y)=5^{x+y}-5^{-(x+y)}\ne f(x)\cdot f(y)=(5^x-5^{-x})(5^y-5^{-y})$
    $\therefore f$ is NOT an exponential function.

(d) $f(x + y) = 7^{2(x+y)} = 7^{2x} \cdot 7^{2y} = f(x) \cdot f(y)$
∴ f is an exponential function.

7. (a) No.
(b) No, odd function $\Rightarrow f(-x) = -f(x)$. If $f(x) = a^x$, $a > 0$ is an exponential function, then
$f(-x) = a^{-x} = \dfrac{1}{a^x}$ which does not equal $-f(x) = -a^x$.

9. (a) and (b)   (c)

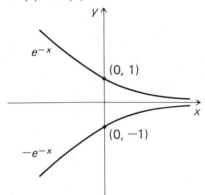

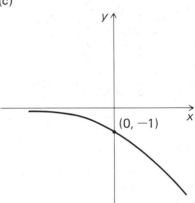

4.3 Exercises

1. (a) $A = 1000 \left(1 + \dfrac{.07}{365}\right)^{(365)(20)}$ (b) $A = 1000 \left(1 + \dfrac{.07}{12}\right)^{(12)(20)}$

(c) $A = 1000 \left(1 + \dfrac{.07}{4}\right)^{(4)(20)}$ (d) $A = 1000 \left(1 + \dfrac{.07}{2}\right)^{(2)(20)}$

(e) $A = 1000(1 + .07)^{(20)}$ (f) $A = 1000e^{(.07)(20)}$

3. $N = (2{,}500) \cdot 5^{13}$ 5. $N = (10) \cdot 2^{-15/25}$

7. $I = 100e^{-(.01)(2)}$ 9. $T_c = 60 + 40 \cdot 10^{-(.01)(10)}$

4.4 Exercises

1. (a) $\log_3 9 = 2$ (b) $\log_5 25 = 2$ (c) $\log_4 64 = 3$ (d) $\log_3 27 = 3$

(e) $\log_{1/2} \left(\dfrac{1}{8}\right) = 3$ (f) $\log_{1/4} \left(\dfrac{1}{16}\right) = 2$ (g) $\log_{27} \left(\dfrac{1}{3}\right) = -\dfrac{1}{3}$ (h) $\log_{32} \left(\dfrac{1}{2}\right) = -\dfrac{1}{5}$

(i) $\log_{10} 1000 = 3$ (j) $\log_{10} (.1) = -1$ (k) $\log_{13} 1 = 0$ (l) $\log_{49} 7 = \dfrac{1}{2}$

(m) $\log_e 1 = 0$ (n) $\log_8 4 = \dfrac{2}{3}$ (o) $\log_{3/5} \left(\dfrac{9}{25}\right) = 2$ (p) $\log_6 6 = x$

3. (a) $x = 32$ (b) $x = 0$ (c) $x = 256$ (d) $x = -2$

(e) $x = \dfrac{1}{64}$ (f) $x = 10$ (g) $x = 100$ (h) $x = \dfrac{1}{10}$

(i) $x = 16$

4.5 Exercises

1. (a) $\dfrac{1}{2} \{\log_b x + \log_b y\}$ (b) $\log_b 3 + 2 \log_b r$

(c) $\dfrac{1}{7} \{\log_b a + 2 \log_b c\}$ (d) $2 \log_b r + \log_b s - 3 \log_b t$

(e) $-\dfrac{4}{5} \log_b u$ (f) $\log_b 2 + \log_b \pi + \dfrac{1}{2} \log_b l - \dfrac{1}{2} \log_b g$

(g) $\dfrac{1}{3} \{2 \log_b (x + y) + 2 \log_b (x - y)\}$ (h) $\dfrac{1}{4} \log_b x + \dfrac{1}{3} \log_b y$

(i) $\dfrac{1}{6} \{\dfrac{1}{3} \log_b x + 5 \log_b z - 2 \log_b y\}$

3. (a) $x = 25$ (b) $x = 5$ (c) $x = 12$
5. (a) $x = \log_2 5$ (b) $x = (\log_5 13) - 1$ (c) $x = -\log_4 10$
   (d) $x = \pm\sqrt{\log_3 7}$ (e) $x = 1 - \log_5 15$ (f) $x = (\log_e 4) + 2$

4.6 Exercises

1. (a) 2 (b) 1 (c) 0 (d) -1
   (e) 4 (f) 0 (g) -2
3. (a) 9.58 (b) 70.6 (c) .603
   (d) .0506 (e) 3090 (f) 1.01
5. (a) $N = 3544$ (b) $N = .8026$ (c) $N = 6.934$
7. (a) $N = 92.0$ (b) $N = .0302$

9. (a) $\log_9 10 = \dfrac{1}{\log_{10} 9} = \dfrac{1}{0.9542} = 1.0480$

   (b) $\log_4 5 = \dfrac{1}{\log_5 4} = \dfrac{1}{\log_5 2^2} = \dfrac{1}{2} \cdot \dfrac{1}{\log_5 2} = \dfrac{1}{2}\log_2 5$

4.7 Exercises

1. (a) 0.2777 (b) 20490 (c) .002402
3. $T = 2.074$ sec 5. $k = 0.0277$

Chapter 4 Exercises

1. (a) $\log_3 81 = 4$ (b) $\log_b 1 = 0$ (c) $\log_8 4 = \dfrac{2}{3}$

   (d) $\log_a 16 = x$ (e) $\log_4\left(\dfrac{1}{16}\right) = -2$ (f) $\log_5 x = y$

3. (a) $3^{-x} = 3^3$ (b) $x = \dfrac{\log 7}{\log 10}$ (c) $x = \dfrac{\log 4}{\log 1.52}$
   $x = -3$

5. (a) $\log\left(\dfrac{x^2 z}{y^3}\right)$ (b) $\log\sqrt{\dfrac{(x+y)}{(x-y)}}$

7. $x = 4 + \dfrac{1}{\log 2}$ 9. 145.5

11. (a) $N = 8.27$ (b) $N = 3000$

13. $t = \dfrac{\log 3}{\log 1.03} \approx 37$ years

CHAPTER 5

5.1 Exercises

1. (a) $P(3\pi) = (-1, 0)$ (b) $P\left(\dfrac{-5\pi}{2}\right) = (0, -1)$ (c) $P(14\pi) = (1, 0)$

3. (a)

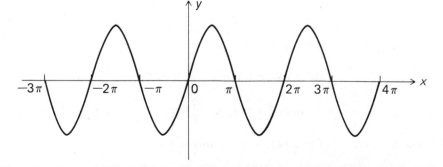

(b)

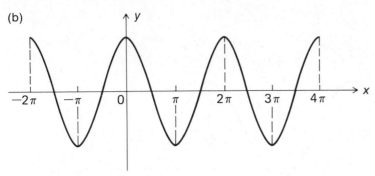

5. All are periodic functions.

5.2 Exercises

1. (a) $\theta = \dfrac{\pi}{3}, \dfrac{5\pi}{3}$ (b) $\theta = \dfrac{\pi}{2}, \dfrac{3\pi}{2}$ (c) $\theta = \dfrac{\pi}{6}, \dfrac{11\pi}{6}$

    (d) $\theta = \dfrac{3\pi}{4}, \dfrac{5\pi}{4}$ (e) $\theta = \dfrac{\pi}{6}, \dfrac{5\pi}{6}$ (f) $\theta = \dfrac{\pi}{2}$

    (q) $\theta = \dfrac{4\pi}{3}, \dfrac{5\pi}{3}$ (h) $\theta = \dfrac{\pi}{4}, \dfrac{3\pi}{4}$

3. (a) $\cos \theta = \dfrac{4}{5}$ (b) $\sin \theta = -\dfrac{\sqrt{3}}{2}$

5. (a) $\dfrac{\sqrt{2}}{4}$ (b) $\dfrac{\sqrt{6}}{4}$ (c) $\dfrac{1}{4}$ (d) $\dfrac{3}{4}$

    (e) 3   (f) 1

7. Odd function   9. No inverses, not $1 - 1$

5.3 Exercises

1. (a) Amplitude = 1
     Period = π
     Phase shift = 0

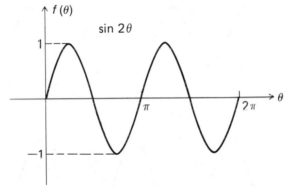

  (b) Amplitude = 1
     Period = π

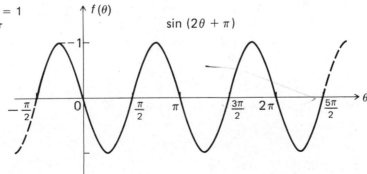

Phase shift = $\dfrac{\pi}{2}$ to left

(c) Same as in (b) except amplitude $= \dfrac{3}{4}$.

(d) The graph of (c) shifted up one unit.

3. (a) $f(\theta) = \cos\left(\dfrac{\theta}{2} + \pi\right)$

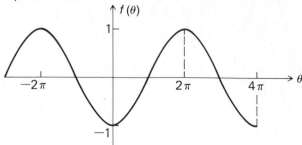

(b) $f(\theta) = -3\sin\left(2\theta - \dfrac{\pi}{2}\right)$

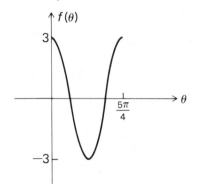

(c) $f(\theta) = \left(\dfrac{1}{3}\right)\sin\left(3\theta + \dfrac{\pi}{2}\right)$

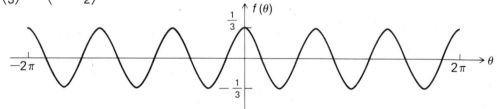

(d) $f(\theta) = 2\cos(2\theta - \pi)$

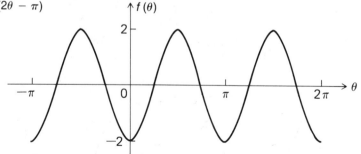

5. Amplitude $= 10$, period $= \dfrac{1}{30}$, phase shift $= 0$, frequency $= 30$

5.4 Exercises

1. (a) 0   (b) $\dfrac{\pi}{3}$ (c) $\dfrac{\pi}{6}$ (d) $-\dfrac{\pi}{2}$

   (e) $\dfrac{5\pi}{6}$ (f) $\dfrac{\pi}{3}$ (g) π (h) $\dfrac{\pi}{4}$

5. $\sin^{-1}$ is symmetric with respect to origin.

7. (a) True   (b) False   (c) True   (d) True

9. (a) $\theta = \dfrac{\pi}{2} + \sin^{-1}(x + 1)$ (b) $\theta = \dfrac{\pi}{2} - \cos^{-1} x$

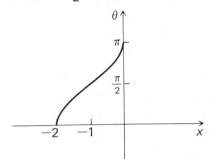 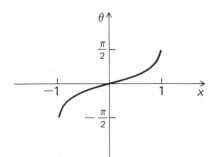

5.5 Exercises

1.

θ	0	$\dfrac{\pi}{6}$	$\dfrac{\pi}{4}$	$\dfrac{\pi}{3}$	$\dfrac{\pi}{2}$	$\dfrac{-\pi}{6}$	$\dfrac{-\pi}{4}$	$\dfrac{-\pi}{3}$
$\tan \theta$	0	$\dfrac{\sqrt{3}}{3}$	1	$\sqrt{3}$	undef.	$\dfrac{-\sqrt{3}}{3}$	-1	$-\sqrt{3}$

3. (a) $y = \tan 2\theta$ (b) $y = 2 \tan \left(\theta + \dfrac{\pi}{2} \right)$

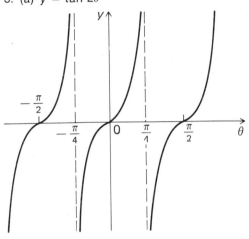 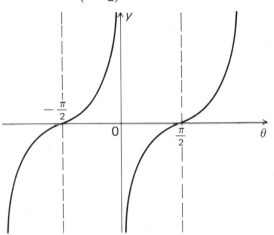

5. (a) $y = \cot 2\theta$ (b) $y = 2 \cot \theta$

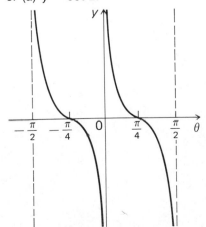 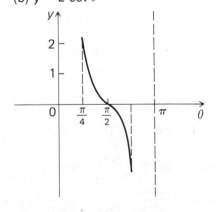

7. (a) $y = \sec\left(\theta + \dfrac{\pi}{2}\right)$

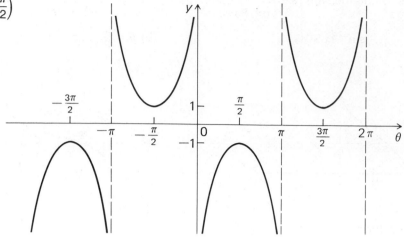

(b) $y = 1 + \csc \theta$

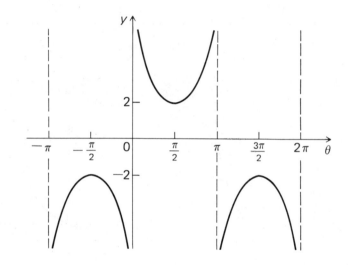

9. $y = \tan^{-1}(-x) \Leftrightarrow \tan y = -x$ and $\dfrac{-\pi}{2} < y < \dfrac{\pi}{2}$

$\Leftrightarrow -\tan y = x$ and $\dfrac{-\pi}{2} < y < \dfrac{\pi}{2}$

$\Leftrightarrow \tan(-y) = x$ and $\dfrac{-\pi}{2} < y < \dfrac{\pi}{2}$

$\Leftrightarrow -y = \tan^{-1}(x)$ and $\dfrac{-\pi}{2} < y < \dfrac{\pi}{2}$

$\Leftrightarrow y = -\tan^{-1}(x)$ and $\dfrac{-\pi}{2} < y < \dfrac{\pi}{2}$

Hence $\tan^{-1}(-x) = -\tan^{-1} x$

Chapter 5 Exercises

1. $P\left(-\dfrac{3\pi}{4}\right) = \left(-\dfrac{\sqrt{2}}{2}, -\dfrac{\sqrt{2}}{2}\right)$

3. (a) $\sin \theta \cot \theta = \sin \theta \cdot \dfrac{\cos \theta}{\sin \theta} = \cos \theta$

(b) $\cot \theta \sec \theta = \dfrac{\cos \theta}{\sin \theta} \cdot \dfrac{1}{\cos \theta} = \dfrac{1}{\sin \theta} = \csc \theta$

5. (a) $1 + \tan^2 \theta = 1 + \dfrac{\sin^2 \theta}{\cos^2 \theta} = \dfrac{\cos^2 \theta + \sin^2 \theta}{\cos^2 \theta} = \dfrac{1}{\cos^2 \theta} = \sec^2 \theta$

(b) $1 + \cot^2 \theta = 1 + \dfrac{\cos^2 \theta}{\sin^2 \theta} = \dfrac{\sin^2 \theta + \cos^2 \theta}{\sin^2 \theta} = \dfrac{1}{\sin^2 \theta} = \csc^2 \theta$

7. (a) Odd function   (b) Odd function

9. x

CHAPTER 6

6.1 Exercises

1. (a) 60°   (b) 150°   (c) 216°   (d) 570°
   (e) 18°   (f) 0°   (g) 40°   (h) 300°

3. (a) $\dfrac{7\pi}{6}$ inches   (b) 5π inches   (c) $\dfrac{5\pi^2}{180}$ inch   (d) $\dfrac{3}{4}$ inch

5. 100π inches/minute

6.2 Exercises

1. (a) 1, 0, undefined   (b) $\dfrac{-2}{\sqrt{5}}, \dfrac{-1}{\sqrt{5}}, 2$ (c) $\dfrac{-\sqrt{5}}{\sqrt{14}}, \dfrac{3}{\sqrt{14}}, \dfrac{-\sqrt{5}}{3}$

   (d) 1, 0, undefined   (e) $\dfrac{12}{13}, \dfrac{-5}{13}, \dfrac{-12}{5}$ (f) $\dfrac{-3}{\sqrt{13}}, \dfrac{-2}{\sqrt{13}}, \dfrac{3}{2}$

3. $\sin 60° = \dfrac{\sqrt{3}}{2}$, $\cos 60° = \dfrac{1}{2}$, $\tan 60° = \sqrt{3}$, $\cot 60° = \dfrac{1}{\sqrt{3}}$, $\sec 60° = 2$, $\csc 60° = \dfrac{2}{\sqrt{3}}$

5. Since θ is in standard position, $\sin \theta = \dfrac{y}{r}$ and $\cos \theta = \dfrac{x}{r}$, which gives $y = r \sin \theta$ and $x = r \cos \theta$.

6.3(A) Exercises

1. (a) $\sin^2 \theta - \cos^2 \theta$ (b) $\cos \theta$ (c) 1   (d) $\sec^2 \theta$

3. $\dfrac{1 + \cos \theta}{\sin \theta} = \dfrac{(1 - \cos \theta)(1 + \cos \theta)}{(1 - \cos \theta)\sin \theta} = \dfrac{1 - \cos^2 \theta}{(1 - \cos \theta)\sin \theta} = \dfrac{\sin^2 \theta}{(1 - \cos \theta)\sin \theta}$

$= \dfrac{\sin \theta}{1 - \cos \theta}$

5. $\sec^2 \theta + \csc^2 \theta = \dfrac{1}{\cos^2 \theta} + \dfrac{1}{\sin^2 \theta} = \dfrac{\sin^2 \theta + \cos^2 \theta}{(\sin^2 \theta)(\cos^2 \theta)} = \dfrac{1}{(\sin^2 \theta)(\cos^2 \theta)}$

$= \dfrac{1}{\sin^2 \theta} \cdot \dfrac{1}{\cos^2 \theta} = \sec^2 \theta \csc^2 \theta$

7. $\tan \theta + \cot \theta = \dfrac{\sin \theta}{\cos \theta} + \dfrac{\cos \theta}{\sin \theta} = \dfrac{\sin^2 \theta + \cos^2 \theta}{\cos \theta \sin \theta} = \dfrac{1}{\cos \theta \sin \theta} = \dfrac{1}{\cos \theta} \cdot \dfrac{1}{\sin \theta}$

$= \sec \theta \csc \theta$

9. $(a \cos \theta + b \sin \theta)^2 + (-a \sin \theta + b \cos \theta)^2 = a^2 \cos^2 \theta + 2ab \cos \theta \sin \theta + b^2 \sin^2 \theta$
$+ a^2 \sin^2 \theta - 2ab \sin \theta \cos \theta + b^2 \cos^2 \theta$
$= a^2 \cos^2 \theta + a^2 \sin^2 \theta + b^2 \sin^2 \theta + b^2 \cos^2 \theta$
$= a^2(\sin^2 \theta + \cos^2 \theta) + b^2(\sin^2 \theta + \cos^2 \theta)$
$= a^2 + b^2$

11. $\dfrac{1 - \cos \theta}{\sin \theta} = \dfrac{(1 - \cos \theta)}{\sin \theta} \cdot \dfrac{(1 + \cos \theta)}{(1 + \cos \theta)}$

$= \dfrac{1 - \cos \theta + \cos \theta - \cos^2 \theta}{\sin \theta(1 + \cos \theta)}$

$= \dfrac{1 - \cos^2 \theta}{\sin \theta(1 + \cos \theta)}$

$= \dfrac{\sin^2 \theta}{\sin \theta(1 + \cos \theta)}$

$= \dfrac{\sin \theta}{1 + \cos \theta}$

13. $(\csc\theta - \cot\theta)^2 = \csc^2\theta - 2\csc\theta\cot\theta + \cot^2\theta$

$$= \frac{1}{\sin^2\theta} - \frac{2}{\sin\theta}\cdot\frac{\cos\theta}{\sin\theta} + \frac{\cos^2\theta}{\sin^2\theta} = \frac{1 - 2\cos\theta + \cos^2\theta}{\sin^2\theta}$$

$$= \frac{(1 - \cos\theta)^2}{(1 - \cos^2\theta)} = \frac{(1 - \cos\theta)(1 - \cos\theta)}{(1 - \cos\theta)(1 + \cos\theta)} = \frac{1 - \cos\theta}{1 + \cos\theta}$$

15. $\dfrac{\tan^3\theta - \cot^3\theta}{\tan\theta - \cot\theta} = \dfrac{(\tan\theta - \cot\theta)(\tan^2\theta + \tan\theta\cot\theta + \cot^2\theta)}{(\tan\theta - \cot\theta)}$

$$= \tan^2\theta + 1 + \cot^2\theta = \tan^2\theta + 1 + \cot^2\theta$$

$$= \tan^2\theta + \csc^2\theta$$

6.3(B) Exercises

1. $\sin\theta$ 3. $-\sin\theta$ 5. $\cos\theta$

7. $\dfrac{1}{4}(\sqrt{6} - \sqrt{2})$ 9. $2 + \sqrt{3}$ 11. $\sqrt{\dfrac{2 - \sqrt{3}}{2 + \sqrt{3}}} = 2 - \sqrt{3}$

13. $\cot(\alpha + \beta) = \dfrac{\cot\alpha\cot\beta - 1}{\cot\alpha + \cot\beta}$

15. (a) $\sin 3\theta = 3\sin\theta - 4\sin^3\theta$

     (b) $\tan 3\theta = \dfrac{3\tan\theta - \tan^3\theta}{1 - 3\tan^2\theta}$

     (c) $\cot 3\theta = \dfrac{\cot^3\theta - 3\cot\theta}{3\cot^2\theta - 1}$

17. (a) $\dfrac{1}{2}(\sqrt{2 - \sqrt{2}})$ (b) $\sqrt{2} - 1$ (c) $\sqrt{2} + 1$ (d) 1

     (e) $2 - \sqrt{3}$ (f) $\dfrac{-\sqrt{2 - \sqrt{2}}}{2}$

19. $\left[\sin\left(\dfrac{\theta}{2}\right) + \cos\left(\dfrac{\theta}{2}\right)\right]^2 - \sin\theta$

$$= \sin^2\left(\frac{\theta}{2}\right) + 2\sin\left(\frac{\theta}{2}\right)\cos\left(\frac{\theta}{2}\right) + \cos^2\left(\frac{\theta}{2}\right) - \sin\theta$$

$$= 1 + 2\sin\left(\frac{\theta}{2}\right)\cos\left(\frac{\theta}{2}\right) - \sin\theta$$

$$= 1 + \sin\theta - \sin\theta = 1$$

22. $\dfrac{\sin\alpha - \sin\beta}{\cos\alpha + \cos\beta} = \dfrac{2\cos\left[\frac{1}{2}(\alpha + \beta)\right]\sin\left[\frac{1}{2}(\alpha - \beta)\right]}{2\cos\left[\frac{1}{2}(\alpha + \beta)\right]\cos\left[\frac{1}{2}(\alpha - \beta)\right]} = \dfrac{\sin\left[\frac{1}{2}(\alpha - \beta)\right]}{\cos\left[\frac{1}{2}(\alpha - \beta)\right]} = \tan\left[\dfrac{1}{2}(\alpha - \beta)\right]$

6.4 Exercises

1. $\{2k\pi, k = \text{any integer}\}$ 3. $\left\{\dfrac{\pi}{4} + 2k\pi\right\} \cup \left\{\dfrac{3\pi}{4} + 2k\pi\right\}$ 5. $\left\{\dfrac{\pi}{3} + 2k\pi\right\}$

7. $\left\{\dfrac{\pi}{6} + \dfrac{2k\pi}{3}\right\}$ 9. $\left\{\dfrac{\pi}{12} + \dfrac{k\pi}{3}\right\}$

11. $\{k\pi\}$ 13. $\left\{\pm\dfrac{\pi}{3} + 2k\pi\right\}$

15. $\left\{\dfrac{\pi}{6} + 2k\pi\right\} \cup \left\{\dfrac{5\pi}{6} + 2k\pi\right\} \cup \left\{\dfrac{3\pi}{2} + 2k\pi\right\}$ 17. $\{2k\pi\} \cup \left\{\dfrac{\pi}{2} + 2k\pi\right\}$

19. $\{2k\pi\} \cup \left\{\pm\dfrac{2\pi}{3} + 2k\pi\right\}$ 21. $\varnothing$

23. $\left\{\dfrac{k\pi}{2}\right\}$

25. $\left\{\pm\dfrac{\pi}{6} + k\pi\right\}$

6.5 Exercises

1. (a) $\beta = 30°, a = 10\sqrt{3}, c = 20$ (b) $\beta = 59°, b \approx 121, c \approx 142$
   (c) $a = 13, c = 19.9, \beta = 49°$ (d) $c = 13, \alpha \approx 23°, \beta \approx 67°$
   (e) $a \approx 4.9, \alpha \approx 44.5°, \beta = 45.5°$ (f) $\alpha \approx 56°, \beta \approx 34°, b = 23$
   (g) $\alpha = 47°, a \approx 7.8, c \approx 11$ (h) $\alpha = 41°, c \approx 18.3, b \approx 18.3$
3. Approximately 46 feet
5. (a) $\beta = 54°, a = 7.3, c = 5.3$ (b) $\gamma = 70°, b = 4.52, c = 4.91$
   (c) $\gamma = 78°, b = 726, c = 752$ (d) $\alpha = 44.4°, \beta = 78.5°, \gamma = 57.1°$
7. (a) $\beta = 134°10', \gamma = 10°50', c = 2.62$
   (b) $\beta = 45°50', \gamma = 99°10', c = 13.8$
9. Largest $= 100°57'$, smallest $= 24°9'$
11. (a) 90   (b) 148   (c) 1188   (d) $60\sqrt{6}$
13. 109 feet   15. 65 million miles

Chapter 6 Exercises

I. (a) $\dfrac{7\pi}{12}$ (b) $\dfrac{3\pi}{4}$ (c) $\dfrac{11\pi}{6}$ (d) $\dfrac{2\pi}{3}$

   (e) $\dfrac{7\pi}{6}$ (f) $-\dfrac{7\pi}{4}$ (g) $\dfrac{5\pi}{3}$ (h) $\dfrac{\pi}{5}$

   (i) $\dfrac{3\pi}{5}$

3. (a) $s = 2\pi$ inches, $A = 4\pi$ sq. inches
   (b) $s = 40\pi$ inches, $A = 2000\pi$ sq. inches

   (c) $s = \dfrac{3\pi}{2}$ inches, $A = \dfrac{3\pi}{2}$ sq. inches

   (d) $s = \pi$ inches, $A = \dfrac{\pi}{2}$ sq. inches

5. (a) $\sin\theta = \dfrac{4}{5}, \cos\theta = \dfrac{3}{5}, \tan\theta = \dfrac{4}{3}, \cot\theta = \dfrac{3}{4}, \sec\theta = \dfrac{5}{3}, \csc\theta = \dfrac{5}{4}$

   (b) $\sin\theta - -\dfrac{1}{2}, \cos\theta - \dfrac{\sqrt{3}}{2}, \tan\theta - \dfrac{-1}{\sqrt{3}}, \cot\theta - -\sqrt{3}, \sec\theta - \dfrac{2}{\sqrt{3}}, \csc\theta - -2$

   (c) $\sin\theta = \dfrac{2}{\sqrt{5}}, \cos\theta = \dfrac{1}{\sqrt{5}}, \tan\theta = 2, \cot\theta = \dfrac{1}{2}, \sec\theta = \sqrt{5}, \csc\theta = \dfrac{\sqrt{5}}{2}$

   (d) $\sin\theta = 0, \cos\theta = -1, \tan\theta = 0, \cot\theta = $ undefined, $\sec\theta = -1, \csc\theta = $ undefined

   (e) $\sin\theta = \dfrac{-3}{\sqrt{13}}, \cos\theta = \dfrac{-2}{\sqrt{13}}, \tan\theta = \dfrac{3}{2}, \cot\theta = \dfrac{2}{3}, \sec\theta = \dfrac{-\sqrt{13}}{2}, \csc\theta = \dfrac{-\sqrt{13}}{3}$

   (f) $\sin\theta = \dfrac{2\sqrt{39}}{13}, \cos\theta = \dfrac{1}{\sqrt{13}}, \tan\theta = 2\sqrt{3}, \cot\theta = \dfrac{1}{2\sqrt{3}}, \sec\theta = \sqrt{13}, \csc\theta = \dfrac{13}{2\sqrt{39}}$

7. (a) $\sin 75°$ (b) $\cos 105°$ (c) $\sin\left(-\dfrac{\pi}{8}\right)$ (d) $\cos 0° = 1$

9. $\cos 20\theta = 1 - 2\sin^2(10\theta) = 1 - 2(2\sin 5\theta \cos 5\theta)^2$
   $$= 1 - 8\sin^2 5\theta \cos^2 5\theta$$
   $$= 1 - 8\sin^2 5\theta(1 - \sin^2 5\theta)$$
   $$= 1 - 8\sin^2 5\theta + 8\sin^4 5\theta$$

11. (a) $\theta = 30°, 150°, 210°, 330°$ (b) $\theta = 0°, 48°35', 131°25', 180°$
    (c) $\theta = 90°, 120°, 240°, 270°$ (d) $\theta = 210°, 270°, 330°$

13. $\dfrac{129.4}{\sqrt{3}} \approx 74.7$ ft

15. $74°38'$

CHAPTER 7

7.1 Exercises

1. (a) $x^2 + y^2 = 25$ (b) $(x - 3)^2 + (y - 2)^2 = 16$

  (c) $\left(x + \dfrac{1}{2}\right)^2 + y^2 = 1$ (d) $x^2 + y^2 - 2ry = 0$

  (e) $x^2 + y^2 - 8x - 2y - 3 = 0$

3. (a) $x^2 + y^2 = 49$ (b) $(x - 3)^2 + (y - 1)^2 = 16$

  (c) $(x - 5)^2 + y^2 = 36$ (d) $(x + 2)^2 + \left(y - \dfrac{5}{2}\right)^2 = \dfrac{25}{4}$

  (e) $x^2 + \left(y - \dfrac{2}{3}\right)^2 = \dfrac{25}{9}$ (f) $\left(x + \dfrac{7}{2}\right)^2 + y^2 = 9$

5. (a) $x^2 + y^2 - 6x + 4y - 12 = 0$ (b) $x^2 + y^2 + 6x - 8y = 0$
  (c) $x^2 + y^2 - x - 3y - 10 = 0$ (d) $8x^2 + 8y^2 - 79x - 32y + 95 = 0$

7. $x^2 + y^2 - 8x - 4y - 12 = 0$

7.2 Exercises

1. (a) $(1, -3)$ (b) $(-2, -4)$ (c) $(-2, 0)$ (d) $(-6, 0)$
  (e) $(-3, 2)$ (f) $(3, -4)$ (g) $(-4, -7)$ (h) $(3, 3)$

3. (a) $x' = x - 1, y' = y; y' = (x')^2$
  (b) $x' = x - 2, y' = y + 1; y'x' = 3$
  (c) $x' = x + 4, y' = y - 1; (y')^2 = (x')^3$
  (d) $x' = x - 2, y' = y + 1; (x')^2 + (y')^2 = 49$

5. (a) $4x'y' + 7 = 0$ (b) $x'y' = 8$ (c) $x'y' = -7$

7.3 Exercises

1. (a) $(0, 0), \left(\dfrac{3}{2}, 0\right), x = -\dfrac{3}{2}$ (b) $(0, 0), (0, 2), y = -2$

  (c) $(0, 0), \left(\dfrac{-3}{16}, 0\right), x = \dfrac{3}{16}$ (d) $(5, 1), (3, 1), x = 7$

  (e) $(3, 2), (5, 2), x = 1$ (f) $(2, -3), \left(2, -\dfrac{9}{4}\right), y = \dfrac{-15}{4}$

3. (a) $y^2 + 4y - 8x + 36 = 0$ (b) $(x - 2)^2 = -20(y - 3)$
  (c) $y^2 - 6y - 12x + 21 = 0$ (d) $x^2 - 6x - 2y + 10 = 0$

5. (a) $x^2 - 9y - 36 - 0$ or $x^2 + 9y + 36 = 0$
  (b) $y^2 - 2y - 8x + 9 = 0$ or $y^2 - 2y + 8x - 39 = 0$

7. 32.5 feet

7.4 Exercises

1.

Center	Foci	Vertices	Major Axis
(a) $(0, 0)$	$(2\sqrt{7}, 0), (-2\sqrt{7}, 0)$	$(0, \pm6), (\pm8, 0)$	along x-axis
(b) $(6, -4)$	$(6 + 2\sqrt{5}, -4)$ and $(6 - 2\sqrt{5}, -4)$	$(6, 0), (6, -8)$ $(0, -4), (12, -4)$	parallel to x-axis
(c) $(1, -2)$	$(1, -2 + \sqrt{5}), (1, -2 - \sqrt{5})$	$(1, 1), (1, -5),$ $(-1, -2), (3, -2)$	parallel to y-axis
(d) $(-3, 1)$	$(-3, 1 + 2\sqrt{3}),$ $(-3, 1 - 2\sqrt{3})$	$(-3, 5), (-3, -3),$ $(-5, 1), (-1, 1)$	parallel to y-axis

3. $\dfrac{(x - 1)^2}{45} + \dfrac{(y - 2)^2}{20} = 1$

5. Circle

7.5 Exercises

1. (a) Foci $(\pm 5\sqrt{2}, 0)$; vertices $(\pm 5, 0)$; $y = \pm x$

  (b) Foci $(\pm 7, 0)$; vertices $(\pm 3\sqrt{5}, 0)$; $y = \pm \dfrac{2\sqrt{5}}{15} x$

  (c) Foci $(0, \pm 3\sqrt{2})$; vertices $(0, \pm 3)$; $y = \pm x$

  (d) Foci $(\pm 5, 0)$; vertices $(\pm 4, 0)$; $y = \pm \dfrac{3}{4} x$

3. $\dfrac{(x - 1)^2}{16} - \dfrac{(y + 2)^2}{9} = 1$

Chapter 7 Exercises

1. (a) Center $(-2, 1)$; radius $= 3$ (b) Center $\left(\dfrac{1}{2}, 1\right)$; radius $= \dfrac{\sqrt{5}}{2}$

  (c) Center $(-2, 0)$; radius $= \dfrac{1}{2}$ (d) Center $\left(-\dfrac{3}{4}, -\dfrac{5}{4}\right)$; radius $= \dfrac{3\sqrt{2}}{4}$

3. (a) $x' = x + 1, y' = y - 3$; $3(x')^2 - 4(y')^2 = 102$

  (b) $x' = x - 2, y' = y + \dfrac{1}{2}$; $3(x')^2 + 4(y')^2 = 0$

  (c) $x' = x - 3, y' = y + 1$; $2(x')^2 + 5(y')^2 = 40$

5. (a) $y^2 = 8x$ (b) $(x - 1)^2 = 4y$ (c) $y^2 + 4y + 6x - 17 = 0$

7. (a) $\dfrac{(x + 3)^2}{16} + \dfrac{(y - 5)^2}{4} = 1$ (b) $\dfrac{(y - 3)^2}{61} + \dfrac{(x - 2)^2}{25} = 1$

9. (a) $3x^2 - y^2 = 12$ (b) $4x^2 - y^2 = 16$

  (c) $3y^2 - 4x^2 = 12$ (d) $\dfrac{(y - 1)^2}{20} - \dfrac{(x - 2)^2}{5} = 1$

INDEX

A

Abscissa, 21
Absolute value, 17–20
  definition, 17, 18, 28
  of complex number, 186
  function, 61
  properties, 18
Addition formula
  of cosine, 301, 326
  of sine, 301, 326
  of tangent, 303, 326
Addition of two functions, 57, 85
Algebra
  of functions, 57, 85
  fundamental theorem, 182
Ambiguous case, in solution of triangles, 317–319
Amplitude, 263
Angle
  of depression, 315
  of elevation, 315
  initial side, 287
  measure, in degrees, 289
    in radians, 288
  negative, 287
  standard position, 287
  terminal side, 287
  vertex, 287
Angular speed, 292
Antilogarithm, 236
Arc length, of circle, 290
Arccosine function, 270
Arcsine function, 268
Arctangent function, 275
Area of circular sector, 291
Associative Law of functions, 59
Asymptotes
  of cosecant functions, 281
  of cotangent functions, 277
  horizontal, 188
  of hyperbola, 366
  of rational functions, 189
  of secant functions, 278
  of tangent functions, 275
  vertical, 189
Augmented matrix of system of linear
    equations, 133
Axis(es)
  major, 357
  minor, 357
  real, 1, 20
  translation of, 340, 370
  x-, 20
  y-, 20

B

Base, 204
  e, 238
  of exponential function, 212
  logarithmic, 238
  ten, 233
Binomial
  coefficients, 375, 376
  expansions, 377
  theorem, 374

C

Cartesian coordinate system, 20, 28
Cartesian product, 40, 85
Center
  of circle, 331
  of ellipse, 356
Characteristic of logarithm, 235
Circle
  area of sector, 291
  definition, 331
  degenerate case of, 333
  equation of, 331, 335
  unit, 247
Circular function
  definition, 247
  graphs, 249
  period, 252
Coefficients of polynomial, 89
Combinations $\binom{n}{k}$, 376
Commutative Law of functions, 59
Complex numbers, 178, 201
  absolute value, 186
  addition, 184
  conjugate, 180
  division, 185
  equality, 184
  form, 179
  imaginary part, 179
  multiplication, 185
  real part, 179
  roots, 178
  subtraction, 184
Complex zeros, 178
Composite functions, definition, 76, 86
Conic section, 329
Conjugate of complex number, 180
Coordinates
  of points on number line, 1, 28
  of points in plane, 20, 29

Cosecant
   circular function, 280
   graph, 281
   inverse, 281
   trigonometric function, 296
Cosine(s)
   circular function, 257
   evaluation, 258
   graph, 262
   inverse, 270
   law of, 319
   properties, 257
   trigonometric functions, 295
Cotangent
   circular function, 276
   graph, 277
   inverse, 277
   trigonometric function, 295
Cramer's rule, 149, 200

D

Decreasing functions, 71
Degree, as measure of angle, 289
Degree of polynomial, 90
Dependent system of linear equations, 122, 199
Determinant(s)
   application of, 151
   and cofactor of element, 147
   definition, 143
   expansion, 143
   function, 142
   minor, 147
   properties, 146
   use in solutions of system, 149
   value of, 143, 148
Difference
   formula of cosine, 326
   formula of sine, 326
   formula of tangent, 326
Directrix, of parabola, 344, 371
Discriminant, 161
   of quadratic equation, 161
Disjoint sets, 39
Distance formula, 22, 24, 29
Distributive Law of functions, 59
Dividend, 96
Division
   of complex numbers, 185
   of polynomials, 96
   of two functions, 57
Divisor, 96
Domain
   of cosecant function, 280
   of cosine function, 257
   of cotangent function, 276
   of exponential function, 204
   of function, 44, 48
   of logarithmic function, 223
   of relation, 41
   of secant function, 278
   of sine function, 257
   of tangent function, 274
Double angle formula
   for cosine function, 304
   for sine function, 304
   for tangent function, 305

E

e, 5, 215
Element of set, 30, 32
Elimination method, 124
Ellipse
   center, 356
   definition, 355
   eccentricity, 361
   foci, 355
   vertices, 357
Empty set, 33
Equality
   of ordered pairs, 20
   of polynomials, 91
   of sets, 33, 35, 84
Equations
   exponential, 209
   linear, 103
   logarithmic, 224
   quadratic, 156
   systems of, 120
   trigonometric, 308
Even functions, 67
Expansion
   binomial, 377
   of determinant, 143
Exponential function
   definition, 212, 244
   graph, 211, 212
   inverse, 222, 223
Exponential equations, 209
Exponents
   definition, 204–207
   properties, 205, 208, 244
   rational, 208
   zero, 205

F

Factor, of polynomial, 99
Factor theorem, 99
Factorial, 375
Factorization theorem, 182
Finite sets, 32
First degree equation, 103
Focus
   of conic, 329
   of ellipse, 355
   of hyperbola, 364
   of parabola, 344
Function(s)
   absolute value, 61
   algebra of, 57, 85
   circular, 247
   composite, 76, 87
   constant, 62
   cosecant, 280
   cosine, 256
   cotangent, 276
   decreasing, 71
   definition, 44, 85
   even, 67, 85
   exponential, 212
   geometric representation, 44, 51, 85
   graphs, 51
   greatest integer, 62
   identity, 61

Function(s) (*Continued*)
  increasing, 70, 85
  inverse, 80, 86
  linear, 103, 198
  logarithmic, 223
  mapping, 44, 85
  notation, 44, 85
  odd, 67, 85
  one-to-one, 72, 85
  periodic, 251
  polynomial, 89
  quadratic, 160
  rational, 187
  reciprocal, 62
  secant, 278
  sine, 256
  step, 62
  tangent, 274
  trigonometric, 294
  value of, 46
  winding, 249
Fundamental period
  of cosine function, 256
  of sine function, 256
  of tangent function, 274
Fundamental theorem of algebra, 182, 201

G

Geometry of lines, 120–121
Graph
  of absolute value function, 61
  of cosecant function, 281
  of cosine function, 262
  of cotangent function, 277
  of exponential function, 211
  of greatest integer function, 62
  of inverse cosine function, 270
  of inverse function, 80
  of inverse sine function, 268
  of inverse tangent function, 275, 276
  of linear function, 103
  of logarithmic functions, 223
  of quadratic function, 156
  of rational function, 187
  of secant function, 278
  of sine function, 262
  of systems of linear equations, 120–122
  of tangent function, 275
"Greater than," symbol for, 6, 28

H

Half-value formulas
  for cosine function, 305
  for sine function, 306
  for tangent function, 306
Homogeneous system, 132
Horizontal asymptotes, 188
Horizontal stretchings (contractions), 54, 85
Horizontal translations, 52, 86
Hyperbola
  definition, 364
  foci, 364
  vertices, 365

I

Identities
  summary of, 226
  trigonometric, 299
Identity function, 59, 61
Imaginary part of complex number, 179
Inconsistent system of linear equation, 120–122
Increasing function, 70
Induction, mathematical, 379
Inequalities, 11, 28
  linear, 13
  nonlinear, 15
Infinite set, 32
Integers, 2, 27
  negative, 1, 27
  positive, 1, 27
Intercept of a graph, 112
Intercept form of line, 113
Interpolation, linear, 232
Intersection of sets, 38
Intervals, 11, 28
  closed, 11
  half-, 11
  open, 11
Inverse
  of circular functions, 268–273, 275
  of cosine, 270
  of exponential function, 222
  of function, 80, 86
  of sine, 268
  of tangent, 275
Irrational numbers, 3, 27

L

Law of cosines, 319
Law of sines, 316
Law of tangents, 226
Length of circular arc, 290
"Less than," symbol for, 6, 28
Line segment, 11
Linear equation
  point-slope form, 111, 199
  slope-intercept form, 113, 199
  two-point form, 110, 199
Linear function, definition, 103, 198
Linear inequalities, 13
Linear interpolation, 237
Linear systems, 120
  consistent and dependent, 121
  consistent and independent, 122
  Cramer's rule, 149
  elimination method, 124
  equivalent systems, 123
  inconsistent and independent, 121
Logarithmic function(s)
  definition, 223, 245
  domain, 223
  graph of, 223
  properties, 226, 245
  range, 223
Logarithms, 223, 245
  applications, 242
  base e, 238, 245
  base ten, 233, 245
  change of base, 240, 245
  characteristic, 235, 245

Logarithms (*Continued*)
   common, 232, 245, 383–384
   computations with, 232, 245
   mantissa, 235, 245
   natural, 232, 245, 385–386
   properties, 226, 245

M

Major axis of ellipse, 357
Mantissa, 235
Mapping, 44
Mathematical induction, 379
Matrix, 133, 199
   augmented, 133
   coefficient, 133
   column, 133
   definition, 133
   determinant, 143
   entry, 133
   reduced-echelon form, 135
   row, 133
   row-equivalent, 136
Maximum value of quadratic function, 162
Measure of an angle, 287
Minimum value of quadratic function, 162
Minor axis of ellipse, 357
Monomial, 89
Multiplication
   associative law of, 59
   commutative law of, 59

N

n factorial, 375
nth term, 377
Negative angle, 287
Negative numbers, 1, 28
Null set, 33
Numbers
   complex, 178
   integers, 2, 27
   irrational, 3, 27
   natural, 1, 27
   negative, 1, 27
   positive, 1, 27
   rational, 2, 27
   real, 1, 5, 27
Number line, 1, 28

O

Odd function, 67, 85
One-to-one function, 72, 75, 85
Ordered pairs, 20, 29
   as functions, 44, 85
   as relations, 41, 85
   in Cartesian product, 40, 85
Order relation
   definition, 6, 28
   notation, 6, 28
   properties, 7, 8, 9, 28
   transitive, 8
   trichotomy, 8
Ordinate, 21, 29
Origin, of Cartesian coordinate system, 20, 28
   of real number line, 1, 27

P

Parabola
   axis of symmetry, 345
   definition, 157, 344
   directrix, 344
   focus, 344
   vertex, 157, 345
Parallel lines, 115, 199
Partial fractions, 194
Pascal's Triangle, 374–375
Period
   fundamental, 252
   of circular functions, 252, 283
   of function, 251
   of winding function, 252
Perpendicular lines, 117, 199
Phase shift, 266
Point-slope form of linear equations, 111
Polynomial(s)
   coefficients, 89, 198
   complex, 181
   constant function, 93
   definition, 89, 198
   degree, 90
   division, 96
   equal, 91, 198
   factor theorem, 99, 198
   fundamental theorem of algebra, 182
   rational root theorem, 168
   rational zeros, 168
   remainder theorem, 98, 198
   zeros, 93, 198
Positive angle, 287
Positive numbers, 1, 27
Product, Cartesian, 40, 85
Proper subset, 35, 84

Q

Quadrants, 22
Quadratic equations, 160
   completing square, 158
Quadratic formula, 161, 200
Quadratic functions, 156, 200
Quadratic inequalities, 15
Quotient, 96

R

Radian measure of angle, 288
Radius of circle, 331
Range
   of cosecant function, 280
   of cosine function, 256
   of cotangent function, 276
   of exponential function, 204
   of function, 44, 48, 85
   of logarithmic function 223
   of relation, 41
   of secant function, 278
   of sine function, 256
   of tangent function, 274
Rational numbers, 2
Rational root theorem, 168, 201
Real line, 1, 28
Real numbers, 1, 5, 27

Real part of complex number, 179
Reflection across $y = x$, 83
Relations, definitions, 41, 85
Remainder theorem, 98
Repeating decimal, 3
Right triangle trigonometry, 312
Roots
  complex, 178
  irrational, of polynomials, 168
  principal, 207
  rational, of polynomials, 168
Row-equivalent matrices, 136

S

Scientific notation, 234
Secant
  circular function, 278
  trigonometric function, 295
Set(s), 30, 84
  complement, 37, 84
  description, 30, 84
  disjoint, 39
  element, 30, 32, 84
  empty, 33
  equal, 33, 35, 84
  finite, 32
  infinite, 32
  intersection, 38, 84
  null, 33
  operations, 36
  proper subset, 35, 84
  relations, 41, 85
  subset, 33, 84
  union, 37, 84
  universal, 36
Set builder notation, 31
Sine
  circular function, 257
  evaluation, 258
  graph, of function, 262
  inverse of, 268
  trigonometric function, 295
Sines, law of, 316
Slope, 105, 198
Slope-intercept form of line, 113
Solution
  of linear inequalities, 13
  of linear systems, 120
  of quadratic equations, 161
  of quadratic inequalities, 15
  by graphic method, 15
Standard form of logarithm, 235
Standard position of angle, 287
Subset(s), 33
Symmetry of graphs, 63, 85
Synthetic division, 100

T

Tangent, 274
  circular function, 274
  graph, 275
  inverse, 275

Tangent (*Continued*)
  period, 275
  trigonometric function, 295
Terminal point, 11
Terminal side of angle, 287
Terminating decimals, 3
Transitive property of order, 8
Translation of axes, 340
Triangle trigonometry, 312
Trichotomy, 8
Trigonometric equations, 308
Trigonometric functions, 294
Trigonometric identities, 226

U

Union of sets, 37, 84
Unit circle, 249
Universal set, 36

V

Value, absolute, 17
Variables
  dependent, 46
  independent, 46
Venn diagrams, 34
Vertex
  of angle, 287
  of ellipse, 357
  of hyperbola, 365
  of parabola, 162
Vertical asymptotes, 189
Vertical stretchings (contractions), 53, 85
Vertical translations, 51, 86

W

Winding function, 249

X

x-axis, 20

Y

y-axis, 20
y-intercept, 112

Z

Zero(s)
  factorial, 376
  function, 59
  of polynomial, 93
  of quadratic function, 157, 160
  rational, 168